More Praise for *The Handbook of Nanotechnology*

"As someone who has successfully transitioned nanotechnology from the university lab to the commercial world, I can recommend this book as a 'must-read' for nascent academic entrepreneurs, those wishing to commercialize technologies they have developed at the university. As a faculty member at a major research university that is comprehensively involved in nanoscience and nanotechnology research, I come into continual contact with such persons: I can now point them to the first book they should study. The chapters on business development are particularly valuable; I wish I had had this book 5 years ago."

—Max G. Lagally, founder nPoint,
and professor of Surface Science,
at the University of Wisconsin-Madison

"In order for our society to realize the awe-inspiring potential for revolutionary change which nanotechnology promises in every industry from transportation to pharmaceuticals, complex issues of business, public policy and law must be managed at the highest levels of leadership in both the public and private sector. This Handbook provides an invaluable guide for that leadership."

—Rodney E. Slater,
former U.S. Secretary of Transportation
and partner at the law firm of Patton Boggs

"This is an excellent work that is both comprehensive and practical across a wide range of perspectives. It is great to have solid, detailed analysis of and advice on the science, business, and policy aspects of nanotechnology. This is one of the few publications to identify both the commonality of nanotechnology with existing industry and regulatory structures as well as its unique characteristics that have implications for policy, law, and running a business. It is a book that intelligently anticipates future developments in policy and intellectual property, as well as mergers and other financing activity."

—Randy Levine, Ph.D.,
president and CEO, ZettaCore, Inc.

"Indispensable. I can't imagine an attorney or policy maker not having this book on their shelf. It is an extraordinarily insightful and thorough book that delves into the intricacies of the emerging nanotechnology field in an accessible and easy-to-understand manner."

—F. Mark Modzelewski,
founder of the NanoBusiness Alliance
and managing director, Lux Research, Inc.

THE HANDBOOK OF NANOTECHNOLOGY

Business, Policy, and Intellectual Property Law

John C. Miller
Arrowhead Research Corporation

Ruben M. Serrato
Canon U.S.A.

Jose Miguel Represas-Cardenas
Stanford University

Griffith A. Kundahl
NanoBusiness Alliance

with an editorial contribution by
Mark Graffagnini

John Wiley & Sons, Inc.

Copyright © 2005 by John C. Miller, Ruben Serrato, Jose Miguel Represas-Cardenas, and Griffith Kundahl. All rights reserved.

Published by John Wiley & Sons, Inc., Hoboken, New Jersey.
Published simultaneously in Canada.

No part of this publication may be reproduced, stored in a retrieval system, or transmitted in any form or by any means, electronic, mechanical, photocopying, recording, scanning, or otherwise, except as permitted under Section 107 or 108 of the 1976 United States Copyright Act, without either the prior written permission of the Publisher, or authorization through payment of the appropriate per-copy fee to the Copyright Clearance Center, Inc., 222 Rosewood Drive, Danvers, MA 01923, 978-750-8400, fax 978-646-8600, or on the web at www.copyright.com. Requests to the Publisher for permission should be addressed to the Permissions Department, John Wiley & Sons, Inc., 111 River Street, Hoboken, NJ 07030, 201-748-6011, fax 201-748-6008.

Limit of Liability/Disclaimer of Warranty: While the publisher and author have used their best efforts in preparing this book, they make no representations or warranties with respect to the accuracy or completeness of the contents of this book and specifically disclaim any implied warranties of merchantability or fitness for a particular purpose. No warranty may be created or extended by sales representatives or written sales materials. The advice and strategies contained herein may not be suitable for your situation. You should consult with a professional where appropriate. Neither the publisher nor author shall be liable for any loss of profit or any other commercial damages, including but not limited to special, incidental, consequential, or other damages.

For general information on our other products and services, or technical support, please contact our Customer Care Department within the United States at 800-762-2974, outside the United States at 317-572-3993 or fax 317-572-4002.

Wiley also publishes its books in a variety of electronic formats. Some content that appears in print may not be available in electronic books.

For more information about Wiley products, visit our web site at www.wiley.com.

Library of Congress Cataloging-in-Publication Data

The handbook of nanotechnology business, policy, & intellectual property law / John C. Miller... [et al.].
   p. cm.
  Includes bibliographical references and index.
  ISBN 0-471-66695-5
  1. Microelectronics industrty. 2. Nanotechnology—Industrial applications. 3. Intellectual property. I. Title: Handbook of nanotechnology business, policy, and intellectual property law.
II. Miller, John C.
   HD9696.A2H36 2004
   620.5′068—dc22
                                                  2004009884

Printed in the United States of America

10 9 8 7 6 5 4 3 2 1

This book is dedicated to the Nomads, who take risks to follow their dreams.

Contents

Acknowledgments	ix
Foreword	xi
Introduction	1
Part I: Introduction to Nanotechnology	**11**
Chapter 1: Understanding Nanotechnology	13
Chapter 2: The Industrial Structure Giving Rise to Nanotechnology	33
Part II: Nanotechnology Policy and Regulation	**39**
Chapter 3: Societal and Ethical Implications	41
Chapter 4: Environmental Regulation	51
Chapter 5: The Patent and Trademark Office	65
Chapter 6: FDA Regulation	83
Chapter 7: National Security and Export Controls	103
Chapter 8: Federal Funding	115
Chapter 9: Conclusions	131
Part III: Nanotechnology Business	**137**
Chapter 10: Starting a Nanotech Company	139
Chapter 11: Business Plans and Strategy	161
Chapter 12: Early Stage Financing	189
Chapter 13: Intellectual Property	209
Chapter 14: Corporate Partnering and Globalization	235
Chapter 15: Consolidation and Standardization	251
Chapter 16: Exit Opportunities	267
Chapter 17: Conclusions	281
Notes	287
About the Authors	355
Index	357

Acknowledgments

Special thanks to all of our families and friends: Caroline Campbell, Mike, Scotti, and Brian Miller, Wayne and Ila Harris, Bob Gilliland, Ruben, Rafaela, Linda, and especially Gabriel Serrato, Jose Represas-Perez and Ana Beatriz Cardenas, Chuck Ballingall, Bill Southworth, Byron Arthur, Kim, Kate, and Caroline Kundahl, and Cheryl and Joseph Graffagnini.

John also wishes to thank the Campbell family for providing hospitality in Australia and Palo Alto during the writing of this book and Professor Hank Greely for providing inspiration to see the book to completion.

Jose Miguel also wishes to thank the Department of Electrical Engineering at Stanford University for his fellowship award, as well as the many professors that over the years have been examples of generosity, intellectual independence, and scientific integrity.

We would also like to thank all of the people that provided ideas, comments, and assistance:

Mark Modzelewski (NanoBusiness Alliance); K.J. Cho (Stanford University); Mike McGehee (Stanford University); Hari Manoharan (Stanford University); Leon Radomsky (Foley & Lardner); Ken Barovsky (Quantum Dot Corp.); Steve Maebius (Foley & Lardner); Rita Colwell (National Science Foundation); Larry Bock (Nanosys); Margaret Radin (Stanford Law School); John Barton (Stanford Law School); Deborah Hensler (Stanford Law School); Joe Grundfest (Stanford Law School); Rich Wolf (Caltech Office of Technology Transfer); Max Lagally (nPoint); Katharine Ku and Linda Chao (Stanford Office of Technology Licensing); Veronica Lanier and Carol Mimura (Berkeley Office of Technology Licensing); Rebecca Goodman and Robert Nidever (UCLA Technology Licensing Office); Joyce Brinton and Robert Benson (Harvard Technology Licensing Office); Craig Zolan (Uventures); Randy Levine (ZettaCore); Rodney Slater (Patton Boggs); Randy Bell (Nanotechnologies Inc.); J. Kevin Gray (Fish & Richardson P.C.); John Belk (The Boeing Corporation); Chad Mirkin (Northwestern University); Mark R. Wiesner (Rice University); Vicki Colvin (Rice University); Jess Milbourn and Thomas Schults (PST); Christine Peterson (Foresight Institute); Nadrian Seeman (NYU); Joe Mauderly (Lovelace Respiratory Research Institute); Kathy Jo Wetter (ETC Group); Elisabeth Lutanie (Institute of Physics); Terry Lowe (Metallicum); Robert Bradbury (Robobiotics); Bruce Stewart (Arrowhead Research Corporation); Peter Grubstein (NGEN Partners); Raj Bawa (Bawa Biotechnology Consulting and Rensselaer

Polytechnic Institute); Donald Marlow (FDA); Jim Hurd (Nanoscience Exchange); Chad Wieland and Nhat D. Phan (Burns, Doane, Swecker & Mathis, L.L.P.); Jeffrey Weinshenker (FDA); Eric Werwa (Rep. Honda's Office); Ben Boyer (Lehman Brothers); Doug Jamison (Harris & Harris Group, Inc.); Phil Sayre (EPA); Bill Goddard (Caltech); Edward Rashba (IEEE); and Daniel Gamota (Motorola).

Finally, special thanks to Dr. Barbara Karn (EPA) for facilitating the peer review process with federal agencies and Dr. David Reisner (Inframat) for his extensive, detailed review and comments.

Foreword

On January 21, 2000, President Clinton unveiled the National Nanotechnology Initiative (NNI) in a major policy address at Caltech. In his speech, he announced that his budget would propose almost doubling the federal investment in nanoscale science and engineering, from $270 million in FY2000 to $495 million in FY2001. He asked his audience to imagine "materials with 10 times the strength of steel and only a fraction of the weight; shrinking all the information at the Library of Congress into a device the size of a sugar cube; detecting cancerous tumors that are only a few cells in size." The next week, with 51 million Americans watching, Clinton again referred to the promise of nanotechnology in his State of the Union address. (His speech writers tried to take this section out to shorten the 89-minute speech, but Clinton insisted on leaving it in!)

As a strong supporter of the NNI, I was thrilled that President Clinton had decided to embrace it as one of his top science and technology priorities. I had the privilege of working for President Clinton and Vice President Gore for eight years, and eventually served as the Deputy Director of the White House National Economic Council and the Deputy Assistant to the President for Technology and Economic Policy.

I was convinced that there is a strong intellectual case for increasing the federal government's investment in nanoscale science and engineering. First, nanotechnology has the potential to be what economists call a "general purpose" technology—similar in the size and scope of its economic and societal impact to the steam engine, electricity, the transistor, and the Internet. Second, long-term, high-risk research will be needed to realize the potential of nanotechnology. Some of this research is beyond the time horizons of individual firms, and government support for research is critical when firms cannot fully capture the benefits of investing in research and development. Third, the NNI can help address the growing imbalance between biomedical research and the physical sciences and engineering by increasing support for critical disciplines such as condensed matter physics, chemistry, materials science, and electrical engineering. Fourth, the NNI will help create the workforce of the twenty-first century, since most of the government funds support university research. Furthermore, as Nobel Laureate Rick Smalley has observed, nanotechnology might get our young boys and girls excited about science and engineering, in the same way that Sputnik or the space race captured the public's imagination in previous generations. Finally, global leadership in nanotechnology is up for grabs, and increased federal investment will help strengthen the U.S. position in this key area.

The development of the NNI began in earnest in September 1998, when the White House created a working group on Nanoscience, Engineering, and Technology under the auspices of the National Science and Technology Council (NSTC). I served as the White House Co-Chair, and Mike Roco, the point person on nanotechnology for the the National Science Foundation (NSF), served as the Chair. In January 1999, the NSTC convened a workshop with experts from industry and academia. University researchers such as UC Berkeley's Paul Alivisatos and industrial researchers such as Hewlett-Packard's Stan Williams helped identify the most important and promising R&D opportunities in nanoscale science and engineering.

Throughout 1999, dedicated public servants like Mike Roco (NSF), Jim Murday (Naval Research Laboratory), Iran Thomas (Energy), Meyya Meyyapan (NASA), Jeff Schloss from the National Institutes of Health (NIH), and Kelly Kirkpatrick from the Office of Science and Technology Policy (OSTP), worked tirelessly to develop a concrete proposal for the President's FY2001 budget. My colleagues at the OSTP and I met with senior officials from the science agencies; we convinced them that we would fight to protect any increases in nanotechnology research that they proposed as part of their budget submission to the Office of Management and Budget. We also asked the President's Council of Advisors on Science and Technology to review our proposal, which they strongly endorsed.

I also began to ask federal program managers and leading researchers in the field to identify potential "grand challenges"—ambitious but plausible outcomes from increased research in nanoscale science and engineering. Although I knew that it was impossible to predict what might eventually come out of the NNI, my time in the White House had taught me that it was essential to identify some exciting possibilities that could be easily understood by politicians, reporters, and the general public. Armed with these examples (several of which wound up in Clinton's Caltech speech and State of the Union address), I started briefing the most senior White House staff about nanotechnology—people like Gene Sperling, the head of the National Economic Council, John Podesta, the President's Chief of Staff, and David Beier, Vice President Gore's Chief Domestic Policy Adviser.

In the fall of 1999, the White House staff began to identify possible initiatives for consideration by President Clinton. I convinced Gene Sperling that nanotechnology should be one of the priorities in the President's FY2001 budget, as part of a larger increase in support for science and technology that we called the "21st Century Research Fund." Neal Lane, the President's Science Advisor, was also a staunch advocate for the NNI. Working together, the National Economic Council and the Office of Science and Technology Policy made a compelling case to President Clinton to support a large increase for the NNI in his budget. In a December 1999 meeting in the White House Cabinet Room, President Clinton approved the NNI.

Although hardly an impartial observer, I believe that President Clinton's decision to launch the NNI served as a catalyst for increased investment by universities, large companies, venture capitalists, and state governments. Federal expenditures have continued to rise during the Bush administration, and will reach nearly $1 billion in the FY2005 budget. Media coverage of nanoscience and nanotechnology has exploded, and Congress has passed legislation that authorizes the NNI for four years. Many foreign governments have also increased their investments in nanotechnology research.

Of course, only time will tell whether these increased public and private investments in nanotechnology will lead to revolutionary advancements in computing and communications, clean energy, health care, transportation, advanced materials, environmental applications, national security, and space exploration. As President Clinton noted in his Caltech speech, "some of these research goals will take 20 or more years to achieve." There is always the risk that advocates of nanotechnology, whether in government, industry, finance, or academia, will overpromise and underdeliver.

What is clear is that we must now address the ethical, legal, policy, regulatory, and business issues associated with the commercialization of nanotechnologies. This is why *The Handbook of Nanotechnology* is so timely. John Miller, Ruben Serrato, Jose Miguel Represas-Cardenas, and Griffith Kundahl have done a terrific job of analyzing the key economic and societal issues facing nanotechnology. How should the EPA regulate nanomaterials, which may have different environmental and human health effects than the same materials in bulk? Will the Patent and Trademark Office be able to handle the rush to file nanotechnology patents without slowing down the rate of innovation? How can the government most effectively manage its investment in nanoscale science and engineering? How can entrepreneurs successfully launch nanotechnology start-ups? *The Handbook of Nanotechnology* is invaluable for anyone who is seeking to move nanotechnology from the lab to the marketplace in an ethical and responsible fashion.

—THOMAS A. KALIL
July 2004

Introduction

Congress, Washington D.C.: 2000

Congress is a long way from the Texas laboratory where Richard Smalley made his Nobel Prize–winning discovery. Nevertheless, the gray-bearded scientist speaks with more than just confidence—he has a fiery passion. Perhaps he is inspired by an ardent desire to cure a vicious disease; perhaps he wants to express to the world the importance of his life's work. Dr. Smalley declares:

> But twenty years from now, nanotechnology will have given us specially engineered drugs...Cancer—at least the type that I have—will be a thing of the past.[1]

The politicians are persuaded.

Silicon Valley, CA: 2003

The investors take their seats in the polished conference room on Sand Hill Road. Although still visibly weathered by the dot-com storms, they are somewhat upbeat. The sky is clear, and there is something new in the air. Veteran entrepreneur Larry Bock dims the lights and starts the tape. The screen lights up, and it is Mr. Robinson from the film *The Graduate*. "Benjamin, can I have a word with you?" The tall, handsome college graduate nervously consents and walks down the hall with Mr. Robinson. "Benjamin, I just want to say one word...are you listening?" "Yes sir," replies the graduate. "Nanotechnology..." says the older man. The investors are intrigued.

Pentagon, Washington D.C.: 2002

In one of the deepest, darkest rooms in the Pentagon, five-star generals and White House officials meet. In light of recent international events, there is a great deal of unease—it seems the future has never been more uncertain. But today the generals are confident as they present their vision of the military battle suit of the twenty-first century, a suit enhanced by nanotechnology. It can change color to blend in with the surrounding environment and can transform itself from a soft fabric into bulletproof armor. Sensors in the suit detect when the soldier is wounded, tiny devices transmit vital signals to a distant medic, and antidotes are released. It even enables a soldier to jump twenty feet in the air. The presentation ends and the president is encouraged.

IBM Headquarters, Armonk, NY: 2002

When IBM's CEO Louis Gerstner hears from Phaedon Avouris, he knows it is time to listen. Avouris, head of IBM's nanoscience and technology group, is a skeptical scientist. Reporters turn to him for a "realistic assessment" of semiconductor technology. Today Gerstner is intrigued by Avouris' unusual excitement. Avouris hands the CEO an image of what appear to be thin threads laid out in a crisscross fashion. On closer examination, Gerstner identifies the microscopic image as a logic gate—the fundamental computer component responsible for selectively routing electrical signals and transforming them into meaningful ones and zeros. Comprised of carbon nanotubes, this transistor is one of the first examples of molecular-scale electronics. Avouris boldly remarks,

> [I]t is no longer a question of whether nanotubes will become useful components of the electronic machines of the future but merely a question of how and when.[2]

Gerstner is convinced.

While advances in biotechnology and the rise of the Internet dazzled investors and made headlines in the final years of the twentieth century, a quiet revolution was taking place in the field of nanotechnology. In 2000, the National Science and Technology Council (NSTC) observed that nanotechnology, which involves the manipulation of matter at the atomic and molecular levels, "could be at least as significant as the combined influences of microelectronics, medical imaging, computer-aided engineering, and man-made polymers" developed in the last century.[3] Nanotechnologies may speed cures for cancer and AIDS, make possible solar and hydrogen fuel cells to eliminate reliance on fossil fuels, or enable computers with magnificent processing power. Dr. K. Eric Drexler, one of nanotechnology's most controversial figures, goes so far as to predict that nanotechnology will result in self-replicating nanorobots capable of doing everything from assembling automobiles to unleashing weapons of mass destruction.

As with any promising new technology, the difference between hyperbolized rhetoric and scientific reality may not be immediately apparent. It is clear, however, that nanotechnology is already a significant factor in the nation's long-term strategy for continued scientific and industrial leadership.

At the federal level, government funding has grown dramatically since President Clinton launched the $422 million National Nanotechnology Initiative (NNI) in 2000. For fiscal year (FY) 2004, the Bush administration doubled NNI funding to $847 million. In December 2003, Congress passed the 21st Century Nanotechnology Research and Development Act, which authorized $3.7 billion for nanoscale science and engineering projects between FY 2005 and FY 2008. The goal of this program is to galvanize the field by allocating funds for fundamental research, new science centers, a new national research infrastructure, workforce education and training, and further study of the ethical, legal, and societal implications of nanotechnology. At the state level, legislatures have also begun to channel funds toward nanotechnology and many such as California, Illinois, and Massachusetts have created task forces to attract new nanotech companies and federal research dollars.

Government funding of nanotechnology is also being complemented by private investment. Venture capitalists are actively focused on finding nanotechnology investments and have poured more than $1 billion in to new companies over the last few years. Eager individual investors are attending conferences, creating investment groups and buying nanotech investment reports. Many of the world's largest private companies are also very active nanotech investors. Corporations like Hewlett-Packard, IBM, and Intel are allocating substantial portions of their research budgets to nanotechnology. Other large companies like Canon and the Boeing Corporation are making direct investments in nanotech companies.

Outside of the United States, many countries now recognize that their global trade competitiveness will soon be dictated by their competency in nanoscale research and development. Japanese and European Community government investments in nanotech currently rival U.S. expenditures. Their leading universities such as Osaka University and Oxford University have nanotech development programs modeled after the best universities in the United States. China, South Korea, Canada, Taiwan, Russia, Germany, and the Netherlands also have significant government programs designed to attract and retain scientists. The international community is actively gearing up for the day when nanotech leadership could be the most important factor in determining global economic and political leadership.

Despite the rapid progress in nanoscale research and the massive investments taking place, little attention has been devoted to the legal, policy, regulatory, and business issues associated with this new era of technological power. This book is the first attempt to fully explore these issues and prepare industry, government, and society for the revolution in nanotechnology. The book is divided into three parts. Part I provides an overview of nanotechnology, describes the industrial structure giving rise to its commercial development, and identifies areas where nanotech is having the greatest impact. Part II focuses on the regulatory and policy issues confronting those in the field. It explores how federal agencies and Congress should prepare for nanotechnology. As such, it is directed to an audience of regulators, policymakers, and industry leaders. Part III explores the legal and business issues confronting nanotech companies and is directed to individual managers, lawyers, investors, and scientists. In the following pages, we provide a brief overview of the contents of each section.

PART I: INTRODUCTION TO NANOTECHNOLOGY

In order to walk through the legal, policy, regulatory, and business issues in this field, it is necessary to first understand the scientific underpinnings and potential applications of the technology. In Chapter 1, we provide a clear and technical description of nanotechnology. Chapter 2 builds on this scientific foundation by creating a model of the industrial structure giving rise to nanotechnology. It is our hope that these two chapters will enable readers to distinguish between likely applications in the near future and long-term visions that may or may not be realized.

PART II: NANOTECHNOLOGY POLICY AND REGULATION

As of this writing, there has been no coordinated framework for regulation of nanotechnology. In September 2000, the National Science and Technology Council, Committee on Technology, Subcommittee on Nanoscale Science, Engineering, and Technology (NSET)—the federal interagency group coordinating the NNI—brought together the first group of nanotechnology researchers, social scientists, and policymakers to address "how nanotechnology will change society and the measures to be taken to prepare for these transformations."[4] Although the resulting work, *Societal Implications of Nanoscience and Nanotechnology Report*, was useful as a springboard for policy discussion, it did not provide an organized or prioritized analysis of the relevant issues. Instead, the report was simply a collection of different people's thoughts on the societal impacts of nanotechnology. The report was not comprehensive, and many of the issues identified in this book were not even mentioned. Since the report took place in 2000, much of the discussion was also necessarily an "inherently speculative exercise."[5]

The lack of well-informed and rigorous academic analysis of the policy issues associated with nanotechnology is problematic. As we argue in the following chapters, the history of law and technology warns of the consequences of the failure by policymakers and regulators to adequately prepare for rapid technological progress. For example, misperceptions concerning the environmental risks of nuclear power and agricultural biotechnology have precluded these domains from realizing their full potential. Patent decisions made by courts and the Patent and Trademark Office have created substantial barriers to progress in drug discovery and software, and mistakes made by the Food and Drug Administration have plagued biotechnology for years. But the greatest policy mistakes involve funding. In several instances, poor preparation and insufficient communication with industry have resulted in government funding of failed technologies.

Despite oxygen deprivation from political and regulatory mishaps, however, the flames of past emerging technologies have continued to burn. History may still write a different story about the ability of government to adequately prepare for the advent of nanotechnology. Its far-reaching potential to radically transform the world renders it almost entirely dependent upon government nurturing for its survival. Never before has the flame of a technological movement relied as much on the oxygen supplied by government officials as it does the wood provided by scientists. Unfortunately, while the scientists are forging ahead, their counterparts in government are lagging behind. Thus, a rigorous analysis of the legal, political, and regulatory issues associated with nanotechnology is urgently needed.

Chapter 3

The first policy topic that we discuss is whether a "nano" world is desirable. Bill Joy of Sun Microsystems launched the public debate about the wisdom of pursuing nanotechnology research and development by highlighting the field's long-term threats. According to Joy, the convergence of nanotechnology, robotics, and genetic engineering could produce "spiritual machines" that ultimately replace humans. Further, accidental or intentional misuse of self-replicating nanorobots could result in a catastrophe of cataclysmic proportions. Thus far, the debate has been mostly limited to a shouting match between doomsayers and scientists. Michael Crichton, author of *Jurassic Park*, recently released a new thriller depicting nanorobots that invade and take control of human bodies.

A public backlash could shatter the emerging nanotechnology industry. Before myths and irrational fears impede technological progress, a more scholarly analysis is needed of whether in the long run the risks justify the benefits. Chapter 3 begins this discussion. Even if nanotechnology results in the blending of humans and machines and self-replicating nanorobots present substantial environmental and security risks, nanotechnology research and development should not and cannot be prohibited. The most prudent course of action is to cultivate nanoscience while promulgating regulations to prevent potential harms. Scientists and policymakers must actively work to convince the public that this is an appropriate strategy.

After concluding that nanotechnology is a worthy endeavor for humanity, we investigate how government can pave a smoother path for nanotechnology. Because nanostructures and nanodevices represent whole new classes of materials and products, they present challenges to the fundamental organization of the regulatory state. The Environmental Protection Agency (EPA), the Patent and Trademark Office (PTO), the Food and Drug Administration (FDA), and agencies responsible for administering export control laws are struggling to deal with these challenges.

Chapter 4

Chapter 4 explores the complex challenges that nanotechnology poses for the EPA. Environmental groups have voiced concerns that nanosized particles, because they have different properties than bulk materials of the same substance, could present a host of new health and environmental hazards. However, since environmental laws require regulation of new chemicals and nanomaterials are just smaller versions of existing chemicals, these substances are starting to enter the environment with minimal regulatory review. We engage in a comprehensive analysis of the toxicity risks associated with nanomaterials and weigh these risks against the benefits. Ultimately, we conclude that, while more data is needed, the current data dictate that EPA should not subject nanomaterials to prohibitively stringent regulations. The EPA should first garner more data and consider subjecting nanomaterials to additional review procedures.

Chapter 5

Chapter 5 describes how the Patent and Trademark Office has encountered problems in reviewing nanotechnology patents. Due to the absence of a core group of examiners well versed in nanoscience, the agency has issued broad and overlapping patents on the building blocks of nanotechnology. Further, a compulsion to patent has swept nanotechnology researchers and companies; the number of patents and different patent holders is large and rapidly growing. A chaotic and fragmented intellectual property base will make it more difficult for start-up companies to research and develop commercial products downstream. The chapter concludes by analyzing tools that might be used by government to alleviate the intellectual property quagmire.

Chapter 6

Chapter 6 turns to the issue of regulation of nanomedicine. The FDA classifies products as either drugs, devices, or biologics, and it regulates each category differently. Because nanomedical products are often a combination of drugs, devices, and biologics, classification will become increasingly difficult and confusing. Similarly, because nanotechnology will primarily improve upon existing products, the agency will undoubtedly encounter complicated issues associated with clinical data necessary for product approval. The agency has done little to acquire the expertise necessary for effective review of the safety and efficacy of nanomedical products. The FDA should consider implementing several reforms now to ensure that it is adequately prepared to regulate nanomedicine.

Chapter 7

In Chapter 7, we review challenges posed by nanotechnology to the export control sector. Navigating through the various export control laws can be a confusing and frustrating endeavor. We show that, while most nanotechnology companies will encounter export control laws, it is unclear how these laws should be applied. We urge agencies responsible for administering such laws to clearly address how they apply to nanotechnology.

Chapter 8

We explore the issue of federal funding in Chapter 8. A survey of the financing landscape reveals that large corporations and governments are making massive investments in nanotechnology. At this stage, an enormous federal commitment is necessary to foster basic research and establish an infrastructure to sustain an economy based on nanotechnology. The history of public R&D programs suggests that such programs are prone to failure if they are not managed and implemented effectively. Thus, we engage in a thorough and detailed review of the NNI and the new legislation. We conclude by offering several broad themes that policymakers should keep in mind in the coming years.

Chapter 9

We conclude Part II by calling for government to carefully nurture nanotechnology. Although some initial preparation for the Nano Age has begun, regulators, policymakers, and the public will be required to take additional measures in coming years. Chapter 9 lays out several broad themes for policymakers and regulators to consider as they address the future of nanotechnology.

PART III: NANOTECHNOLOGY BUSINESS

Prior to this writing, few efforts have been made to identify and analyze the complicated legal and business issues confronting nanotech companies. Most of the analysis of the nanotechnology business environment to date has been confined to industry journals, newsletters, and accounts by the popular press. While these publications provide useful information about the latest products, companies, and investments, they do not provide the comprehensive analytical guidance to aid entrepreneurs, lawyers, business executives, and investors.

A discussion of the business and legal issues confronting individual companies is needed because professionals will face unique challenges in facilitating the commercialization of products based on developments in the field. For example, drafting claims of nanotech patents involves different technical and strategic considerations than drafting claims of biotech patents. Similarly, because nanotechnology is primarily an enabling technology for a variety of different industries, there are unique considerations for writing business plans and engaging in corporate partnerships. The advent of nanotechnology will take place in a very different climate than past technological waves. From choosing an organizational form to seeking venture financing and planning for an initial public offering, nanotechnology companies will encounter a very different legal and business environment than technology companies of the past. Part III of this book is the first attempt to more fully explore these issues from the perspective of individual nanotech companies. It is our hope that the issues identified here will aid all members of the nanotech business community in creating better companies.

Chapter 10

We begin with the issues facing a newly born company. From patenting and licensing the necessary patents to choosing the name and organizational form, starting a nanotechnology company is a daunting task. Many scientists and managers launching companies in this exciting new field have dreams of turning their projects into the next Intel or Hewlett-Packard. Some successful start-ups will be acquired and others will make initial public offerings. But many will also fail. Ultimately, the success of a start-up turns on strategic, legal, and business decisions made in the earliest days of the company. Chapter 10 attempts to assist entrepreneurs and investors by exploring these initial considerations.

Chapter 11

We devote Chapter 11 to writing the nanotech business plan and identifying areas of important consideration for nanotech companies. Writing a compelling business plan and creating a sound strategy are crucial for companies to maximize their access to capital and potential for success. This chapter discusses how business plans are made and then identifies specific considerations for nanotech companies. We provide suggestions for entrepreneurs in writing the executive summary and choosing which markets to focus on. We also discuss risk and review issues related to operations of the company. The chapter concludes with a discussion of financial modeling, valuation of nanotech companies, and business plan outsourcing.

Chapter 12

In Chapter 12, we take up the issue of financing nanotech companies. We first describe how early-stage companies should seek capital from friends and family, angel investors, and the federal government. We then discuss venture capital. Although investors are increasingly willing to bet on nanotechnology, most start-ups are starving for capital. We detail the current state of venture investing and analyze data on nanotechnology investments. We also profile ten important venture funds that actively invest in nanotechnology companies, identify issues that companies will encounter in seeking venture financing, and provide an overview of the terms and conditions that accompany venture financing.

Chapter 13

While Chapter 5 analyzes broad patent issues facing the PTO, courts, and industry as a whole, Chapter 13 focuses on specific intellectual property issues facing individual companies. We first discuss drafting nanotech patents and obstacles that applicants may encounter in obtaining nanotech patents. We provide a roadmap for companies as they wade through the fragmented patent landscape—and describe how companies should approach patent disputes through different licensing and litigation strategies. Finally, the chapter surveys trademark, trade secret, and copyright issues that might arise for a start-up company.

Chapter 14

Chapter 14 engages in a detailed analysis of corporate partnering and globalization trends in nanotechnology. Because nanotech is primarily an enabling technology, finding a corporate partner is essential for many start-ups. With so much change and uncertainty in this field, developing successful corporate alliances is challenging. We describe current partnerships between start-ups and large companies in nanotech and draw lessons on risks and opportunities. We also provide a brief overview of the terms of partnership deals for nanotech companies. The last part of the chapter explores issues that nanotech companies will encounter in the global marketplace.

Chapter 15

In Chapter 15, we explore the case for consolidation and standardization in nanotech. Similar to the biotech and Internet waves, nanotech development will give rise to highly competitive industries. Competition between start-ups could stifle commercialization. In some cases, nanotech start-ups should consider using mergers in order to reduce duplicative research, avoid costly patent litigation disputes, and improve their negotiating position with suppliers and buyers. Like consolidation, standardization would aid companies engaged in nanoscale research. Standards are needed to allow different players to consistently characterize nanomaterials and develop materials, devices, and systems that are interoperable.

Chapter 16

Chapter 16 analyzes the exit opportunities available for nanotech companies. The field is so new that, at the time of this writing, it is unclear exactly what the exit strategies for nanotechnology start-ups will be—it is unknown if many of these companies will go public, be acquired, or go bankrupt. We first survey the considerations involved in planning for an IPO or acquisition. We then provide a brief overview of the issues that must be tackled for a start-up to successfully execute its strategy.

Chapter 17

The book concludes with a summary of the multitude of issues facing nanotech start-ups and offers some general strategies for companies to adopt. We identify five main lessons that can help guide prospective entrepreneurs, lawyers, and managers in dealing with the new world of nanotechnology.

PART I

Introduction to Nanotechnology

Part I of this book provides a description of nanotechnology and how it will be commercialized. In Chapter 1, we engage in a technical discussion of nanotechnology and its different applications. In Chapter 2, we construct a model of the industrial structure that is giving rise to nanotechnology. Understanding the materials in these chapters is crucial for readers who wish to clearly digest the concepts and ideas presented in the remainder of the book.

CHAPTER 1

Understanding Nanotechnology

Everything should be made as simple as possible—but not simpler.[1]
—Albert Einstein

In order to explore the business, policy, and legal issues associated with an emerging technology, it is necessary to have a grasp of the scientific underpinnings and potential applications of the technology. This is especially important in the context of nanotechnology, where rhetoric in the popular press has blurred the line between fact and science fiction. This chapter attempts to define what nanotechnology is, explore the history of the field, and then provide a lucid but technical description of the science and some of its potential applications. We hope that it is specific enough to serve as a reference for existing technology and yet general enough that readers may apply its overarching framework in the coming years.

DEFINING NANOTECHNOLOGY

Nanotechnology involves the investigation and design of materials or devices at the atomic and molecular levels. One nanometer, a measure equal to one-billionth of a meter, spans approximately 10 atoms. Formulating a precise definition of nanotechnology, however, is a difficult task. Even scientists in the field maintain that it "depends on whom you ask."[2] Biophysicist Steven M. Block notes that some researchers "reserve the word to mean whatever it is they do as opposed to whatever it is anyone else does."[3] For example, some researchers use the term to describe almost any research where some critical size is less than a micron (1,000 nanometers) while other scientists reserve the term for research involving sizes between 1 and 100 nanometers. There is also debate over whether naturally occurring nanoparticles, such as carbon soot, fall under the rubric of nanotechnology. Finally, some reserve the term "nanotechnology" exclusively for manufacturing with atomic precision whereas others employ the term to describe the use of nanomaterials to construct materials, devices, and systems. According to the Foresight Institute, a nonprofit organization dedicated to preparing society for nanotechnology, molecular nanotechnology "will be achieved when we are able to build things from the atom up, and we will be able to rearrange matter with atomic precision."[4] The National Science Foundation, on the other hand, defines nanotechnology as "research and technology development at the atomic, molecular or macromolecular levels, in the length scale of approximately 1–100 nanometer range, to provide a fundamental understanding of phenomena and materials at the nanoscale and to create and use structures, devices and systems that have novel properties and functions because of their small and/or intermediate size."[5] Rather than adopt one of the preceding definitions to confine our discussion of nanotechnology, we survey nearly anything and everything that has been described as nanotechnology and provide a new and useful framework for understanding different types of nanotechnology.

UNDERSTANDING NANOSCIENCE

The dawn of the journey into the nano world can be traced back to 1959, when Caltech physicist Richard Feynman painted a vision of the future of science. In a talk titled "There's Plenty of Room at the Bottom," Feynman hypothesized that atoms and molecules could be manipulated like building blocks.[6] The first "proof-of-principle" that atoms could be precisely positioned by a manmade tool (living cells have, of course, been positioning atoms since time immemorial) took place in 1989 when scientists at IBM manipulated 35 xenon atoms to form the letters IBM (Figure 1.1). In the last

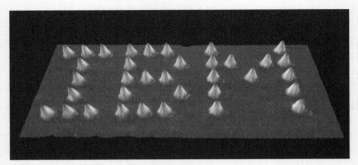

Figure 1.1 Xenon atoms on a nickel substrate positioned by STM. Courtesy: IBM Research, Almaden Research Center. Unauthorized use not permitted.

few years, exploration within the field of nanotechnology has ramped up substantially.

The nano world is full of surprises and potential. In this realm, the disciplinary boundaries between chemistry, molecular biology, materials science, and condensed matter physics dissolve as scientists struggle to understand new and sometimes unexpected properties. Although these professionals are only on the first leg of the journey, they have made significant progress in synthesizing and understanding the "building blocks" of nanotechnology. In the coming years, the ability to utilize these building blocks for practical purposes will greatly increase. Let us first survey the building blocks of nanotechnology before turning to the potential applications.

The Building Blocks

Throughout this book, we use the term "building blocks" to describe the nanomaterials that can be positioned and manipulated for a variety of different applications. The analogy of building a house is appropriate to understanding nanotechnology. Houses can be comprised of a variety of materials: wood, nails, sheet rock, bricks, and so on. Just as a builder puts together different shapes and pieces of these materials to construct a home, nanotechnologists experiment with a variety of different nanomaterials to build complex materials, devices, and systems.

Atoms are the most basic units of matter. They can be combined to form more complex structures such as molecules, crystals, and compounds. Nanomaterials are arrangements of matter in the length scale of approximately 1 to 100 nanometers that exhibit unique characteristics due to their size. Fabrication, or the making, of nanomaterials falls into one of two categories: *top-down* or *bottom-up*.

The top-down method involves carving nanomaterials out of bulk materials.[7] Approaches in this category are referred to as different forms of lithography. Lithography can be understood through the concepts of *writing* and *replication*.[8] Writing involves designing a pattern on a negative (usually a mask), and replication involves transferring the pattern on the negative to a functional material. There are several types of lithography. Photolithography, which uses different kinds of electromagnetic radiation, is currently used to manufacture computer chips and other microelectronic devices.[9] Photolithography, as currently used, is not an effective tool for fabricating structures with features below 100 nanometers. E-beam lithography, a technique that employs beams of electrons to write, can produce some nanostructures with high resolution, albeit in an essentially serial fashion.[10] Soft lithographic techniques, such as printing, molding, and embossing, involve the physical or chemical deformation of the functional material to yield the desired structure. While soft lithography can be used to construct less planar nanostructures, it may be less precise than other techniques. A novel approach, which

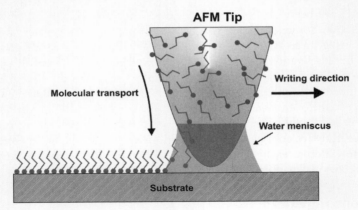

Figure 1.2 "Dip Pen" nanolithography.
Reprinted with permission, Chad Mirkin, Northwestern University.

is conceptually different from conventional lithography, is "dip pen" lithography, a technique developed by Chad Mirkin's lab at Northwestern University. As seen in Figure 1.2, different types of molecules can be placed on a nano-sized probe. Water molecules between the probe and a gold substrate act as a bridge over which the molecules are transferred from the probe to the substrate, thus creating a pattern.

A second method of producing nanomaterials, known as the bottom-up approach, describes techniques for coaxing atoms and molecules to form nanomaterials. One bottom up technique, referred to as "positional assembly," involves using a probe to move atoms into certain arrangements. The use of an atomic force microscope to individually position xenon atoms to spell "IBM" is an example of this approach. Although positional assembly allows control over individual atoms, it is time-consuming, cannot presently be used to create complex nanostructures, and does not represent an efficient means for commercial production. Positional assembly, as realized today, is largely a *serial* process: Each step is performed after the previous one is completed. Photolithography, by way of contrast, is a massively *parallel* procedure—a very large number of features are created in each step. Both methods are, however, largely restricted to *planar* constructions or stacks thereof.

Another bottom-up approach is chemical self-assembly. Different atoms, molecules, or nanomaterials are mixed together and, because of their unique geometries and electronic structures, spontaneously organize into stable, well-defined structures. Because self-assembly methods are based on chemical reactions, they are simple and relatively inexpensive. However, they do not offer the precision necessary for constructing designed, interconnected patterns that top-down approaches currently do. Different categorization schemes have been used to describe the building blocks of nanotechnology. For example, some scientists categorize the building blocks into "soft" and "hard" categories. We describe two different, popular ways of classifying nanomaterials. Nanomaterials are often classified in the literature based on dimensionality. Crucial to this classification is the concept of *confinement*, which may be roughly interpreted as a restriction in the ability of electrons to move in one or more spatial dimensions. 0-D nanomaterials, such as quantum dots and metal nanoparticles, are confined in all three dimensions. 1-D nanomaterials are confined in two directions and extended in only one: electrons flow almost exclusively along this extended dimension. Examples of one-dimensional nanomaterials are nanotubes and nanowires. Finally, 2-D nanomaterials, which are confined in one dimension and extended in two, include thin films, surfaces, and interfaces. Interestingly, material structures currently used as elementary semiconductor devices fall under this category. Nanomaterials can also be divided into inorganic and organic classes.

Inorganic Nanomaterials

The term *inorganic nanomaterials* describes nanostructures in which carbon is not present and combined with some other element. We discuss four types of inorganic nanostructures: fullerenes and carbon nanotubes, nanowires, semiconductor nanocrystals, and nanoparticles.

Fullerenes and carbon nanotubes are the most well-known inorganic nanostructures.[11] Fullerene, formally known as buckministerfullerene, was a new form of carbon discovered by Richard Smalley in 1985.[12] "Buckeyballs," as they are called, are molecules comprised of 60 carbon atoms and have the symmetry of soccer balls. The discovery of fullerenes sparked a raging fire of enthusiasm in the scientific community. It was predicted that the unique properties of fullerenes could be leveraged in everything from windshields to medicine. Although buckeyballs still hold great promise in nanotechnology, the spotlight has shifted to a relative of the fullerene molecule: carbon nanotubes.

Carbon nanotubes, first observed by Sumio Iijima in 1991, are tubular structures that can be thought of as "rolled-up" layers of interconnected carbon atoms. The arrangement of such atoms, because of the electronic structure of carbon, is graphically depicted as a network of hexagons: The lines that form the hexagons represent bonds between adjacent carbon atoms. There are two main types of carbon nanotubes. Multi-walled nanotubes (MWNTs), discovered in 1991, contain a number of hollow cylinders of carbon atoms nested inside one another.[13] Single-walled nanotubes (SWNTs), first synthesized and observed in 1993, consist of a single layer of carbon atoms and a hollow core.[14] (See Figure 1.3.)

Both types of nanotubes are narrow and long and exhibit unique electrical, mechanical, and thermal properties. For example, depending on size and

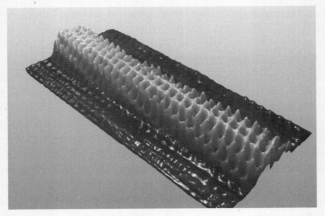

Figure 1.3 STM image of a single-walled carbon nanotube. Reprinted with permission, Cees Decker, Technical University Delft.

shape, nanotubes can display a range of different conducting properties between metallic and semiconducting.[15] Further, nanotubes have a current carrying capacity of one billion amps per square centimeter while copper wires burn out at one million amps per square centimeter.[16] They also have more than 20 times the tensile strength of high-steel alloys,[17] but are lighter than aluminum.[18] Finally, it is estimated that nanotubes can transmit nearly twice as much heat as pure diamond[19] and are likely to remain stable in higher temperatures than metal wires.[20] At the time of this writing, researchers do not have substantial control over the synthesis of carbon nanotubes. The potential of nanotubes to serve as reliable building blocks is largely contingent on the ability to precisely engineer their size and properties.

Nanowires, also known as "nanorods" or "nanowhiskers," are another potential inorganic building block in nanotechnology. Nanowires are solid wires made from silicon, zinc oxide, and various metals. While their diameters are in the nanometer range, they can have lengths in the tens of micrometers. Nanowires have unique optical and electrical properties that, like those of nanotubes, emerge primarily from their low dimensionality. For example, they can emit laser light, act like optical fibers, and change conductance when bound to different molecules.[21] Unlike carbon nanotubes, researchers have today substantial control over the growth of nanowires.

Semiconductor nanocrystals (Figure 1.4), which are sometimes referred to as quantum dots, are fabricated by both lithography and several different self-assembly methods. Researchers are currently exploring the electrical and optical properties of quantum dots.[22] Altering the sizes of quantum dots can alter the wavelengths of light they can be made to emit.

Other types of inorganic nanoparticles, such as metals, oxides, glass, and clay, are also being developed and researched. They have been produced

Figure 1.4 Vertical quantum dots of different shapes.
Reprinted with permission, Leo Kouwenhoven, Technical University Delft.

using both top-down and self-assembly methods.[23] These nanomaterials can have superior properties to their bulk counterparts. For example, nanostructured alloys can be designed to exhibit a greater toughness and creep resistance than conventionally-manufactured alloys.

Organic Nanomaterials

Organic nanomaterials are compounds containing the element carbon. Chemists have long been able to synthesize small complex molecules. Recent advances enable researchers to create organic nanomaterials with specific atoms, geometries, and electronic arrangements. Several different types of organic nanostructures are being tested as potential building blocks. First, some researchers are experimenting with deoxyribonucleic acid (DNA).[24] The molecule is relatively rigid, and several strands can be combined to increase its stiffness. Artificial, repeatable DNA sequences can self-assemble into geometric structures (see Figure 1.5). Researchers have engineered cubic and truncated eight-sided structures from DNA[25] and, more recently, a complete eight-sided structure that is responsive to cloning and, thus, fast replication.[26]

Proteins, which are basic materials of living organisms, might also serve as building blocks. DNA contains the blueprints for proteins. Some scientists are experimenting with altering the DNA of cells to produce proteins that incorporate amino acids not found in nature.[27] Other researchers are experimenting with modified proteins that can form different nanostructures. For example, a group at NASA has shown that "heat shock protein 60" can be induced to self-assemble into tubes, after which the tubes associate to form filaments.[28] Dip pen lithography has also been used to generate protein nanoarrays.[29] Protein-based nanostructures can be made to conduct electricity and might be used in a number of different applications.[30]

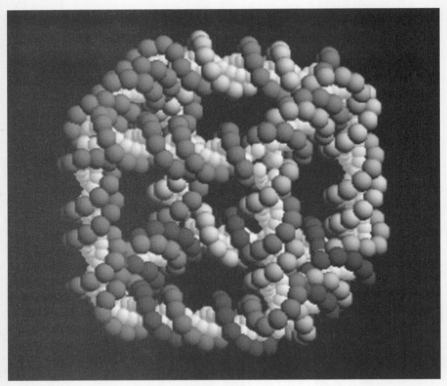

Figure 1.5 Representation of a DNA cube.
Reprinted with permission of Nadrian C. Seeman, New York University.

Researchers also have begun experimenting with viruses and virus fragments as potential building blocks for nanotechnology. Viruses are readily available in very large quantities and possess the three-dimensional structure and chemical reactivity to make them suitable templates for building nanoscale devices.[31] Figure 1.6 shows electrically conducting clusters of virus particles (the large, grayish circles) stuck together with small gold particles (the small, bright circles). The viruses are genetically engineered to have sulfur atoms on their surfaces, which stick very well to metallic gold. The different clusters are formed spontaneously when gold and virus are mixed in different amounts.

Other researchers are working with a variety of different types of polymers. For example, block copolymers are formed by combining chemically different polymer species. Altering parameters such as temperature or pressure can cause the copolymers to spontaneously self-assemble into different morphologies.[32] One class of polymers that has received a great deal of attention is dendrimers. They are treelike molecules that can be made to function like a variety of biological structures.[33] They have surface properties that allow them to bind to other molecules and can carry molecules internally.

The Tools

Fullerenes, nanotubes, nanowires, semiconductor nanocrystals, nanoparticles, and polymers are examples of building blocks in nanotechnology. However, returning to the earlier analogy, a builder who has the necessary raw materials (wood, bricks, and so on) is helpless without tools to put together the materials in a fashion that results in a home. Blueprints are necessary, as well as the physical equipment such as hammers, saws, drills, tape measurements, and so on. Similarly, developing materials and devices based on nanomaterials requires the ability to model, observe, and position nanomaterials. Nanotechnologists employ computational tools as well as laboratory tools.

Computational nanotechnology involves designing and modeling nanomaterials and devices. As computational models enable researchers to model experimental results and predict new phenomena, this enterprise plays a critical role in nanotechnology.[34] There are several different computational methods being developed and integrated in different software packages.[35]

There are a variety of tools used by experimentalists to prepare, characterize, manipulate, and test nanostructures. For example, the scanning tunneling microscope (STM) allows researchers to view nanostructures by

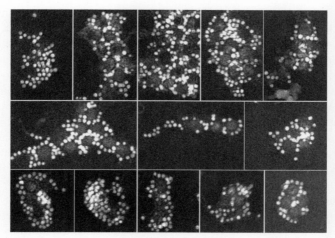

Figure 1.6 Electrically Conducting Clusters of Virus Particles.
Reprinted with permission, M. G. Finn, The Scripps Research Institute.

measuring small currents passing between the microscope's tip and the sample under evaluation.[36] Both the STM and the atomic force microscope (AFM) can be utilized to position nanostructures;[37] however, as mentioned before, they are limited on the vertical dimension and also are relatively slow and impractical for large-scale production. Scientists are developing "nanotweezers" to enable researchers to grab nanomaterials, while researchers can move nanomaterials using a movable platform.[38] Rays of light, known as "optical tweezers," are also used to manipulate nanoparticles.[39]

Applications of Nanotechnology

We identify and describe three general classes of nanotechnology applications based on the degree of control over the synthesis, characterization, and positioning of nanomaterials. First, we use the term *simple nanotechnology* to describe applications involving mass production of nanomaterials. Commercial products based on simple nanotechnology do not involve precise fabrication and positioning of nanostructures. We describe the second class of nanotechnology applications as "building small." This category refers to the use of nanomaterials to build advanced materials, devices, and systems. Within the next 5 to 15 years, "building small" nanotechnology could have a major impact on a number of different products in a range of different industries. We term the final class of nanotechnology applications "building large." This category describes the as-yet-unrealized vision of self-replicating nanorobots.

"Simple Nanotechnology"

We refer to products as "simple nanotechnology" when they are manufactured through mass production and dispersion of nanomaterials in a random fashion. In other words, there is no precise fabrication and positioning of nanostructures. Examples of simple nanotechnology are the use of nanomaterials as catalysts and coatings and in composites and textiles.

Catalysts are substances that regulate the rate at which chemical reactions proceed. When the catalytic rate varies with the surface area of catalysts, nanoparticles that present a large surface area can serve as excellent catalysts in certain reactions. For example, nanoscale metal oxides are currently being utilized in catalytic converters.[40] Aluminum nanoparticles can be found in energetics to enhance the performance of rocket propellants and as lead-free primers in explosives.[41] Finally, researchers are utilizing nanoparticles to develop technology that converts coal directly into liquid fuels.[42]

Coatings and films, traditionally composed of epoxies and paints, are put on objects to make them durable and give them other qualities. Nanofilms serve as invisible coatings that are more durable and cost-effective than traditional coatings. Coatings comprised of nanoparticles can be extremely rough or slippery, or exhibit unusual properties, such as altering color when an electric current is applied. Recent applications of nanoparticles include coatings on walls that make them more resistant to graffiti, a wax used by skiers to increase traction, and transparent suntan lotion.[43] Glassmakers have even begun selling self-cleaning windows coated with dirt-repelling nanoparticles. It is important to distinguish these "macroscopic" thin films from sophisticatedly engineered surfaces or layers—called thin films in a different context—used in certain devices in which the atomic structure is extremely well controlled, such as thin layers of semiconductor materials in devices.

Composites are combinations of materials differing in combination and form. The use of nanomaterials in composites can increase their mechanical properties, decrease their weight, enhance their chemical and heat resistance, and alter their interaction with light and other radiation. As such, they are likely to enhance metals, plastics, textiles, and so on. For example, ceramic composites made of nanoparticles that afford superior performance may be applied as protective coatings in environments subject to harsh thermal and mechanical conditions.

"Building Small" Nanotechnology

We describe the second class of nanotechnology as "building small," because it primarily involves using nanomaterials to construct novel materials, devices, and systems. Unlike "simple" nanotechnology, "building small" requires the ability to precisely fabricate and position nanostructures. The ability to leverage the unique mechanical, electrical, chemical, and optical properties of different nanomaterials could have a major impact on a number of different products in a range of industries. The following discussion provides a brief explanation of some of the different types of products that will be impacted by "building small" nanotechnology. The products we discuss can loosely be grouped into six different classes: sensors and measurement, electronics, communications, energy, life sciences, and aerospace and defense.

Sensors and Measurement: Some of the initial applications of "building small" nanotechnology that will hit the market in the next three to ten years will be a range of different sensing and measurement devices. First, nanostructures are being used to develop better chemical sensors. Such devices can be used for leak detection, medical monitoring, environmental hazard monitoring, and industrial control. Nanotubes and nanowires can serve as the basis for these sensors, because they change their electrical resistance when exposed to alkalis, halogens, and other gases. Several start-up companies are racing to bring nanotechnology sensors to market that are smaller, more sensitive, and use less power.

Nanostructures are also being used to improve biological detection. For example, different-sized quantum dots can be put together in various combinations in latex beads to produce a large number of distinct labels. Each bead can be attached to a different gene. When the "library" of gene-bead structures is exposed to a sample of DNA, the complementary genes bind, and researchers can determine which genes are present in the sample.[44]

In the long run, nanostructures might be used in advanced spectroscopic devices that measure minute concentrations of molecules in different settings. For example, airborne particulates absorb electromagnetic radiation at specific wavelengths. By using quantum dots, researchers could tune a laser to the wavelength at which a certain particle absorbs radiation. Shining the beam through the air would enable researchers to measure the amount of radiation absorbed.

Electronics: A second general category of products that are likely to be substantially improved by "building small" nanotechnology are electronic materials, devices, and systems. We survey the impact of nanotechnology on computer processing, memory, data storage, and display technologies.

Moore's law—an empirical observation rather than actual physical law—holds that the number of transistors that can be fabricated on a silicon-integrated circuit doubles every 18 to 24 months. Microelectronics has progressed along this path for nearly forty years. Modern chips, which are a few square centimeters in size, hold approximately 100 million transistors. Within the next 10 to 12 years, however, silicon electronics will be unable to increase computing speed at the current rate. Stray signals on the chip, thermal instability caused by densely packed transistors, and excessive fabrication costs are predicted to crash the silicon wave.[45]

Semiconductor and computer companies such as Hewlett-Packard (HP), Intel, and IBM have begun to research the possibility of using nanotechnology to build chips in the future. Researchers have demonstrated a field-effect transistor based on a semiconducting single-walled nanotube.[46] In July 2001, a research team at UCLA and HP revealed an electronic switch consisting of a layer of several million rotaxane molecules.[47] Charles Lieber's team at Harvard University announced in January of 2002 a transistor comprised of a silicon nanowire and a gallium nitride nanowire.[48] Cees Dekker's group at Delft University of Technology has observed diode-like properties on a single carbon nanotube that has imperfections in the atomic network.[49] (See Figure 1.7.) And not long before this book went to press, researchers at Stanford University and UC Berkeley created the first integrated silicon circuit with nanotube transistors.[50] The size, cost, and ease of fabrication of nanoscale building blocks make it possible to engineer system architectures with redundancy, an engineering quality that allows for the failure of several individual components without impairing the overall performance of the system.[51]

In addition to making more advanced forms of conventional electronic devices, nanostructures could spawn the creation of entirely novel devices. For example, in a certain regime of transport, electrons maintain their angular momentum, or "spin," as they travel through a nanotube. Spin, much like charge, is a physical quantity that could encode and/or process information. Researchers are attempting to leverage this property by constructing "spintronic" devices that switch on or off in response to electron spin.[52] This approach contrasts with traditional electronic devices, which turn on and off in response to electric charge.

Ultimately, the use of nanotechnology to make smaller and denser circuits could lead to "artificial brains" that have intellectual capabilities comparable—or even superior—to those of human beings."[53] Of course, the obstacles on the path to achieving such a goal are monumental. Indeed, several challenges must be overcome before devices based on nanostructures can have a revolutionary impact on computer processing. Researchers must discover how to interconnect nano-sized transistors into complex circuits—a growing challenge even in silicon technology. Further, they must find a mechanism that can mass-produce complex circuits in an efficient and cost-effective manner.

While nanotechnology is unlikely to be used to develop new devices and systems for computer processing in the near future, it could have a more immediate impact on memory devices. Conventional dynamic random-access memory (DRAM) is the short-term electronic memory that a computer uses to run its software. Because information is only stored as long as there is power, stored information must be loaded from the hard drive onto memory every time the machine is turned on. Nano devices are being tested to create new types of RAM that are nonvolatile—that is, that preserve information

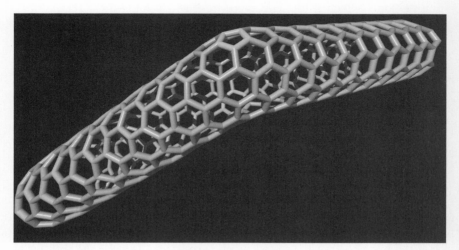

Figure 1.7 Representation of a "kinked" nanotube that exhibits diode-like properties.
Reprinted with permission, Cees Dekker, Technical University Delft.

even when the power is turned off.[54] Such devices could conceivably replace current RAM and Flash technologies in the future.[55]

The very concept of nonvolatile memory and the expected storage densities of these devices blur somewhat the distinction between "memory" and "data storage." Nonetheless, applications of nanotechnology to traditional data storage applications such as hard disks have yielded fruitful results: For example, in 2002, IBM announced that it could pack a trillion bits of data onto a chip the size of a postage stamp.[56]

The arena of electronics in which nanotechnology is likely to have the greatest impact in the nearest future is display technologies. Historically, television screens and desktop computer displays used cathode-ray tubes, in which electrically heated wires shoot electrons onto a phosphor-coated screen. Liquid crystal display technology has led to the introduction of flat panel displays for a variety of different devices. Organic light-emitting diodes (OLEDs) are already being marketed as a superior alternative to liquid crystal displays on account of their lower operation costs, ease of fabrication, and superior usability characteristics. As the screen lives of OLEDs lengthen and the costs of production decrease, OLED-based displays will become more popular.

Communications: A third class of products likely to reap substantial benefits from miniaturization are communications devices. Nanotechnology will enable better optical networks, where information is transmitted by light. Already, a clever combination of self- and directed-assembly is used to fabricate photonic crystals, which are essentially semiconductors of light, as opposed to electrical current.[57] A novel approach to optical devices is the use of plasmonic circuits, which utilize light to induce fluctuations in the charge density of a nanostructure and, therefore, the transport of energy and information. This technology would allow for a dramatic scaling of optical components.[58]

Energy: Traditional energy supplies could reach some significant limits in coming years.[59] "Building small" nanotechnology, though, is likely to revolutionize energy conversion and storage. Technology that would enable exploitation of alternative energy sources to supply electric power on a large scale at a lower cost than oil could play a key role in the future of humanity.

Examples of such technology are photovoltaic and photochemical cells. Solar energy can be used to produce electric power directly through the use of photovoltaic cells. It can also be used to produce fuel—hydrogen—by splitting water molecules in a process that closely parallels that of photovoltaic conversion. Nanostructured photovoltaic cells, which include dye-sensitized and organic cells, essentially divide the tasks of absorbing light and generating current between two different entities.[60] This results not only in greater durability but also in an enormous cost advantage vis-à-vis other technologies. The practical challenge that organic cells face for large-scale production is to achieve efficiencies at least three times greater than are possible today. This task, despite several fundamental phenomena that must be better understood, seems within technological reach.[61]

The storage of hydrogen, whether produced by using solar energy to cleave water molecules or, for example, by recuperating it from biological waste, is an area that could also benefit from nanotechnology. Several nanostructures have been studied[62] and some have been successfully demonstrated in the laboratory.[63] Nanomaterials have also been advocated for lithium-ion batteries,[64] now a ubiquitous technology. It is envisioned that large-scale energy storage—necessary to enable solar technology as a mainstream source of electric power—could be accomplished through the use of superconductors, although many fundamental problems must be solved before such use becomes feasible.[65]

Medicine: Some promising applications of nanotechnology are also likely to bear fruit in the world of medicine. Nanotechnology could spawn new drugs, drug delivery systems, diagnostic devices, materials for tissue engineering, and other devices.

Researchers are experimenting with nanomaterials to develop a variety of new drugs. For example, NanoBio Corporation is developing an antimicrobial material that is effective against a wide range of microbial pathogens. The company is in clinical trials for use of the material to treat herpes and toenail fungus and expects to bring products to market within five years. Another company, C Sixty Inc., is investigating the use of the fullerene molecule as a drug. Fullerene can interact with cells, proteins, and viruses, and can be altered to perform specific tasks. The company hopes to develop several therapies, including a novel treatment for HIV.

Nanotechnology could also give rise to new mechanisms for delivering drugs. The most basic drug delivery systems under development enhance the effectiveness of drugs by targeting certain types of cells, speeding up delivery time, and preventing digestive enzymes from breaking down the medication. Researchers are also experimenting with more advanced delivery systems, such as a dendrimer device that can infiltrate cells and detect premalignant and cancerous changes in the cells, release a chemical substance to kill the cell, and verify destruction of the cell by becoming fluorescent in the presence of enzymes released by fatally wounded cells. Dendrimers might also be used as delivery vehicles for introducing genes into cells in the body to treat different diseases. For years, researchers have been experimenting with gene therapy procedures that involve wrapping genes in viruses or coatings of fat, but these methods often elicit dangerous immune responses. Because dendrimers are so small, they may be able to insert a gene into a targeted cell without provoking an immune reaction. Other researchers are developing implantable devices that can periodically dispense medicines, such as insulin or morphine. These devices, composed of copolymer-nanoshell composites, are capable of holding medicine. When the nanoshells are exposed to infrared light, the drug is released into the surrounding tissue.[66]

In conjunction with yielding better drugs and drug delivery systems, nanotechnology could significantly improve diagnostic capabilities. One way that it could accomplish this is with devices that perform highly parallel analysis of genetic material. High-throughput assaying is likely to be a valuable aid in identifying variations in the sequence of DNA and thus diagnosing diseases of a hereditary nature. Rapid analytical methods for characterization will also be valuable in studying protein structure and function in order to arrive at a molecular-level understanding of cellular behavior. Finally, molecular assaying will also enable the discovery of highly specific drugs. [67]

Alternatively, nanostructures could be used to build noninvasive devices that can enter the body to determine glucose levels, distinguish between normal and cancerous tissue, and provide genetic screening for multiple diseases. For example, researchers are working with a nanoscale needle that can probe cells for carcinogenic chemicals.[68] Exploratory research in this area includes a pill that would travel through the body and provide a comprehensive diagnosis of the patient's health.[69]

Nanotechnology could also enhance tissue engineering and cell therapy, which involve the use of living cells and other natural or synthetic compounds to develop implantable parts for the restoration, maintenance, or replacement of the body's tissues and organs. To treat patients whose pancreatic cells do not produce enough insulin, researchers have experimented with implanting insulin-producing cells from a pig. The primary problem associated with such a procedure is that the immune system attacks the foreign pig cells. Researchers are conducting clinical trials using a silicon capsule with nano-sized pores that prevents the immune system from identifying the foreign cells. The pores, which are only a few nanometers wide, are small enough to screen out the antibodies employed by the immune system while large enough to allow insulin molecules to exit into the bloodstream.[70] Nanoporous fabrication technology could also be used to direct the growth of tissue[71] and facilitate the integration of synthetic materials into the human body.[72]

Nanomedical research could also result in an array of different medical devices. In the short run, it seems likely that surgical tools will be enhanced by nanotechnology. For example, nanotechnology has resulted in a surgical scalpel based on a nanostructured diamond that slices more neatly into eyeballs.[73] In the more distant future, nanotechnology could result in miniature devices that can be implanted to correct auditory, visual, and sensory impairment. For example, researchers are working on a tiny film that can be implanted on the retina of a blind patient, where it absorbs light and delivers electrical signals that are relayed to and interpreted by the brain.[74] This might emulate to some extent the sensory input from eyesight.

Aerospace and Defense: A final class of products that will be improved by "building small" nanotechnology are products in the aerospace and defense sector. To some extent, the impact of nanotechnology on this class of products will primarily be the culmination of the impact of nanotechnology on other industries. NASA perceives miniaturization as the key to exploring new frontiers in space. The agency envisions that future spacecraft will be comprised of ultrasmall sensors, advanced electronic and photonic systems for communication and navigation, lightweight and strong surface materials, and highly efficient power sources. Meyya Meyyappan, director of the Center for Nanotechnology at the NASA Ames Research Center, declares that nanotechnology "presents a whole new spectrum of opportunities to build device components and systems for entirely new, bold space architectures."[75]

The Defense Department is also making substantial investments in nanotechnology. For example, the Army contracted with Massachusetts Institute of Technology (MIT) to design futuristic combat gear for American soldiers.[76] The envisioned battle suits will purportedly change color to blend in with the surrounding environment and transform from a soft fabric to bulletproof armor. Sensors in the suit would detect when the soldier is wounded, devices would transmit vital signals to a distant medic, and antidotes would be released as needed.

"Building Large" Nanotechnology

We refer to the final class of nanotechnology as "building large," because it attempts to revolutionize macroscopic manufacturing capabilities.[77] Molecular manufacturing, as this type of nanotechnology is sometimes called, was conceived by K. Eric Drexler at MIT in the late 1970's.

Drexler has predicted that molecular nanotechnology will eventually allow scientists to prevent death by cellular repair, build everything from computers to space shuttles, eliminate pollution, rebuild extinct plants and animals, and efficiently produce food to end hunger on the planet.[78] Others have envisioned nanorobots that will travel throughout the human body (Figure 1.8) using molecular motors and computers, treat pathogens, eliminate cancer and HIV as life-threatening conditions, reverse trauma and injury from burns and accidents, enhance mental capabilities and physical abilities, and slow down aging.[79] A common idea in these applications is the concept of *mechanosynthesis*, which is to have positional control over the site of chemical reactions.[80]

Given a description of a structure's atomic constituents and their arrangement, mecanosynthesis would require a tool that could be programmed to perform specific reactions at specific sites, namely binding each atom to its neighbors as per the specification. This tool, called an *assembler*, should be able to position atoms to within a fraction of a nanometer—a rough estimate

Figure 1.8 Artist's impression of a futuristic nanorobot.
Reprinted with permission, (c) ConeyI Jay 2004 www.coneyljay.com.

for the average size of an atomic diameter.[81] As there are by this estimate on the order of 1024 atoms in a cubic inch, a large number of rapid assemblers must be employed to assemble anything of macroscopic proportions. To gain a sense of the scales involved, an example might be useful. If one were to employ a billion such assemblers, and if each took, on average, a billionth of a second per positioning/binding operation, then a rather trivially structured and lightweight cubic inch of matter would result after about eleven days.[82] Employing a thousand assemblers that take a thousandth of a second per operation for the same task would take longer than *twice* the current age of the universe.[83]

To be feasible, then, molecular nanotechnology would need to demonstrate precise, small, fast, inexpensive, and plentiful positioning/synthesizing assemblers. A characteristic that cuts across the last two categories is that of *replication*, the notion that some degree of exponential growth in assembler number—growth that is proportional to the quantity present at any given time—may compensate for the astronomical number of operations that must be performed to assemble a macroscopic product.[84] Indeed, Drexler has claimed that an object weighing one kilogram could be assembled in less than three hours.[85]

Assemblers, as defined by Drexler, are yet to be demonstrated. Certain proponents of manmade molecular machinery also point to living cells as a proof-of-principle for the central idea of building complex microscopic structures from elementary precursors. The way in which biological systems accomplish this, however, is fundamentally different from the mechanical, "dry" approach advocated by Drexler.[86]

Nobel-Prize-winning scientist Richard Smalley has made a vocal case against self-replicating nanorobots, noting that they "will never become more than a futurist's daydream."[87] Smalley's central argument, presented as the "fat fingers" and "sticky fingers" problem, is that the attempt to position individual atoms to guide chemical reactions will fail because unwanted reactions—with other neighboring atoms—will occur. Drexler has countered that his approach involves molecules, not single atoms; furthermore, he argues that if such unwanted reactions were to happen under particular conditions, then it would be necessary to turn to other configurations of building blocks. Smalley finalizes his argument by saying to Drexler that, "in all of your writings, I have never seen a convincing argument that this list of conditions and synthetic targets that will actually work reliably with mechanosynthesis can be anything but a very, very short list."[88] In the end, however, the feasibility of assemblers is not a philosophical issue. Scientific questions are only settled by experiment, and the physical realization of Drexler's ideas will be the ultimate test of their validity.

Meanwhile, there are other obstacles in the path to molecular nanotechnology. On one hand, a number of influential researchers seem to be hostile to Drexler's ideas. If this indifference is shared by a large group of scientists, it might significantly stall technological progress towards the eventual construction of an assembler. On the other hand, even if an assembler is technically possible, other factors might hinder its development. The sum total of the decisions society makes will determine the incentives, opportunities, public interest, and risks for "build large" nanotechnology. In the end, if mechanosynthesis is ever demonstrated, social forces will likely have a substantial hand in its development.

CHAPTER 2

The Industrial Structure Giving Rise to Nanotechnology

We don't make the products you use, we make the products you use better.
—BASF Commercial

The phrase is a familiar one. "We don't make the products you use, we make the products you use better." Although used in commercials by BASF, a company that adds value to existing products with its chemicals and plastics, this phrase could also be used to describe nanotechnology companies. Nanotechnology will serve to improve a number of different types of products in different industries. As we saw in Chapter 1, nanotechnology could yield stronger and lighter materials, faster and smaller computers, and more effective drug delivery devices. One venture capitalist argues that nanotechnology "is not an industry, but an enabling technology that will impact a number of industries."[1] At some level of abstraction, this might be a fair statement. But we are hesitant to embrace this characterization. Although nanotechnology is certainly an enabling technology, there is also an identifiable nanotechnology industry that develops and markets this enabling technology. In this chapter, we define and explain the nanotechnology industry.

In our first attempt to write this chapter, we created a database of several hundred companies investing in nanotechnology and divided them into different categories based on business model and product. However, such a database would become outdated soon after publication. (For readers wishing to obtain a comprehensive list of companies investing in nanotechnology, we recommend consulting *NanoInvestorNews*.[2]) As an alternative to publishing a comprehensive database of companies involved in nanotechnology, we propose an elementary model for understanding the industrial structure giving rise to nanotechnology. As shown in Figure 2.1, the model distinguishes between established companies that are seeking to integrate nanotechnology into their existing technology platforms and companies focused on commercializing nanotechnology.

ESTABLISHED COMPANIES INTEGRATING NANOTECHNOLOGY

There are a number of large companies in established industries that are investigating nanotechnology through internal R&D efforts and partnerships with other companies. They include companies in the materials, computing, life sciences, energy, communications equipment, consumer electronics, and aerospace and defense industries. DuPont, for example, is working with fabrics based on nanotechnology and nanotubes for flat panel displays. In the defense and aerospace sector, Raytheon has launched a nanotechnology initiative. Companies in the energy industry such as Mobil have been experimenting with nanoporous zeolites as catalysts for some time. Roche, a large health care company, is actively researching diagnostics based on nanotechnology and nanoparticles for drug delivery. Perhaps the greatest expenditures on nanotechnology by large companies are taking place in the computing, communications equipment, and consumer electronics industries where companies like IBM, Hewlett-Packard, Intel, Lucent, and Hitachi are pouring substantial sums of money into exploring the viability of different nanostructures, building scanning probes, and developing optical and electronic devices.

Many of these companies have developed fundamental technologies necessary for the advancement of nanoscale science and engineering. However, unlike many nanotech start-ups which are focused on developing and marketing nanoscale materials, devices, and systems, established companies are primarily seeking to integrate nanotechnology into their existing technology platforms. Thus, while these companies are important players in nanotechnology, it is possible to distinguish them from a so-called "nanotechnology industry."

COMPANIES COMMERCIALIZING NANOTECHNOLOGY

The nanotechnology industry can be roughly defined as the group of companies focused on bringing nanotechnology processes, tools, and first-generation materials, devices, and systems to market. Most of these companies today are start-ups. They have sprung up over the past few years, and new ones are incorporating every month. Different people are throwing around various estimates of the number of start-ups in the field. At one conference we attended, the projections ranged from a couple of hundred to 1200.[3]

COMPANIES FOCUSED ON COMMERCIALIZING NANOTECH			COMPANIES INTEGRATING NANOTECH
Tools	Simple	Building Small	
Lithography/ Production *NanoInk* **Manipulation/ Characterization** *Npoint* *Veeco* **CAD Tools** *General Nanotech*	**Bulk Nanomaterials** *Nanophase* **Industrial Coatings** *Inframat* **Catalysts** *Catalytic Solutions* **Textiles** *Nanotex*	**Life Sciences** *Nanosphere* *C Sixty* *NanoBio* **Integrated Systems** *Molecular Nanosystems* *Nanosys* *Nantero* *Nanomix* *Konarka*	**Materials** *DuPont* **Computing** *IBM* *Hewlett-Packard* *Intel* **Life Sciences** *Roche* **Energy** *Exxon Mobil* **Communications Equipment** *Lucent* **Consumer Electronics** *Hitachi* **Aerospace and Defense** *Raytheon*

Figure 2.1 A model of the industrial structure giving rise to nanotechnology.

As nanotechnology progresses and becomes more attractive for investors, start-ups will continue to proliferate.

Most start-ups are spin-offs from research conducted at universities and government labs. The research is usually funded or subsidized by the federal government. When a patentable invention arises from such research, it is often licensed to the inventors to further develop it for commercialization. (In Chapter 10, we explain this process in detail.) Thus, most nanotechnology companies are formed based on pioneering breakthroughs at research institutions such as Stanford, Berkeley, MIT, Harvard, Northwestern, Caltech, and Rice.

Drawing on our description of nanotechnology in Chapter 1, we divide nanotechnology companies into three categories: (1) companies developing "tools"; (2) companies seeking to commercialize "simple" nanotechnology; and (3) companies focused on "building small" nanotechnology. In some instances, determining which category a company belongs to is arbitrary. For example, a company that focuses on developing a research tool in the near term may change its business model to work on producing a downstream product in the future. To the extent that the examples provided appear to neatly fit within one of the two categories, it is because they were chosen to do so. While recognizing that this categorization scheme may greatly oversimplify the current reality, it serves as a useful conceptual guide to the different categories of companies being started.

The "tools" category includes companies whose business plan focuses on the tools to synthesize, manipulate, observe and characterize nanomaterials as well as computational and modeling tools. Several examples are illustrative. NanoInk arose out of Chad Mirkin's work at Northwestern University. It concentrates on developing dip-pen lithography, a technique used to precisely draw nano-sized objects. Veeco, which is a publicly traded company, manufactures tools to measure at the nanoscale and engineer nanoscale devices. Npoint, a company founded by Max Lagally from the University of Wisconsin, produces "nanopositioners" that place and move nano-sized objects. An example of a company focused on producing computational tools is General Nanotechnology.

The "simple" nanotechnology category includes companies marketing bulk quantities of nanomaterials and companies developing novel powders, dispersions, coatings, catalysts, composites, lubricants, chemicals, and textiles. Start-ups in this category have been around longer and, in some instances, already have products. For example, Nanophase, a company that produces nanocrytalline particles that can be dispersed in products such as suntan lotion, was founded in 1989 and was able to complete an initial public offering in 1997. Other examples of start-ups in this category include Inframat, which focuses on industrial coatings, Catalytic Solutions, which uses nanomaterials as catalysts, and Nanotex, which works on improving fabrics.

The "building small" nanotechnology category is comprised of companies seeking to develop devices and systems at the nanoscale. We further subdivided this category into life sciences companies and integrated systems companies. Examples of companies in the life sciences category are Nanosphere, C Sixty, and NanoBio Corporation. Nanosphere uses gold nanoparticles for gene detection and disease diagnosis. C Sixty is experimenting with fullerenes as treatments for cancer and HIV, and the NanoBio Corporation is working with dendrimers for targeted drug delivery.

The integrated systems category includes companies that are attempting to develop optical, electronic, and chemical devices. We grouped all of these companies together, because the devices they are developing might be used in a range of different industries. For example, Nanosys, founded by renowned entrepreneur Larry Bock, focuses on acquiring intellectual property on the synthesis of inorganic nanostructures as well as developing sensors, photovoltaics, and electronic devices. Molecular Nanosystems, a Stanford spinoff, is attempting to use the chemical vapor deposition process for nanotube synthesis to produce sensors and field emission devices.

Nanomix is using nanostructures to develop gas sensors and hydrogen storage materials, and Zia Laser uses quantum dots to build better lasers for communications networks and medical applications. Nantero, a spinoff from MIT, is using carbon nanotubes to develop nonvolatile random access memory. Finally, Konarka uses proprietary nanostructures to create new photovoltaic cells that can more efficiently convert light into energy.

Most nanotechnology start-ups will not attempt to develop and market their own commercial products. Rather, they will seek to partner with large companies in industries that can utilize nanotechnology to improve their commercial products.

IMPLICATIONS OF THE MODEL

Although elementary, this model can be used to make some generalizations about the industrial structure giving rise to nanotechnology. There are a web of complex linkages between academic institutions, start-up companies, and large companies in a variety of different industries. In some instances, start-up companies are competing with each other as well as with large companies attempting to develop similar products. In other instances, start-up companies can form mutually beneficial partnerships with large ones. Understanding this intricate and complicated web through our model will be useful throughout the book.

PART II

Nanotechnology Policy and Regulation

Part I of this book provided a description of nanoscience and its potential applications. In Part II, we explore the policy and regulatory challenges posed by nanotechnology. We examine the societal and ethical implications of nanotech, environmental regulation, the Patent and Trademark Office, the Food and Drug Administration, export control laws, and federal funding. This section is intended for legislators, government officials, and industry groups.

CHAPTER 3

Societal and Ethical Implications

In the Nevada desert, an experiment has gone horribly wrong. A cloud of nanoparticles—micro-robots—has escaped from the laboratory. This cloud is self-sustaining and self-reproducing. It is intelligent and learns from experience. For all practical purposes, it is alive. It has been programmed as a predator. It is evolving swiftly, becoming more deadly with each passing hour. Every attempt to destroy it has failed. And we are the prey.

—Michael Crichton, *Prey*

Before exploring how government can pave a smoother path for nanotechnology, we must ask whether society should travel down this road in the first place. Is a "nano world" desirable? This question does not refer to "simple" nanotechnology in the near term. Rather, it is directed to the long-term visions of a world where any embodiment of nanotechnology is allowed to flourish. This chapter attempts to answer this fundamental question. It first identifies some of the societal and ethical issues posed by nanotechnology, explains the argument for relinquishing it, and points out why policymakers must immediately devote some attention to this issue. It then engages in a careful policy and legal analysis of the call to relinquish nanotechnology. Even if the harms articulated by opponents of nanotechnology are likely to materialize, nanotechnology research and development should not and cannot be prohibited. The most prudent course of action is to cultivate nanoscience while promulgating regulations to prevent the harms. The chapter concludes by offering several specific recommendations in conformity with this general framework.

SOCIETAL AND ETHICAL IMPLICATIONS

Nanotechnology will spawn an array of societal and ethical issues. In the short term, where "simple" nanotechnology will primarily be used to develop new materials applications, there are no real ethical issues. (However, as will be seen in Chapter 4, "simple" nanotechnology does pose health and safety issues). The progression of "building small" nanotechnology in the next several years will raise some difficult questions. For example, how powerful should computer processors be? What are the implications of neurobiochips that could be implanted in humans to stimulate brain function? Will the economic gains generated by nanotechnology further entrench existing social inequalities? The simmering public debate over genetic engineering could explode if nanotechnology enables effective diagnostics of genetic conditions and gene therapy. Perhaps the thorniest issues in the short term involve the effect of nano-sized devices on individual privacy. As explained in Chapter 1, nanotechnology could yield microscopic sensors for different uses. Invisible sensors could be placed in settings and on products to accumulate all sorts of different information. Sensors might also be developed that can probe the body, detect a range of different diseases, and broadcast the information to an external agent. While these issues are almost certain to occupy scholars and policymakers in coming years, they have not yet assumed a prominent role in the societal and ethical debates about nanotechnology. As such, we leave them for a future edition of this book.

In this chapter, we focus on the long-term societal and ethical issues posed by nanotechnology. Although these issues will not have any real impact on society for many years—if ever—they are currently being used in public debates to advocate an immediate ban on nanotechnology research and development. Critics have identified nanotechnology as fueling the march toward mechanization and repression of humans. Bill Joy, of Sun Microsystems, has launched a campaign to illustrate the dehumanizing effects of advanced nanotechnology. In an article titled "Why the Future Doesn't Need Us," Joy argues that the convergence of nanotechnology, robotics, and genetic engineering will yield an age of "spiritual machines."[1] These intelligent machines will ultimately blend with and replace humans. As nanotechnology progresses, this critique of technological progress will grow louder.

In addition to normative concerns regarding the societal changes brought about by nanotechnology, there are also safety concerns associated with nanorobots. Self-replicating nanorobots might be used as weapons of mass destruction. Even worse, a technical error could cause a batch of nanorobots to begin self-replicating out of control and demolish everything in their path.[2] The horrific visions of the nano world have led Joy and others to call for a "relinquishment" of all nanotechnology research and development.

THE NEED FOR IMMEDIATE ACTION

Thus far, the debate over the desirability of a nano world has been largely confined to a handful of participants. The speed at which the field is developing creates an urgent need for well-informed policy analysis. If the opposition to nanotechnology is compelling, the next few years provide a window of opportunity to stop the technology in its tracks. If nanotechnology takes off and becomes an engine of economic growth, it will be virtually impossible to contain it.

If, however, the case against nanotechnology is feeble, policy action is still required in the immediate future. The arguments promulgated by nanotechnology opponents must be shattered before they gain momentum and threaten the development of the technology. Michael Crichton, author of *Jurassic Park*, recently released a new thriller depicting nanorobots invading and taking control of human bodies.[3] Entitled *Prey* and advertised with the slogan "Humanity, get down on your knees," the novel may be turned into a blockbuster film. With media attention like this, nanotechnology could soon be the subject of a public frenzy.

The lesson learned from other emerging technologies is that grassroots opposition can substantially impair an industry. Nuclear power, once lauded as the solution to the energy conundrum, is now seen by many as a technological pariah.[4] Indeed, the introduction of new nuclear power facilities "has been brought to a standstill without any legislative prohibitions or deterrents, but rather by harassment, agitation, and litigation spawned by opposition groups whose efforts have made nuclear power 'too hot to handle' in the political arena."[5] Similarly, the public fear of agricultural biotechnology has crippled that industry. Europe has imposed stringent regulations on DNA research, and European scientists have faced threats of physical violence from militant groups.[6] Although the United States has maintained a favorable regulatory environment, there have been vociferous calls to boycott bioagricultural products.[7] The hostile public and uncertain political environment have deterred investment in the industry.[8] The remainder of this chapter attempts to provide the policy and legal analysis needed to convince policymakers and the public of the need for allowing nanotechnology to flourish.

POLICY ANALYSIS

There are several ways to respond to the call for relinquishment. First, some argue that nanotechnology will never give rise to machines capable of such drastic societal change and safety concerns. As noted in Chapter 1, scientists such as Richard Smalley avidly maintain that self-replicating nanorobots are nothing "more than a futurist's daydream." Further, even if manufacturing self-replicating nanorobots is possible, others have attempted to prove that they could not intentionally or unintentionally cause mass destruction.[9] Scientists have also argued that there is no basis for asserting that computing power will ever be translated into genuinely human capabilities.[10] We refrain from weighing in on this debate. We note, as we did in Chapter 1, that no experimental evidence thus far supports either the claim that manmade nanorobots are feasible or the expectation that nanotechnology will enable artificial computing ability to surpass that of human beings.

Second, a normative defense of technological progress can be offered. Dehumanization is not an inevitable consequence of advances in nanotechnology. Some are still hopeful that humans will continue to be distinct from and dominate the machines that they create.[11] Further, even if the blending of man and machines is inescapable, it is not necessarily insidious. In *The Age of Spiritual Machines*, Ray Kurzweil depicts the synthesis of people and machines as a dignified and even welcome eventuality.[12] According to Kurzweil, the intimate connection between humans and technology could be a natural expansion of human intelligence and capability. We are reluctant to endorse this argument.

While the first two responses to the case for relinquishment are tenuous, the third response is more compelling. Even if there are grave moral and safety hazards associated with advances in nanotechnology, they pale in comparison to the benefits. Technological progress is a double-edged sword—the power to make substantial improvements in the health, welfare, and capabilities of humans is inevitably accompanied by the potential for destruction. If the most imaginative predictions of advanced nanotechnology come true, the world could come closer to eliminating disease, poverty, famine, and ecological crisis. The most ardent proponents of nanotechnology have even suggested that these benefits would decrease the likelihood of violence and international conflict.[13] Throughout history, the benefits of technological progress have far exceeded the harms. As Kurzweil writes:

> If we imagine describing the dangers that exist today (enough nuclear explosives to destroy all mammalian life, just for starters) to people who lived a couple of hundred years ago, they would think it mad to take such risks. On the other hand, how many people in the year 2001 would really want to go back to the short, brutish, disease-filled, poverty-stricken, disaster-prone lives that 99 percent of the human race struggled through a couple of centuries ago?[14]

Time and again, humans have done a decent job of exploiting the benefits of technology while minimizing the harms. We are optimistic that this will hold true for nanotechnology.

Finally, the most powerful argument against relinquishment is that it cannot succeed. Relinquishment would be wholly ineffective without an international agreement. The economic and strategic gains associated with nanotechnology make it extremely unlikely that most nations would agree to a development ban. If several major players in nanotechnology research and development opted not to participate in such a ban, researchers and capital would flood to these countries.[15] Since development of advanced nanotechnology could represent a global threat regardless of where the technology is developed, the ban would fail.

Even if support for an international agreement could be mustered, such an agreement would likely fail. History is replete with examples of international arms control regimes failing to arrest research and development of weapons.[16] The evidence on the inefficacy of international arms control is especially disheartening in light of the particular characteristics of nanotechnology that will make the developments in the industry more difficult to control. There would be a greater incentive to disregard a ban on nanotechnology than other arms control agreements, because the rewards of cheating are greater. Whereas the incentive for violating traditional arms control regimes is primarily strategic, developing nanotechnology could provide a state with economic, strategic, environmental, and health benefits. Further, it would be more difficult to supervise and enforce a ban on nanotechnology research and development. Nanotechnology research "does not require large specialized facilities and multibillion dollar budgets the way nuclear or even chemical research does."[17] Rather, such research can sometimes be "done in basements."[18] Terrorist groups and states ignoring the ban would be able to effectively conceal their research and development efforts. Difficulties with defining prohibited conduct would further undermine enforcement efforts. Nanotechnology has been broadly defined to include a wide variety of research and development and products ranging from photolithography to advanced biotechnology to nanomaterials. To avoid restricting research in these areas, the ban would have to be narrowly tailored by prohibiting, for example, research involving nanorobots. Whatever specific types of conduct are explicitly prohibited, researchers could reclassify their efforts to avoid restrictions.[19]

To the extent that a ban is unenforceable in preventing development of advanced nanotechnology, it could increase the likelihood of an accidental catastrophe or intentional misuse of the technology. Because the technology could be developed on foreign soil under a veil of secrecy, policymakers would be uncertain about the risks of an accident or intentional misuse. Washington would be left guessing whether surreptitious efforts to construct self-replicating nanorobots were succeeding—and if they were, what threat they could pose. Knowledge about the risks of nanotechnology would lead to better policymaking. If policymakers could be assured that self-replicating nanorobots will never be realized, then they would not expend unnecessary time and energy speculating about the potential risks. If, however, scientists could demonstrate that nanorobots are feasible, then policymakers could take appropriate actions to reduce the likelihood of intentional misuse or an environmental catastrophe. With an awareness of the risks of an environmental catastrophe, policymakers could legislate strict guidelines governing research and development of nanorobots. Further, incorporating nanotechnology into American military capabilities could deter others from using nanoweapons.[20] Finally, policymakers could fashion emergency plans and develop technological solutions to mitigate accidents or intentional misuse of nanotechnology. In contrast to a nanotechnology policy fashioned by policymakers armed with the knowledge of the dangers of the industry and the technological base to combat its dangers, uncertainty would leave the United States vulnerable to intentional or unintentional misuse.

LEGAL ANALYSIS

In addition to being ineffective and counterproductive, a policy of relinquishment might be constitutionally impermissible. It is arguable that a ban on nanotechnology research is unconstitutional under the First Amendment.[21] No courts have directly addressed the issue of the constitutionality of government bans on research. The Supreme Court has suggested that scientific expression is speech protected by the First Amendment.[22] Conduct is afforded the same protection as pure speech when conduct is "sufficiently imbued with the elements of communication."[23] Such protected conduct is referred to as expressive conduct or symbolic speech.[24] For example, in *Texas v. Johnson*,[25] the Court held that flag burning is protected by the First Amendment, because it conveys the message of opposition to the government.[26] A ban on nanotechnology research would be unconstitutional if: (1) such research qualifies as speech; and (2) the ban does not serve a sufficiently important governmental interest.

Does Nanotechnology Research Constitute Speech?

Conduct will only be protected under the First Amendment when there is "an intent to convey a particularized message" that is likely to be understood by those receiving the message.[27] Expressing the message need not be the sole or even primary reason for engaging in the conduct.[28] It is arguable that researchers intend to convey two distinct messages when conducting nanotechnology research. First, nanotechnology research is an expression of scientific ideas. Since a scientific concept cannot be expressed scientifically without an experiment to subject the concept to falsification,[29] experimental research likely constitutes the expression of the *scientific* idea.[30] Nanotechnology research is the scientific expression of theories about how the laws of physics operate at the atomic and molecular levels. Second, nanotechnology research is an expression of a particular philosophy. Modern philosophical debates over the wisdom of advancing scientific research pit postmodernism's rejection of technological discourse against liberal humanism's embracing of scientific progress.[31] Scientists researching nanotechnology are expressing their fundamental belief in the ability of science to improve the human condition.

Does the Ban Serve a Sufficiently Important Governmental Interest?

Even if nanotechnology research is protected as symbolic speech under the First Amendment, government regulation is permissible when there is a sufficiently important governmental interest. In *United States v. O'Brien*,[32] the Court outlined a four-part test to determine whether symbolic speech could be restricted without violating the First Amendment:

> (1) if it is within the constitutional power of the government; (2) if it furthers an important or substantial governmental interest; (3) if the governmental interest is unrelated to the suppression of free expression; and (4) if the incidental restriction on alleged First Amendment freedoms is no greater than is essential to the furtherance of that interest.[33]

Banning nanotechnology research would be permissible under the goverment's power to institute regulations to protect public health, safety, and ethics. Such a ban furthers important government interests of preventing environmental harm and intentional misuse of weapons enhanced by nanotechnology, which are unrelated to expression of speech. It is arguable, however, that the restrictions on expression are greater than is essential to furtherance of such interests. A more narrowly tailored regulatory regime that places restrictions on certain types of research or banned the production of certain types of progress could prevent the alleged harms. Ultimately, the constitutional validity of a ban on nanotechnology research and development is uncertain and could turn on how the policy is crafted.

RECOMMENDATIONS

Setting aside the questions of whether a world of intelligent and self-replicating machines is likely or desirable, the ineffectiveness of and legal complications associated with relinquishment dictate that policymakers should not impose severe regulations on nanotechnology. The conclusion that nanotechnology should be allowed to flourish is tempered by the need to regulate the harmful implications of developments in the industry. A prudent course of action might proceed as follows.

First, additional social science research must be conducted to flesh out the societal and ethical implications of nanotechnology. The 21st Century Nanotechnology Research and Development Act requires a study "to assess the needs for standards, guidelines, or strategies for ensuring the development of safe nanotechnology" to be conducted within six years.[34] In addition to this effort, the establishment of nanoethics centers could provide an infrastructure to effectively and continually address new issues posed by nanotechnology. Such centers would conduct studies, make policy recommendations, and serve as a bridge between scientists and engineers, the media, and policymakers.[35]

Second, mechanisms must be established to enable policymakers and industry to maintain a dialogue with the public regarding the safety and ethical implications of nanotechnology. The backlashes against biotechnology and nuclear power were primarily due to the fact that the public was never convinced that the technology was largely benign. As one prominent researcher noted, government must really "be active in educating lay people about nanotechnology."[36] Citizen/scientist panels represent one potential forum for exchanging ideas between the scientific community and the general public. In addition to establishing a climate of trust and openness between technologists and the public, such panels could be used to identify primary public concerns in different arenas of nanotechnology. A working model for a citizen advisory panel is the North Carolina Citizens Technology Forum, which involves online and face-to-face technology forums.[37] Other mechanisms involve education and outreach programs. Middle school, high school, and college students should be exposed to curricula, software, and games that enable them to better understand nanoscience. Additionally, industry groups could sponsor television commercials and IMAX movies that clearly communicate the implications of nanotechnology.[38]

Third, regulations will be needed to address research involving self-replicating nanorobots. Within the next 10 to 20 years, efforts will be made to develop nanorobots. As long as these endeavors do not involve self-replication, they do not present any new risks and should be regulated like any other engineering effort. Few researchers are likely to engage in experimentation involving self-replication. Indeed, influential scientists in the field discount the possibility of self-replicating nanorobots. Nevertheless, in order to quiet public fears of nanorobots, the government should consider establishing regulations on research involving self-replication. In the early days of biotechnology, the National Institutes of Health (NIH) chartered the Recombinant DNA Advisory Committee (RAC) to regulate r-DNA experimentation.[39] Although the RAC guidelines were only binding on NIH-funded research institutions, they were generally followed by private researchers.[40] Something akin to the RAC could be established for regulation of experiments involving self-replicating nanorobots.

In an attempt to preempt a public backlash against nanotechnology, researchers met to promulgate regulations governing research on self-replicating nanorobots. The resulting Foresight Guidelines established development principles and specific design guidelines to prevent accidents as well as intentional misuse.[41] Policymakers could use the Foresight Guidelines as a basis for regulating such research. Thus, regulations might require that experiments only take place in secured, contained facilities.[42] Regulations on the design and construction of self-replicating nano-machines might also be necessary. For example, regulations could ensure that the nano-machines depend on a single artificial fuel source or artificial "vitamins" that don't exist outside of the facility, self-replication only occurs through broadcast transmissions, device instruction is encrypted, and termination dates are programmed into the devices.[43]

Finally, in the long run, if humans are successful in introducing self-replicating nano-machines, a new statutory regime might be needed. The risks posed by self-replicating nanorobots would be similar in magnitude to those posed by the ability to leverage atomic energy and the advancement of chemical and biological capabilities. Policymakers have developed comprehensive statutory responses to ensure that the environmental and security harms associated with these technologies never come to pass. The policies enacted to regulate nanotechnology could be modeled after the chemical, biological, and nuclear regulatory regimes. For example, strict regulatory oversight, international treaties, export controls, and criminal penalties would all be potential tools to prevent the proliferation and accidental misuse of self-replication technology.

CHAPTER 4

Environmental Regulation

More than one CEO has asked "are we sitting on the next asbestos working with all these tiny things?"[1]
—Mark Modzelewski (NanoBusiness Alliance)

As discussed in Chapter 3, most of the opposition to nanotechnology has focused on the long-term risks associated with self-replicating nanorobots. Some environmental groups are highlighting the health and safety risks posed by nanomaterials in the near term. For example, the Action Group On Erosion, Technology, and Concentration (ETC) called for a ban on the commercial production of nanomaterials and the products incorporating them in 2002.[2] As illustrated by a *New York Times* article released in late 2003 entitled "As Uses Grow, Tiny Materials' Safety Is Hard to Pin Down," the environmental concerns are mounting.[3] At the time of this writing, scientists are only beginning to generate answers to the complex toxicological issues presented by nanotechnology.

This chapter engages in a rigorous scientific, normative, and legal analysis of the environmental argument in support of stringent regulations on nanomaterials. Ultimately, it concludes that, even though there are substantial uncertainties concerning the environmental impacts of nanotechnology, there is a strong case for allowing production of nanomaterials without substantial regulations. Further, prohibiting nanomaterials based on the current data may be legally impermissible under statutory and international law. Nevertheless, the Environmental Protection Agency (EPA) should garner more data and subject nanomaterials to additional review procedures.

INCREASING ENVIRONMENTAL CONCERN

ETC argues that mass production of nanomaterials poses a threat to human health for several reasons. First, large amounts of nanomaterials could accumulate in humans. Nanomaterials might be absorbed by bacteria and work their way through the food chain.[4] Second, the substances may bind to common contaminants such as pesticides or polychlorinated biphenyls (PCBs). Since some nanoscale particles can enter cells without eliciting an immune response, they might enable toxins to more effectively penetrate the body's defenses.[5] Third, carbon nanotube fibers may cause serious respiratory problems. Because of their unique, needle-like structure, detached fibers could "wreak havoc" and become "the next asbestos."[6] A casual glance at Figure 4.1 gives the impression that nanotubes resemble asbestos fibers, as viewed under a scanning electron microscope (SEM). (On closer inspection, however, the scale of the photographs reveals that the nanotubes shown are over 60 times smaller than the asbestos fibers.) Finally, when proteins envelop free-floating nanoparticles in the bloodstream, the shape and function of the proteins might change. Altered proteins could cause unintended and dangerous effects, such as blood clotting.[7] Indeed, "[t]he concerns range from cancer or genetic mutation to the risk that materials could get into groundwater and destroy entire ecosystems."[8] The potential environmental risks led ETC to call for a ban on commercial production of nanomaterials and nanoproducts until they are proven safe. The group's position is based on the precautionary principle. According to this maxim, when confronted by scientific uncertainty, government should err on the side of restricting deployment of the technology.

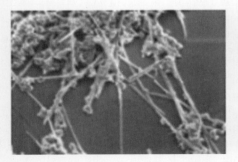
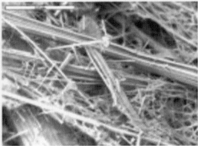

Figure 4.1 Casual resemblance of carbon nanotubes (left) to asbestos fibers (right).[9]
SEM Image Of Randomly Oriented Carbon Nanotubes Grown Using CVD, Reprinted With Permission, Courtesy of J. M. Perez, University of North Texas; Scanning Electron Microscope (SEM) Image of Fibrous Asbestos Insulation Sample, Reprinted with Permission, © 2004 Alion Science and Technology Corporation.

ENVIRONMENTAL REGULATION OF NANOTECHNOLOGY

The Environmental Protection Agency (EPA) is primarily responsible for regulating new substances that may harm the environment. Nanomaterials have begun to appear on the agency's radar screen. Since 2000, the EPA has offered grants to researchers who study potential environmental problems associated with nanomaterials.[10] The agency has also sponsored workshops and resolved to dedicate an increasing amount of energy to nanoscience.[11] EPA officials maintain that the agency is preparing to have a "fleet of people fluent in nanotechnology and its environmental and policy issues."[12]

Despite the increasing focus on the environmental implications of nanotechnology, nanomaterials are beginning to enter the environment with minimal regulatory review. The Toxic Substance and Control Act (TSCA) directs the EPA to regulate "chemical substances" before they are manufactured and released into the environment.[13] Manufacturers are required to file a premanufacturing notice (PMN) and obtain EPA approval if the chemical substance is "new."[14] A "new" chemical substance is one that is not listed on the TSCA Inventory.[15] Because nanomaterials are simply smaller pieces of existing materials that are listed on the TSCA Inventory,[16] manufacturers of nanomaterials do not have to file PMNs. A manufacturer must also file a Significant New Use Notice (SNUN) and obtain EPA approval if it intends to use an existing chemical in a significant new way.[17] In order for a manufacturer to be required to file a SNUN, the EPA must issue a Significant New Use Rule (SNUR) declaring a particular use of a substance as a significant new use.[18] Since the EPA has not issued a SNUR designating nanomaterials a significant new use of existing substances, manufacturers of nanomaterials are not required to file SNUNs. Section 8(e) requires manufacturers to report information relating to substances that present a "substantial risk of injury to health or the environment."[19] Because the risks of nanomaterials are largely unknown at this time, manufacturers have nothing to report. Therefore, nanomaterials are starting to be manufactured without EPA review and approval.[20]

THE NEED FOR ANALYSIS AND ACTION

Similar to the need for a rigorous analysis of the calls to ban nanotechnology research and development is the need to analyze the environmental case against nanomaterials. If ETC's argument is compelling, regulatory action must be taken now. In 2000, a single firm shipped over 250 tons of zinc oxide nanoparticles.[21] Several firms are building facilities to mass-produce carbon nanotubes within the next one to two years.[22] Industries could soon be flooding the environment with nanomaterials. The harms of nanomaterials could be difficult to reverse or control since there are no reliable methods for detecting and remediating hazardous ones.[23]

If, however, the environmental warnings are overstated, policy action is still required in the immediate future. The arguments promulgated by nanotechnology opponents must be addressed before they gain enough momentum to threaten the entire industry. In addition to the arguments set forth in Chapter 3 regarding the risks of public misperceptions stifling the industry, there are also liability risks associated with nanomaterials in the short term. Public perception that nanomaterials are toxic increases the likelihood of plaintiffs' suits as well as favorable jury verdicts.[24] Studies of products liability and toxic tort litigation reveal that a finding of liability for one product can "unleash a firestorm of suits involving the same and related products."[25]

ANALYZING THE DATA

Analyzing the ETC's call for stringent regulations on nanomaterials involves sifting through the data concerning the toxicity of nanomaterials, evaluating the competing normative arguments for and against aggressive regulation, and testing the legality of regulations banning production of nanomaterials. In this section, we analyze the data on the toxicity of nanomaterials. We identify the methods used to measure toxicity of chemical substances, review the available data, and then draw conclusions about the reliability of this data.

Measuring Toxicity

Determining the toxicity of nanomaterials generally involves four steps. First, studies must be conducted to determine the amount of exposure to a substance that causes harm. Epidemiological studies, which use statistical methods in an attempt to quantifiably link a specific factor to a disease that appears within a selected population, can be used. Because such studies are generally unavailable for new chemicals, scientists usually experiment on animals to determine toxicity.[26] Second, when animal studies are used, the dose at which the substances resulted in adverse health effects on animals must be converted to an estimate of the amount of exposure that will cause harmful effects in humans.[27] Third, the level of potential human exposure to the nanomaterials must be estimated.[28] Finally, the dose at which the nanomaterials present a low level of risk must be compared to the estimated exposure dose.[29]

The Data

There are two bodies of research on nano-sized particles. First, there are a number of studies on the health hazards of "ultrafine particles."[30] Ultrafine particles, such as ultrafine carbon black, ultrafine TO_2, and ultrafine metals, are naturally occurring nano-sized particles. Toxicity studies of ultrafine particles were largely conducted in the context of determining the health risks of exposure to ambient particulate matter. Second, there is a limited, but growing, body of research on engineered nanomaterials. Engineered nanomaterials, such as carbon nanotubes and semiconductor nanocrystals, are more consistent in size and composition than ultrafine particles.

Data on Ultrafine Particles

There has been some epidemiological research into the link between ultrafine airborne particles and mortality. For example, Wichmann studied the association of daily mortality data with the number and mass concentrations of ultrafine and fine particles in Erfurt, Germany.[31] The study concluded that ultrafine particles are associated with mortality.[32]

Several toxicology studies have shown that different nanoparticles are toxic when inhaled. For example, exposing human volunteers to nano-sized zinc oxide particles for 30 minutes resulted in bronchoalveolar inflammation.[33] Other studies have shown that titanium dioxide nanoparticles are toxic in rats.[34] Although the precise cause of toxicity is uncertain, researchers predict that the greater surface areas of the nanoparticles facilitate the release of transition metals or generation of free radicals.

Several animal studies have also demonstrated that inhalation of certain nanoparticles does not have toxic effects. For example, studies suggest that cabosil amorphous silica[35] and magnesium oxide particles[36] do not cause bronchoalveolar inflammation.

There is conflicting data concerning the inhalation toxicity of ultrafine carbon. Some studies indicate that ultrafine carbon black particles are associated with physiological changes relevant to pulmonary inflammation, heart attacks, and strokes.[37] Other studies have shown that ultrafine carbon black particles do not induce an inflammatory response in rats.[38]

Data on Engineered Nanomaterials

Although research on the health and environmental risks of engineered nanomaterials is underway, there are few published studies. To our knowledge, there have only been three published studies purporting to test the toxicity of carbon nanotubes. A team of researchers in Warsaw, Poland, concluded that "working with soot containing carbon nanotubes is unlikely to be associated with any health risk."[39] NASA researchers concluded that, on an equal-weight basis, "if carbon nanotubes reach the lung, they can be more toxic than carbon black or quartz."[40] Finally, DuPont found that, when exposed to high doses of single-walled carbon nanotubes, a small percentage of rats died while those that survived showed no sustained inflammatory response.[41]

Analyzing the Data

A careful review of this data reveals that there is substantial uncertainty regarding the health risks of different nanomaterials. The data on ultrafine particles may not be applicable to synthetic nanomaterials, because ultrafine particles are neither a consistent size nor pure in chemical or structural composition. Also, with regard to carbon nanotubes, the data are conflicting and potentially misleading. Studies showing that carbon black particles are toxic do not necessarily prove the toxicity of carbon nanotubes because carbon black may include other organic and metallic substances besides carbon nanotubes. Similarly, the Warsaw study does not necessarily prove that carbon nanotubes are not toxic. The experiment involved giving a test group of pathogen-free guinea pigs an intratracheal instillation of 25 mg of carbon nanotubes in soot while the control group was given soot without carbon nanotubes. Researchers reported no significant differences in the animals' inflammatory reactions. However, even if soot containing carbon nanotubes is not more toxic than soot without carbon nanotubes, this does not prove that carbon nanotubes are not toxic. Finally, the inconsistent results observed in the DuPont study have caused Warheit and his coworkers to question the physiological relevance of their findings.[42]

In addition, the epidemiological data are not reliable. Because personal exposure is seldom determined with any accuracy, it is difficult to establish a conclusive link between ultrafine particles and death.[43] Additionally, although the results demonstrate that ultrafine particles are associated with human deaths, the relative apportionment of causation between ultrafine particles and fine particle mass is not clear. Finally, attempts to determine timing of effect by using the best lag approach can bias the results toward finding positive or negative statistically significant associations.[44]

Further, data concerning the toxic effects of nanomaterials on animals may not be a reliable indicator of the toxic effects of nanomaterials on humans. While the studies involve subjecting animals to a single chemical substance, people are exposed to a wide variety of different compounds at the same time. Synergistic effects can compound the toxicity of a substance by 1,000 times.[45] Additionally, because test animals are genetically similar and usually healthy, young adults, they will often respond to the same substances in the same way.[46] By contrast, because people are genetically diverse, of different ages, and in different states of health, they will respond to the same substances differently.[47] Fundamental differences in human biology and animal biology result in substances having different effects on each.[48] For example, whereas animal tests show that betanapthylamine is relatively harmless, use of the substance in the workplace reveals that workers who have been exposed to it over a five-year period have a "100% bladder cancer" rate.[49] Similarly, years of asbestos studies on animals beginning in the 1930s produced no evidence of risk.[50]

High levels of exposure to a substance over a short period of time is not necessarily an accurate gage for lower levels of exposure over a longer period of time. High-dose studies can exaggerate the toxicity of a substance in several respects. First, inhalation studies on rats can lead to misleading conclusions of inflammation. A phenomenon known as "overload" occurs when a large amount of airborne particles causes a failure of clearance in the lungs of rats, resulting in inflammation.[51] Overload of otherwise nontoxic particles will result in a false-positive pathological outcome.[52] Second, high-dose studies can lead to misleading carcinogenicity estimates. High doses may result in elevated rates of cell reproduction to repair tissue damage, and "[r]apid cell proliferation raises chances for spontaneous mutations leading to cancer."[53] Finally, instillation routes of exposures are unrealistic as compared to dosing, particle distribution, and other factors in realistic inhalation studies.[54]

Humans are likely to be exposed through different routes than the test subjects. The existing data generally focus on the damage done to respiratory tissue as a result of inhalation of nanomaterials and not systemic toxicity resulting from nanomaterials entering the human circulatory system. Nanomaterials are likely to be exposed to other parts of the body than the lungs. The evidence regarding the effects of inhaling nanoparticles may reveal nothing about the toxicity of nanoparticles accumulating in other organs.[55]

Finally, there is an information deficit on the degree humans will be exposed to different nanomaterials. The most hazardous chemicals may not have toxic effects if the exposure is minimal, and relatively inert chemicals may be highly toxic at certain levels.[56] Chemicals can enter the human body through oral ingestion, inhalation, or dermal absorption. Direct exposure involves nanomaterials entering humans during manufacturing or life cycles of the products in which they are used. An example of direct exposure would be workers at manufacturing plants inhaling carbon nanotubes. It is difficult to estimate direct exposure levels. Even when nanomaterials enter the environment, they may not form respirable particles. For example, two recent exposure assessment studies found that individuals working with nanotubes are exposed to very low levels of nanotube dust.[57] Indirect exposure to nanomaterials would take place if nanomaterials got into the water supply or food chain. There are also substantial uncertainties concerning indirect exposure to nanomaterials. For example, it is uncertain whether nanomaterials are biodegradable, whether they can enter the food chain through bacteria, and whether they would pass through water filtration. Research being conducted to resolve these questions has not yet produced any answers. As Dr. Phil Sayre, Associate Director of Risk Assessment at the EPA, notes, "We know very little about the environmental degradation and fate of nanomaterials."[58]

The combined effect of these uncertainties can produce "orders-of-magnitude uncertainties in the assessment of human health risk."[59] Indeed, toxicologists concede that toxicity assessment is "very inexact."[60] Referring to the toxicity of ultrafine particles, one toxicologist noted that this is a field "in which there is much more speculation and legend than there is hard fact."[61] Dr. Vicki Colvin, Director of the Center for Biological and Environmental Nanotechnology at Rice, summarized the uncertainty when she testified to Congress in the spring of 2003:

> [I]f you have used a sunscreen in the last year it is possible that your skin came in contact with nanoscale ceramics. Is this a cause for concern? No one knows. Nanomaterials are valuable in many technologies because they interact quite differently with the body than larger materials. . . . [U]nintended exposures—of research workers, factory workers, and the general public—to nanoscale solids could have more dire consequences than turning skin blue. Or they could turn out to be benign. We just don't know.[62]

NORMATIVE ANALYSIS

The feuding camps of environmentalists and industry apologists could highlight different studies to advance their arguments in the public debate. Generally, however, both sides have refrained from relying on data supporting their positions and recognize that there is substantial uncertainty regarding the toxicological effects of nanomaterials. The ETC report does not even mention studies of ultrafine particles. The authors of the report maintain that introducing such evidence would create the impression that the nanomaterials they seek to restrict are not a new technology.[63] Similarly, industry allies have not offered the Warsaw study as evidence that nanotubes are not toxic. Mihail Roco, the National Science Foundation's senior adviser on nanotechnology, candidly admits that nanomaterials "may have unexpected consequences" and that "some could be toxic."[64]

Rather than focusing on the existing data, the debate is primarily normative. Environmentalists rely on the precautionary principle in arguing that uncertainty warrants banning the production of nanomaterials until they are proven safe. Industry counters that nanomaterials should be regulated according to well-established principles of risk assessment and cost-benefit analysis. A thorough analysis of the competing normative arguments compels a rejection of the call to restrict production of nanomaterials.

The Precautionary Principle

Despite widespread use of the precautionary principle in policy debates, it is an "ill-defined" concept.[65] Generally, it stands for the proposition that "any uncertainty regarding the hazardous properties of a substance or activity ought to be resolved in a manner that favors regulation, with cost considerations of secondary importance."⁶⁶ Extreme application of the principle would permit no tradeoffs between safety and costs, and, therefore, would ban any potentially hazardous substance. A milder form of the principle holds that, while cost considerations are relevant, there should be a greater emphasis on safety than economic costs. Regulation ensures that lives will not be lost while only risking economic gains. As Aaron Wildavsky observes:

> The precautionary principle is a marvelous piece of rhetoric. It places the speaker on the side of the citizen—I am acting for your health—and portrays opponents of the contemplated ban or regulation as indifferent or hostile to the public's health.... The rhetoric seems to present a choice between health and money or even suggests health with no loss whatsoever, for a tangential presumption that industry will find a better and a cheaper as well as safe way.⁶⁷

The Precautionary Principle Should Not Be Used as a Basis for Aggressive Environmental Regulation

Although there is substantial uncertainty surrounding the environmental effects of nanotechnology, the precautionary principle should not compel aggressive environmental regulation for two reasons. First, the underlying rationale supporting the presumption in favor of regulation is flawed. The notion that a higher value should be placed on environmental and safety concerns than economic costs has been subjected to empirical criticism. There is evidence demonstrating that environmental regulation can increase morbidity and mortality by decreasing wealth.⁶⁸ Increasing GNP allows for the development of "better and more diverse research establishment[s], larger markets to stimulate creation of safer products, an infrastructure of health and many opportunities for exercise, and the societal resilience to rapidly and efficiently attack new unforeseen problems threatening our collective health and safety."⁶⁹ Scholars have shown that an economic loss in the range of $5 million to $6 million translates into an additional death.⁷⁰

The economic costs of aggressive regulation of nanotechnology would be considerable. Companies around the world are engaged in races to develop cost-effective methods for the mass production of nanomaterials. Restrictions on commercial production of nanomaterials would cripple American companies' chances of establishing dominance in this field. Beyond its immediate impact on industrial production of nanomaterials, such a policy would also irreparably impair the competitiveness of companies developing products based on nanomaterials. The uncertainty surrounding the regulation of such products combined with the apparent willingness of the government to subordinate the fate of nanotechnology to environmental concerns would evaporate investment in developing such products. Hundreds of nanotechnology start-up companies in the United States, which are currently struggling to obtain financing, would have little chance of weathering through such a regulatory storm. Further, established companies would be likely to reduce the resources devoted to developing products based on nanomaterials; they might shift their research and development and manufacturing facilities overseas. Thus, for example, Nanomed Pharmaceuticals would be unable to obtain financing for further development of drugs and vaccines to treat cancer and neurodegenerative diseases, and IBM could reduce the resources devoted to developing nanoelectronic devices. By drowning part of an industry with significant economic potential, stringent regulation of nanomaterials could indirectly condemn a large number of people to premature deaths.

Alternatively, if the argument that economic costs translate into health and environmental harms is not satisfying, it is arguable that regulations on nanotechnology would directly hinder the introduction of new products with health and environmental benefits. As shown in Chapter 1, nanotechnology is expected to produce cures for diseases such as cancer, technological solutions to environmental pollution, and solutions for other causes of death and suffering. The demise of start-up companies and industrial reorganization that would take place in the aftermath of stringent regulations would prevent such beneficial products from being developed.

Thus, the notion that policymakers should err on the side of stringent regulations simply because there is uncertainty regarding the consequences is flawed. Because economic costs have indirect effects on health and nanotechnology may directly save lives and improve the environment, there should be no presumption in favor of regulation. Policymakers and regulators should engage in cost-benefit analysis to determine whether nanomaterials should be prohibited.[71] Attempting to prove that the environmental and safety harms of regulation are greater than the most severe harms possible absent regulation is beyond the scope of this book.[72] Yet, a cursory weighing of the costs of regulation against the costs of inaction yields a conclusion that nanomaterials should not be prohibited.

LEGAL ANALYSIS

The debate over whether nanomaterials should be prohibited has completely ignored the legal ramifications of a regulatory ban. A thorough legal analysis is imperative. If an EPA ban on the commercial production of nanomaterials is not legally permissible, then there is no bite to the position advocated by ETC. As the following discussion demonstrates, a regulatory ban on nanomaterials until they are proven safe might violate both statutory and international law.

U.S. Law

Domestic environmental laws have adopted cost-benefit analysis rather than the precautionary principle.[73] In the face of uncertainty, regulatory agencies do not have the authority to ban the production of substances until proven safe regardless of the costs of such a ban.[74] Under TSCA, the EPA cannot legally sustain a ban on nanomaterials that have not yet been introduced into commerce or nanomaterials that are already being produced and used in commerce.

In reviewing new chemicals or new uses of existing chemicals under TSCA, the EPA has three regulatory options.

1. If the EPA determines that there is a reasonable basis to conclude that an unreasonable risk would exist from the manufacture or use of the chemical, it can restrict the amounts and uses of the substance, prescribe precautionary measures to be taken in the use and manufacture of a chemical, or prohibit the manufacturing of the substance.[75]
2. If the EPA determines that the chemical does not pose an unreasonable risk, the substance may be manufactured and sold.[76]
3. If the EPA determines that there is insufficient information to make a decision, it can seek an injunction preventing the manufacture of a new chemical or new use of an existing chemical.[77]

A court can grant a temporary injunction on the manufacture of a chemical if it makes two findings: (1) that the available data is "insufficient to permit a reasoned evaluation of the health and environmental effects of a chemical substance" and (2) "the manufacture, processing, distribution in commerce, use, or disposal of such substance, or any combination of such activities, may present an unreasonable risk of injury to health or the environment, or such substance is or will be produced in substantial quantities, and such substance either enters or may reasonably be anticipated to enter the environment in substantial quantities or there is or may be significant or substantial human exposure to the substance."[78] Although no court cases discuss the standards used to decide whether such an injunction should be granted, there are three reasons why courts would be unlikely to enjoin manufacturers of nanomaterials.

First, the EPA would have a difficult time proving that the data is "insufficient to permit a reasoned evaluation." Although there are substantial uncertainties, the EPA routinely makes reasoned evaluations and approves chemicals in the face of minimal data.[79] There is no toxicity data on almost half of all high-production volume chemicals that the EPA allows to be manufactured.[80] Where data is available, it is incomplete.[81] Despite this immense void of information, the EPA uses estimation techniques[82] and generally approves the use of new chemicals.[83]

Second, it is unlikely that the EPA can prove any of the facts under the second component of the test. The available data do not establish that nanomaterials pose an unreasonable risk of injury to health or the environment. Further, there is too much uncertainty surrounding exposure levels to convincingly argue that nanomaterials either "enter or may reasonably be anticipated to enter the environment in substantial quantities or there is or may be significant or substantial human exposure to the substance."

Finally, TSCA directs the EPA to engage in a cost-benefit analysis.[84] Even assuming the worst environmental effects of nanomaterials, courts could find that the benefits outweigh the costs.

Once a chemical has already been manufactured and used, the EPA can mandate testing of the substance.[85] However, the EPA cannot restrict the substance while such tests are taking place.[86]

The litigation over asbestos illustrates the judicial presumption against regulation in the face of uncertainty. Studies have demonstrated that inhalation exposure to asbestos can cause asbestosis, lung cancer, and mesothelioma.[87] Based on the perceived hazards of asbestos, the EPA promulgated the "Asbestos Ban and Phaseout" rule in 1989.[88] The rule banned, at three staged intervals, the future manufacture, importation, processing, and distribution in commerce of approximately 94 percent of the asbestos used in the United States and also imposed labeling requirements on products containing asbestos.[89] In 1991, the a court set aside most of the guidelines established in the policy.[90] The court held that the EPA failed to demonstrate that the products subject to the ban posed an unreasonable risk, and less burdensome actions could have adequately protected public health.[91] Although six asbestos-containing product categories are still subject to the 1989 asbestos ban,[92] there are no regulations on most other asbestos-containing products or uses.

International Law

In addition to violating domestic law, EPA restriction of nanomaterials might also violate international law. In light of the international race to capture the nanotechnology market, it would be nearly impossible for the United States to mobilize support for an international ban. Therefore, in order to be effective in preventing emissions of nanomaterials in the United States, the policy would have to include an import ban on both nanomaterials and the products incorporating such materials. In order to tap into the American market, states that export nanotechnology would likely argue that an import ban is inconsistent with obligations imposed by the General Agreement on Tariffs and Trade (GATT)[93] and associated agreements, including the Agreement on the Application of Sanitary and Phytosanitary Measures[94] and the Agreement on Technical Barriers to Trade.[95] Although the GATT does contain exceptions granting states discretion in health and environmental regulation, GATT panels have consistently weighed in on the side of free trade. For example, when the United States recently challenged the European Union's ban on the import of meat derived from cattle treated with certain growth-producing hormones, the World Trade Organization (WTO) concluded that the health risks did not justify the violation of international trade law.[96]

If an EPA regulation banning importation of nanomaterials or products containing such materials were challenged, the law could be declared invalid. Under the WTO rules, the United States would have the burden of demonstrating that its import ban was consistent with international standards or that nanomaterials are scientifically proven to be unsafe.[97] No existing international standards support such an import ban. Further, the United States would have no scientific basis for proving that nanomaterials are unsafe. The rigorous scientific evidence demanded by the WTO "cannot take account of deep-seated consumer and environmentalist preferences in this sensitive area, and is intolerant of more heightened risk standards that may exist outside the trade law area in customary international law or general principles of law."[98]

RECOMMENDATIONS

A thorough review reveals that the position advocated by ETC is unsound. Although the data show that there are substantial uncertainties, there is a strong normative case for allowing nanomaterials to be manufactured and introduced without substantial regulations. Further, the law supports this general framework. Thus, alternative policies must be considered to shatter the environmental opposition and ensure effective regulation of nanomaterials. Specifically, policymakers should take the following courses of action.

First, nanomaterials should be subject to additional EPA review under the Toxic Substances Control Act. One idea worth exploring is the issuance of a Significant New Use Rule by the EPA declaring that the "nanoscale" form of certain chemicals are a "significant new use of the substances." Such a rule would require manufacturers to submit a Significant New Use Notice to the EPA 90 days before the manufacturer or processor wishes to produce the nanomaterials.[99] The agency would then determine if there is an "unreasonable risk" posed by the substance. Subjecting nanomaterials to TSCA review could have several beneficial results. Based on existing data and the law's emphasis on cost-benefit analysis, EPA review will generally result in regulatory approval. Nevertheless, despite the preference for approval in the face of uncertainty, regulatory review also increases the likelihood that nanomaterials that are clearly toxic will be identified before they are mass-produced. Further, reviewing nanomaterials will quell some of the growing environmental opposition. The lack of a formal regulatory review process provides political ammunition for groups like ETC.

Second, more complete data are needed regarding the environmental implications of nanotechnology. Several projects are currently underway that will provide much-needed information. The most comprehensive studies taking place are being conducted by the Center for Biological and Environmental Nanotechnology (CBEN), established by Rice University with funds from the National Science Foundation.[100] The Center is focusing on testing certain materials that are likely to be produced in large quantities in the near future: silicon nanocrystals, iron nanoparticles, and carbon nanotubes.[101] Experiments conducted by the Center involve monitoring the particles in laboratory-simulated natural microcosms. Specifically, scientists are testing whether nanomaterials would pass through filters in a water treatment plant; whether nanostructures bind to various common contaminants such as pesticides or PCBs; whether bacteria take up the nanomaterials and if so, whether this might open a route for the particles to move into and up the food chain; and how nanomaterials interact with living cells and proteins.[102] NASA is also studying the toxicity of carbon nanotubes in rats.[103] In July 2003, the EPA solicited applications that evaluate the potential impacts of manufactured nanomaterials on human health and the environment.[104] These types of projects must continue to be funded to provide decision-makers at the EPA with accurate information in conducting TSCA review. The EPA should also consider exercising its powers under Section 4 of the TSCA to require industry testing of nanomaterials.

Finally, as suggested in Chapter 3, a mechanism must be established to enable policymakers and industry to maintain a dialogue with the public regarding the environmental implications of nanotechnology.

CHAPTER 5

The Patent and Trademark Office

Patent offices around the world are struggling to evaluate and prosecute nanotechnology patent applications. As the US patent system expands to accommodate nanotechnology-related inventions, the [Patent and Trademark Office] PTO has yet to implement a plan to handle the soaring number of patent applications being filed. The rise of nanotechnology is presenting new challenges and problems to this overburdened agency as it attempts to handle the enormous growth in applications filed and patents granted in a wide range of disciplines encompassing "nanoscience" or "nanotechnology."[1]

—Dr. Raj Bawa, *Patent Agent*

As the compulsion to patent nanotechnology inventions sweeps researchers and corporations around the globe, the Patent and Trademark Office (PTO) will play a crucial role in development of the field. In this chapter, we explore how the PTO is dealing with nanotechnology. We first provide an overview of patent law and the current state of the PTO. We then explain how the PTO has failed to prepare for review of nanotechnology patents. The PTO's lack of expertise combined with euphoria for patenting has resulted in the rejection of valid claims, the issuance of broad and overlapping claims, and a fragmented and somewhat chaotic intellectual property (IP) landscape. These IP roadblocks could severely retard development of nanotechnology. Finally, we consider potential solutions to the intellectual property problems such as PTO reform, judicial action to narrow the scope of broad claims, and patent pools.

PATENT POLICY

Goals of the Patent System

The Constitution empowers Congress to promote the progress of "Science and useful Arts" by granting inventors the exclusive rights to their discoveries.[2] Although academics have wrestled with different justifications for the patent system,[3] the "monopoly-profit-incentive" thesis has become the dominant rationale. According to this theory, temporary monopolies in the form of exclusive patent rights are necessary to encourage socially valuable innovation and product development. Economists have long debated whether the patent system achieves the goal of fostering innovation.[4] Despite the importance of this debate and the need for better data regarding the effects of the patent system on innovation, this chapter will assume that on balance, the system has a positive influence on innovation.

Brief Summary of Patent Law

In order to obtain a patent, an inventor must file an application with the Patent and Trademark Office (PTO) within one year of the first offer for sale or public disclosure of the invention.[5] The application must meet the disclosure requirements in Section 112, which states that the specification must "contain a written description of the invention, and of the manner and process of making and using it, in such full, clear, concise, and exact terms as to enable any person skilled in the art to which it pertains, or with which it is most nearly connected, to make and use the same[.]"[6] The patent will issue if the PTO determines that several conditions are satisfied. First, the invention must be patentable subject matter under Section 101.[7] Second, the invention must display some utility.[8] Third, the invention must be novel. Under Section 102(a), an invention is novel unless the invention was known or used by others in this country or patented or described in a printed publication in this or a foreign country before the date of invention by the applicant.[9] Fourth, the invention must not be obvious.[10] An invention is obvious if the prior art would have suggested[11] to one of ordinary skill in the art that this process should be carried out and would have a reasonable likelihood of success.[12]

Once the PTO issues the application as a patent, the inventor or the inventor's assignee has the right to exclude others from practicing the invention for twenty years from the filing date of the application.[13] Patent claims, which are comprised of elements (or limitations), determine the scope of the patent. Infringement of a claim takes place when there is literal infringement or infringement under the doctrine of equivalents. Literal infringement occurs when every element in a claim is literally found in the accused device.[14] Under the doctrine of equivalents, infringement takes place when an element in a claim is not literally found in the accused device, but its equivalent is present.[15] An element in an accused device can be found to be equivalent if it performs substantially the same function in substantially the same way to obtain substantially the same result as the claim element.[16] If infringement occurs, patent owners may seek enforcement of the patent in a federal district court and obtain preliminary or permanent injunctions, damages as measured by a reasonable royalty or lost profits, attorney's fees, and treble damages.[17]

The State of Patent Law and the PTO

As shown in Figure 5.1, patenting activity is increasing substantially in the United States. In 1991, 177,830 patents were filed.[18] By 2001, the number of filings had nearly doubled to 345,732 patents.[19] In 2003, the PTO maintained a backlog of 450,000 patent applications and patent pendency—the time from application to issuance of patent—was about 27 months.[20] There are a variety of possible explanations for the patent explosion.[21]

Regardless of the cause of the growing patenting activity, the enormous number of filings is stretching the PTO beyond its capacity for effective review. With a stagnant budget, the PTO simply "does not have as many examiners as it should have to issue quality patents in a reasonable time."[22] Also, overworked examiners are reviewing more technically complex applications than ever before.[23] The ultimate impact of less thorough reviews could be increasing chaos in the intellectual property domain. The agency has already received substantial criticism for the issuance of some patents. A well-known example is the patent on the "one-click method" held by Amazon.com.[24] The patent allows Amazon to exclude other online businesses from utilizing customer-identification technology to enable "one-click" purchases. A deeper probe of recently issued patents reveals that even

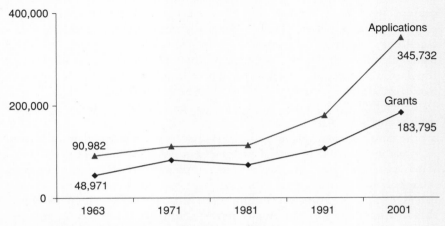

Figure 5.1 Total U.S. patent issuances over time.

more seemingly invalid claims are being granted. As law professor John Thomas notes, "U.S. patent quality appears to be on the decline."[25]

In an attempt to improve the quality and timeliness of review, the PTO unveiled the "21st Century Strategic Plan" in 2002.[26] Substantial controversy surrounds the plan; it remains to be seen how it will be implemented and whether it will succeed in improving the efficacy of the PTO. The Department of Justice Reauthorization Act also contains several provisions intended to enhance PTO functionality.[27] Despite recent attempts at reform, there is little doubt that the PTO is poorly positioned as it embarks on the task of reviewing some of the most technologically complex patent applications in history.

PTO REVIEW OF NANOTECHNOLOGY PATENTS

The compulsion to patent has swept nanotechnology researchers. Thousands of nanotechnology patent applications have been filed over the last several years. Despite the recent flood of nanotechnology patents, the PTO had until very recently done little to acquire the multidisciplinary expertise necessary to effectively review these patents. The PTO maintains several different technology centers focused on reviewing patents in particular fields. For example, the biotechnology and organic chemistry center reviews biotechnology patents, the semiconductor center reviews semiconductor patents, and the chemical and materials engineering center reviews chemical and composite patents. There is as yet no center dedicated to nanotechnology nor is there a group of reviewers specializing in nanotechnology. Since nanotechnology is interdisciplinary, some nanotech patent applications are directed to the biotechnology and organic chemistry center, some to the semiconductor center, and some to the chemical and materials engineering center (Figure 5.2). Examiners in a particular center do not always search the prior art in other centers. There is no classification scheme unique to nanotechnology patents, and the PTO lacks effective automation tools for nanotechnology "prior art" searching.[28] The PTO has not yet demonstrated the flexibility necessary to effectively address the interdisciplinary nature of nanotech, although it has taken some steps in this direction.

The PTO's lack of expertise and disjointed review of nanotechnology patents combined with the euphoria for patenting has created several hurdles for the development of nanotechnology. First, applicants are at risk of having valid claims rejected by unskilled reviewers. Second, the PTO has issued some broad and overlapping claims on upstream inventions. By conferring monopoly power over some of the building blocks of nanotechnology, these patents may stifle downstream innovation. Finally, the sheer number of patents that have and will continue to issue might make it difficult for firms to acquire the licenses necessary for developing downstream products.

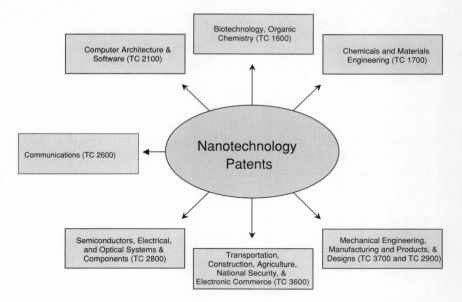

Figure 5.2 PTO review of nanotechnology patents.

Rejection of Claims That Should Issue

The failure of the PTO to prepare for nanotechnology is resulting in the rejection of some valid claims. The complexity of the field may cause some examiners to mistakenly conclude that inventions are obvious or not novel. We will discuss the specific issues faced by applicants in drafting and filing nanotechnology patents in Chapter 13. The denial of valid claims is a wasted expense and reduces the incentive to innovate. Although an applicant can petition for reconsideration, the process is costly, and the delay can be devastating for smaller companies.

Issuance of Broad and Overlapping Claims

In addition to denying valid claims, examiners have also granted broad and overlapping claims on upstream inventions. An exploration of patents on carbon nanotubes and semiconductor nanocrystals reveals several examples of broad and overlapping claims. Hyperion Catalysis International, a materials company, holds a patent with a claim to a "cylindrical discrete carbon fibril." Despite the fact that this patent issued nearly five years before the discovery of carbon nanotubes, the broad language in the patent enables Hyperion to argue that the patent encompasses multi-walled carbon nanotubes.[30] A second example of a broad nanotube patent is Patent 5,424,054, which claims a "hollow carbon fiber having a wall consisting essentially of a single layer of carbon atoms."[31] International Business Machines (IBM), the holder of this patent, can argue that the carbon fiber is a single-walled carbon nanotube.[32] Stanford University holds a patent that claims an "apparatus comprising: (a) a substrate with a top surface; (b) a catalyst island disposed on the top surface of the substrate; and (c) a carbon nanotube extending from the catalyst island."[33] Advanced Technology Materials, Inc. holds a patent that covers a very general architecture for building devices incorporating carbon nanotubes. The patent claims a "microelectronic or microelectromechanical device, comprising: a substrate, wherein the substrate includes an oxide layer and an etch stop layer for the oxide layer; and a fiber formed of a carbon-containing material."[34] Indeed, the patent concedes the breadth of its claims: "[a] wide variety of devices may be formed using a carbon microfiber (nanotube) as a part of the active device."[35] Perhaps the most overt example of the PTO's willingness to issue broad claims on carbon nanotubes is Patent 6,346,189, held by Rice University. The patent, which issued over a decade after single-walled carbon nanotubes were discovered, claims: "A composition of matter comprising at least about 99% by weight of single-wall carbon molecules."[36]

Several broad and overlapping claims have also been granted on semiconductor nanocrystals. For example, Patent 6,268,041 claims silicon nanocrystals "having a size distribution that varies by less than 20 percent of the average particle diameter of between 1 and 30 nanometers" and germanium nanocrystals having "an average particle diameter of between 1 and 8 nm and a size distribution which varies by less than 20 percent of the average particle diameter."[37] This patent provides Starfire Electronic Development and Marketing, Inc. with a monopoly over silicon and germanium nanocrystals having these dimensions. Nanosys and Quantum Dot Corp. have co-licensed a patent from UC Berkeley that claims "particles of III-V semiconductor, said particles being crystalline, being soluble in quinoline or pyridine, and being sized such that at least about 50% are between 1 nanometer and 6 nanometers across."[38] Many patents claim "quantum dots" or "semiconductor nanocrystals," even though the invention only involved a certain type of semiconductor material. For example, Patent 6,322,901 claims:

> A coated nanocrystal capable of light emission, comprising: a core comprising a first semiconductor material, said core being a member of a monodisperse particle population; and an overcoating uniformly deposited on the core comprising a second semiconductor material, wherein the first semiconductor material and the second semiconductor material are the same or different, and wherein the monodisperse particle population is characterized in that when irradiated the population emits light in a spectral range of no greater than about 60 nm full width at half max (FWHM).[39]

Similarly, Patent 5,990,479 claims a "luminescent semiconductor nanocrystal compound capable of linking to an affinity molecule and capable of emitting electromagnetic radiation in a narrow wavelength band when excited..."[40]

A broad patent on an upstream invention provides the patent holder with three options. First, the patent holder can refuse to license the technology in order to prevent competition with its products. For example, Hyperion can refuse to grant a license for multi-walled nanotubes in composites, IBM may not grant a license for single-walled nanotubes in nanoelectronics, and Nanosys may not license semiconductor nanocrystals for different applications. Second, the patent holder can grant nonexclusive licenses to anyone wishing to use the technology. Finally, the patent holder can grant exclusive licenses to particular firms wishing to develop the technology in a particular manner. For example, IBM could grant an exclusive license for single-walled nanotubes to one company developing an energy device, an exclusive license to another company constructing a gas sensor, an exclusive license to another company for an optical device, and so on.

If the holders of broad patents on upstream nanotechnology inventions are willing to widely license their technology, then the patents will not stifle development of products based on nanotechnology. For example, Stanford University held the patent on the fundamental technology used in recombinant DNA research. Licensing the patent on a nonexclusive basis and at reasonable rates fueled the development of the early biotechnology industry.[41] If, however, holders of broad patents refuse to license their patents, the potential commercial impact of nanotechnology could be severely limited.

Issuance of Too Many Patents

While rejection of valid claims and overly broad patents present stumbling blocks, the largest challenge to developers of downstream nanotechnology products is the large number of patents issued by the PTO. The fragmentation of intellectual property creates minefields that firms must cross in bringing products to market. The costs of licensing numerous patents are exorbitant, and alternative arrangements such as cross-licensing and patent pools are unlikely to emerge. As a result, the large number of patents could limit the commercial potential of nanotechnology.

The Patent Thicket

Estimates of the total number of nanotech patents that have issued varies depending on the definition of nanotechnology and search terms used.[42] Based on research conducted in early 2004, we estimate that there are approximately 4,000 patents particularly relevant to nanotechnology (Figure 5.3). The majority of these patents have issued in the past few years. The

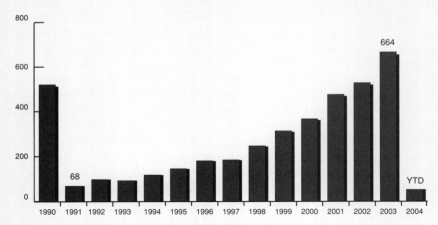

Figure 5.3 Total patent issuances in nanotechnology.

large number of patents is the culmination of the modern obsession with patenting, the incremental nature of technological progress, and ineffective review at the PTO.

A careful review of patents claiming inventions related to carbon nanotubes illustrates the "patent thicket."[43] We searched all patents in the PTO database containing the words "carbon nanotubes" in the claims. A search conducted in early 2004 produced several hundred patents. We identified those patents relevant to companies developing materials, devices, and systems based on carbon nanotubes and constructed the "tree diagram" shown in Figure 5.4.

As shown in Figure 5.4, a company wishing to commercialize a downstream product incorporating carbon nanotubes could be required to license a number of patents. For example, in order to market a hydrogen storage system using nanotubes, a company would be required to license the patents claiming nanotubes as well as patents on the production methods. Further, the company might have to seek licenses from several patent holders on the use of carbon nanotubes in hydrogen storage.[44]

Even if the company's system was revolutionary, the breadth of these issued and pending patents would enable the patent holders to assert a claim of literal infringement. At the very least, they could argue infringement under the doctrine of equivalents.

A similar study of semiconductor nanocrystal patents conducted by the NanoBusiness Alliance in the spring of 2002 describes the intellectual property landscape in this field.[45] The survey identified 238 patents having claims related to "quantum dot" or "semiconductor nanocrystals."[46] The number of licenses required to commercialize a product based on semiconductor nanocrystals is illustrated by the intellectual property portfolio that

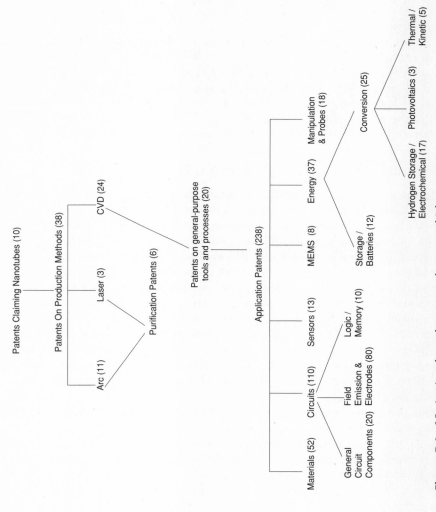

Figure 5.4 Navigating the carbon nanotube patent thicket.

Quantum Dot Corporation has acquired in its quest to launch a commercial product. Quantum Dot is utilizing quantum dots for biomedical applications such as DNA arrays and diagnostics. It has licensed 22 issued patents and owns or has licensed over 90 domestic and international patent applications currently under prosecution.[47] Similarly, Nanosys has assembled a comprehensive intellectual property portfolio of over 130 patent and patent applications covering fundamental discoveries in inorganic semiconductor nanostructures.[48]

Other studies have identified similar trends with other "building blocks." For example, a study conducted by Foley and Lardner shows that over 100 patents related to dendrimers are issuing each year.[49] Indeed, at the time of this writing, several new nanotech patents issue every day. As one young Stanford professor remarked, "I wonder how anyone will ever get licenses to all of the patents that are being generated."[50] As nanotechnology continues to develop, the minefields of patents will become more difficult to traverse.

Difficulties of Navigating through the Patent Thickets

If firms wishing to commercialize products could obtain necessary licenses in an efficient manner, then a fragmented intellectual property environment would not be problematic. There are several reasons, however, why it will be difficult for firms to successfully navigate through the patent thicket.

First, firms holding patents needed by other firms to commercialize their products might use their patents as strategic weapons to strangle competitors. Start-up companies might block other start-ups from bringing products to market. Further, established corporations seeking to prevent start-up companies from disrupting their market dominance with nanotechnology-based products should perceive the intellectual property quagmire as a favorable condition. They can slowly starve their start-up competitors into extinction by waging a protracted battle on the IP front.[51]

Even if all licensors are willing to license their technology, substantial difficulties in concluding license agreements remain. If licensees seek licenses before engaging in research and development of the downstream product, they will be reluctant to disclose sensitive information related to their plans for a commercial product. This reluctance is due to the fact that "at the time it is revealed, this information would be protectable against improper appropriation only through trade secret law, which is a rather weak form of protection."[52] It is difficult to conclude a licensing agreement without both parties having a comprehensive understanding of the downstream product for which the patent would be licensed. Further, even if the licensee would be willing to fully disclose its research and development plans, it will be difficult for firms to value undeveloped, downstream products. Substantial uncertainties abound concerning whether and when nanotechnology will result in commercial products, and evidence suggests that researchers in cutting edge technologies are likely to overvalue their inventions.[53] Moreover, in an industry plagued by environmental opposition and composed of fierce competitors, even an engineering success could be a commercial failure.

If licensees seek licenses after developing the downstream product, there are still significant obstacles to negotiating licenses. Because licensors are in a strategic bargaining advantage, they can demand excessive royalties, and the license seeker must relent.[54] Even holders of patents that contribute little value relative to the product can attempt to appropriate as much of the value of the improvement as possible.

The difficulties associated with concluding a licensing agreement with an individual patent holder are magnified by the fact that companies will require a handful of licenses. Having to conclude separate agreements with several different patent holders presents three problems for the license seeker. First, each patent holder has an incentive to delay concluding an agreement until the licensee has concluded agreements with the other licensors.[55] Holding out to be the final licensor increases the licensor's leverage in negotiations.[56] Second, each patent holder attempts to extract the maximum royalty from the company. When a licensee must seek a number of licenses from different entities which are all trying to extract the maximum value, the business endeavor can quickly lose its value.[57] Finally, there are significant transaction costs associated with concluding such agreements.[58] The high transaction costs associated with such negotiations combined with the expectation that there is a low likelihood of concluding reasonable agreements will deter many firms from attempting to traverse the patent thicket.[59]

Cooperative Licensing Arrangements

Even if individual licensing agreements cannot be concluded, firms can obtain rights to the necessary intellectual property if alternative intellectual property arrangements emerge. In some industries, informal norms serve as a substitute for complicated licensing arrangements.[60] For example, in the semiconductor industry, several firms maintain more than 1,000 patents, and there is a high level of reciprocal infringement.[61] The mutual threat of patent litigation amounts to "a tacit cross-license."[62] Firms find it rational to "forego full enforcement of property rights in exchange for reciprocal forbearance from competitors."[63]

Some industries even resort to formal arrangements to solve the problems associated with concluding individual licensing agreements. In a patent pool, a group of patent holders "assign or license their individual rights to a central entity, which in turn exploits the collective rights by licensing."[64] Typically, participants license into the pool patents on "building block" technologies in exchange for use of any other member's technology for a set fee.[65] Licensing fees are generally determined by the central entity based on the value of the technologies being licensed.[66]

Patent pools have been utilized to facilitate development of pioneering technologies throughout American history.[67] As early as 1856, the Sewing Machine Combination established one of the first patent pools consisting of sewing machine patents.[68] In 1917, a committee formed by the Assistant Secretary of Navy recommended aircraft manufacturers set up a patent pool.[69] More recently, patent pools have been formed on patents that are essential to compliance with the MPEG-2 compression technology standards[70] and patents involved in ensuring compliance with DVD-ROM and DVD-Video formats.[71]

Arrangements such as implied cross-licensing agreements or formal patent pools will be less useful in nanotechnology than in other fields. Such arrangements are "most likely to arise when horizontal competitors who share similar values and are engaged in repeat-play transactions each hold roughly similar portfolios of blocking patents."[72] Where the parties are heterogeneous and have divergent positions and attitudes toward intellectual property, alternative arrangements are less likely to arise.

There are two main characteristics of the industrial structure giving rise to nanotechnology that limit the use of cross-licensing or patent pools. First, there is a significant disparity in size between the competing players. As discussed in Chapter 2, there are hundreds of large corporations and hundreds of start-up companies attempting to commercialize nanotechnology-based products. The asymmetrical sizes and legal arsenals of the different companies preclude the emergence of the mutual threat of costly litigation that is necessary for cross-licensing and patent pools. For example, Nantero is a start-up company using carbon nanotubes to construct a chip for nonvolatile RAM. In order to market its product, Nantero might be required to obtain a number of licenses on nanotube patents from other companies in different industries. Since Nantero does not have a large patent portfolio or, presumably, the stomach for litigation, it has no leverage to seek formal or informal cross-licensing agreements.[73]

Second, divergent business models will hinder cooperation between firms. Some firms are focusing on developing building blocks and applying those building blocks to end-product development while other firms plan to generate revenue by licensing out enabling technology. For example, if a start-up company uses laser ablation to produce single-walled nanotubes for an energy device, it might have to obtain a license from IBM on the nanotube, licenses on the laser ablation process, and licenses on the purification and preparation of nanotubes. Carbon Nanotechnologies holds most of the patents on the laser production method. Since Carbon Nanotechnologies is focused on commercial production of nanotubes and not building downstream devices, the start-up may not have any patents in which Carbon Nanotechnologies would have any interest. Similarly, the start-up may not have any patents in which IBM would have any interest. If the start-up used a chemical vapor deposition process to produce the nanotubes for the energy device, it would have to seek licenses for the patents necessary to the chemical vapor deposition (CVD) process. Several patents covering directed growth of nanotubes on a substrate using the CVD method are held by Molecular Nanosystems. Since Molecular Nanosystems appears to be focused on mechanical probes, field emitters, and chemical and biological sensors, the start-up may not have any intellectual property that could be cross-licensed.

Impact of Too Many Patents on the Commercialization of Nanotechnology

The exorbitant transaction costs associated with concluding individual licensing agreements and the failure of the industry to establish a system for sharing intellectual property rights may limit development of nanotechnology. As economist Carl Shapiro argues, the result will be "that some companies avoid the minefield altogether."[74] Professors Michael Heller and Rebecca Eisenberg have described the proliferation of intellectual property rights on upstream inventions as the "tragedy of the anticommons."[75] They argue that the large number of patents on gene and gene fragments will deter research and development of downstream therapeutic applications.[76]

The resulting uncertainty and large transaction costs will have a particularly significant impact on start-up companies.[77] Attracting venture capital and other forms of financing requires a defensible IP base.[78] Venture capitalists will be reluctant to invest in start-ups if they are concerned that the company will likely be subject to liability for patent infringement. Those start-up companies that can secure financing will spend precious resources on conducting acrimonious licensing negotiations with dubious outcomes and expensive court battles.[79] As the *Venture Capital Journal* declared, "Even if a start-up has the brains and the cash to fuel its research, there are complicated intellectual property issues to throw a wrench into the best-laid business plans."[80] This inhospitable environment is unfortunate, because nanotechnology start-ups could be engines for rapid technology development and wealth creation. The experience in biotechnology demonstrates that start-up companies are better innovators than large pharmaceutical companies.[81]

RECOMMENDATIONS

Nanotechnology is still in its infancy. To enable industry to transform nanoscience into downstream commercial products, government must eliminate obstacles on the intellectual property front. Three courses of action could pave a smoother path on the road to the Nano Age. First, the PTO should adequately equip itself with the knowledge and resources to effectively review applications in nanotechnology. Second, courts should be willing to narrow the scope of overly broad patents when they are wielded as monopolistic weapons. Third, government should consider encouraging the formation of patent pools on the building blocks of nanotechnology.

PTO Reform

As nanotechnology begins to flourish, the PTO will be confronted with an increasing number of patent applications. Although the agency cannot undo its past mistakes, it should take action to prevent the rejection of valid claims, force patentees to narrow overly broad claims, and reduce the total number of patents issuing by ensuring that claimed inventions are not obvious in light of the prior art.

The PTO has faced challenges similar to those presented by nanotechnology with other emerging technologies. For example, in reviewing software patents, the agency "missed the ball for over a decade."[82] The Office failed to hire examiners skilled in the software arts and neglected to effectively classify software prior art.[83] The result has been an intellectual property nightmare in software that has stunted the growth of the industry.[84] In contrast, the PTO has fashioned a more effective response to the flood of business method patents. Although commentators highlight the "1-click" Amazon.com patent as an egregious example of the PTO's failure,[85] a careful analysis reveals that the agency has made considerable efforts to improve review of business method patents. After the *State Street* decision legitimized the patenting of business methods in 1998,[86] the PTO developed new guidelines and started training examiners in 1999.[87] By 2000, the agency had instituted special review procedures, including subjecting applications to a second review by a second patent examiner before issuance.[88] These reforms have led one patent attorney to conclude that "the patent system, as now implemented, is ready and able to meet the challenge of this new technology."[89]

The PTO cannot remain passive or it risks contributing to an IP crisis in nanotechnology similar to the prior crisis in the software industry. Rather, just as it has made an active effort to improve review of business method patents, the agency should take immediate action to prepare for an even greater flood of nanotechnology patent applications.

Some lawyers and industry advocates have argued that the PTO should consider launching a nanotechnology center. A technology center currently comprises about 500 examiners reviewing 40,000 to 80,000 applications. Thus, at this stage, it is probably premature to establish a center devoted to nanotechnology. However, even if the field is not developed to justify the creation of a new center, the PTO should institute a system that directs nanotechnology patents to specific people tutored in nanotechnology within the different technology centers. "Having a set of nanotechnology specialists within the USPTO and in communication with each other could unify prior art searches and ensure more accurate consideration of nanotechnology patents and increased quality of granted patents."[90]

The Patent and Trademark Office

Some initial steps have been taken to educate PTO examiners about nanotechnology. In 2002, a group of nanoscientists gave examiners tutorials on various subjects in nanotechnology.[91] Further, the agency is sending examiners to the Atlantic Nano Forum, which provides detailed, technology-based lectures.[92] There have also been efforts to informally designate certain examiners from each center as Nanotech Specialists.

In September 2003, the PTO launched the first Nanotechnology Customer Partnership Meeting. The purpose of the meeting was to provide the audience with information concerning the PTO review process of nanotechnology patents and to get feedback from attendees—mainly patent lawyers and industry leaders—on possible ways to improve this process. The meeting fostered a valuable dialogue between the attendees and PTO officials.[93] The PTO concluded the meeting by declaring that additional meetings would take place periodically in the future.

Perhaps the most effective mechanism to improve PTO review of nanotechnology patents is for the agency to secure additional resources. Congress "diverts" user fees collected by the PTO from patent applicants to the general budget. Allowing the PTO to keep this money would go a long way toward improving quality and efficiency of review.

Judicial Action

The validity of overly broad claims may ultimately be litigated. When these claims are wielded as monopolistic tools to prevent downstream development, courts should utilize legal tools to narrow their scope. For example, when reviewing claims to carbon nanotubes or semiconductor nanocrystals, courts could limit claims to cover the nanomaterials produced using the methods described in the patents. In Chapter 13, we discuss how the enablement doctrine can be used to narrow the scope of claims.

Government Intervention To Facilitate Licensing

Preparing for effective review of future nanotechnology applications might alleviate some future mistakes and confusion. Litigation may even result in the narrowing of the overly broad patents on nanostructures. However, the large number of patents on the tools needed to produce downstream products will continue to plague the burgeoning industry. Mechanisms to combat the tragedy of the anticommons should be explored.

Compulsory Licensing

Some commentators have proposed compulsory licensing as a solution to the problems of "blocking patents" in gene patents and business method patents.[94] Despite academic support, the courts and legislature have generally resisted compulsory licensing.[95] The absolute right of the patent owner to refuse to license an invention is statutorily protected: "no patent owner otherwise entitled to relief for infringement or contributory infringement of a patent shall be denied relief or deemed guilty of misuse or illegal extension of the patent right by reason of his having... refused to license or use any rights to the patent...."[96] The only two statutory compulsory licensing provisions are for inventions related to atomic energy[97] and air pollution control.[98] Courts have held patents unenforceable when the patent was obtained fraudulently, the litigation was a sham, or there was an illegal tying arrangement.[99]

The resistance to compulsory licensing is well grounded and should be continued in the context of nanotechnology patents. A compulsory licensing scheme would stifle innovation and investment in nanotechnology.[100] Because of the high risks and development costs associated with developing downstream products, companies need patent protection. Venture capitalists would be much more hesitant to invest in start-ups if they perceived that government could shatter the patent barriers that protect their investments. Further, compulsory licensing could decrease public disclosure of technological progress. Companies would be more likely to rely on trade secrets to protect their inventions instead of patents. Since trade secrets are hidden, it is arguable that they slow technology development.

Government Encouragement of Patent Pools

Patent pools on the building blocks of nanotechnology could facilitate timely development of downstream products. Such pools would be limited to complementary technologies and would remain open to future licensees. An example of a useful pool would be a pool on the patents covering the chemical vapor deposition process for producing carbon nanotubes. Pools would eliminate the need for future firms to conclude separate licensing agreements with a number of different entities.[101] In the case of CVD, a company seeking to develop a nanotube device based on the process could obtain a single license rather than having to engage in separate negotiations with a handful of entities holding different patents on the CVD process. Moreover, the constant threat of litigating access to royalties after a downstream product has been developed would be eliminated.[102] Start-up companies could devote substantially fewer resources to licensing transactions and would be more likely to receive financing. The increased commercial development would also generate increased royalties for patent holders. Additionally, reducing transaction costs of licensing would increase the amounts that license seekers are willing to pay for patents. The PTO has endorsed the use of patent pools to facilitate the development of downstream products on gene patents. "[P]atent pools can eliminate the problems associated with blocking patents or stacking licenses in the field of biotechnology, while at the same time encouraging the cooperative efforts needed to realize the true economic and social benefits of genomic inventions."[103]

For the reasons explained earlier in the chapter, such pools are unlikely to arise on their own. The patents on upstream processes and tools are held by universities, start-ups, and large corporations. The divergent goals, size disparities, and business models between the relevant patent holders doom private efforts to establish pooling arrangements. Nevertheless, government might consider encouraging formation of such pools in nanotechnology. Public intervention has been responsible for the formation of successful pools in the automobile, aircraft, and synthetic rubber industries.[104] First, the PTO might publish a white paper encouraging the formation of nanotechnology patent pools. A committee of technology exchange officers could then be formed to design frameworks for useful pools and present them to relevant entities. Government-sponsored patent pools might be more acceptable to industry than privately sponsored pools for three reasons.

First, government-sponsored patent pools would reduce the legal uncertainties associated with pooling arrangements. Patent pools often raise antitrust concerns.[105] Indeed, one of the primary obstacles to the formation of patent pools is the risk that such pools will be struck down by antitrust authorities.[106] Pools designed by the committee of technology officers could be reviewed and approved by the Department of Justice in advance to resolve any antitrust concerns. As one scholar maintains, government-encouraged patent pools "give the government a chance to prevent the most egregious misuses of pooling arrangements."[107]

Second, government involvement could provide the catalyst necessary to form the pools. A free-rider problem plagues the establishment of a pool involving numerous patent holders. No single party has an incentive to expend resources in jump-starting the effort when the group as a whole benefits. Public assistance in designing, coordinating, and implementing the pools might serve as a "visible hand [to help] overcome the collective action problem inherent in group bargaining."[108]

Third, government might be able to leverage its funding commitments to entice participation. As discussed in Chapter 8, the federal government directly funds most of the basic research at universities. Grants are also available for start-up companies, and large corporations even benefit from government aid. Further, there will be an array of lucrative government contracts to develop products. The government could potentially utilize these financial carrots to encourage parties to participate in pooling arrangements.[109]

We are not, however, optimistic that government can facilitate the emergence of comprehensive pools on the building blocks of nanotechnology. Based on interviews and discussions with members of industry and technology licensing offices at universities across the country, we fear that the coordination and implementation difficulties may be insurmountable. An entire article could be written on these complications, but we will identify the most devastating ones. First, government pressure is unlikely to overcome the strong disincentive for participation by companies that wish to use their patents as competitive weapons. As explained earlier, the fragmented intellectual property landscape can be advantageous for large firms vis-à-vis smaller competitors. Second, even if all of the players would be willing to participate, there are complicated problems associated with establishing such pools. As Lita Nelson, director of MIT's licensing office, observed, "the devil is in the details."[110] For example, it is extremely difficult to establish a fair pricing mechanism for the pool of intellectual property. Different licensees may need to license different bundles of intellectual property from the pool. Thus, the entity administering the pool may have to continually determine values of different patents in different packages. Patent holders are certain to disagree on the relative importance of their intellectual property in different packages. Patent holders tend to overvalue their patents and may ultimately conclude that they can extract greater royalties from refraining from participation in the pool. Further, there are complications associated with attempting to include overlapping patents in the packages. Determining which patents are valid and how patents that may be invalid should be priced could give rise to intractable conflicts between patent holders. Finally, government intervention has only succeeded in the past as a result of threatened compulsory licensing.[111] Stopping short of compulsory licensing, government efforts to establish pooling arrangements may be thwarted by the same coordination and collective action difficulties that preclude the emergence of pools through the market. As such, innovative solutions to the fragmentation of the intellectual property landscape are needed.

Chapter 6

FDA Regulation

> *Products on the near horizon will no doubt meld all three: nanorobots that can enter the circulatory system, delivering just the right amount of drug or gene product to the right place. Those who make decisions at the [Food and Drug Administration] FDA about such traditional or complex and high-tech products must be scientifically equal to the intellectual cognitive development that has invented these advanced technologies as we judge which products are ready for the marketplace. If we are not scientifically strong, our decision-making will become risk-averse or, what is worse, simply wrong.*[1]
>
> —Jane Henney (former commissioner, FDA)

As we saw in Chapter 1, nanotechnology could produce an array of new products in medicine, from novel drugs to devices that travel through the body finding and diagnosing illness. This chapter will explore the regulatory problems that the Food and Drug Administration (FDA) will encounter in regulating nanomedical products—specifically, the problems of fitting such products into the agency's classification scheme, determining the appropriate level of regulatory scrutiny to apply to such products, and maintaining adequate scientific expertise in the field. Although the FDA has addressed these issues in regulating biotechnology, nanotechnology will present more difficult regulatory challenges. As of this writing, the agency has only begun to prepare for this burgeoning technology. A failure to effectively regulate nanomedical products could be disastrous for public health, the emerging nanotechnology industry, and the FDA. Thus, the FDA must take further steps to ready itself for the miniaturization of medical technologies.

We first sketch out the current state of the FDA and discuss the consequences of the FDA's failure to effectively prepare for nanomedical products. We then provide a detailed description of the regulatory challenges posed by nanomedicine, highlight what the FDA has done to address these issues in the context of other emerging technologies such as biotechnology, and explain why it will be more difficult to deal with these challenges in regulating nanotechnology. Finally, we propose several courses of action for the agency to effectively prepare for nanomedicine.

STATE OF THE FDA

The FDA is the agency responsible for regulating the safety and effectiveness of most food products, human and animal drugs, therapeutic agents of biological origin, medical devices, radiation-emitting products, cosmetics, and animal feed.[2] The agency, which operates under the Department of Health and Human Services, has a budget of approximately $1.7 billion.[3] The FDA is organized into several Centers that specialize in regulating particular types of products: the Center for Food Safety and Applied Nutrition (CFSAN), the Center for Drug Evaluation and Research (CDER), the Center for Biologics Evaluation and Research (CBER), and the Center for Devices and Radiological Health (CDRH).

In the mid-1990s, the FDA came under attack for unnecessary delays in reviewing applications for new products. Congress passed the FDA Modernization Act (FDAMA) in 1997 to improve the efficiency and effectiveness of FDA regulation.[4] The FDAMA was initially haled as a success in improving regulation. The late 1990s witnessed the FDA significantly decreasing review times despite increasing applications for sophisticated products.[5] In response to increased review times for medical devices in 2000 and 2001, Congress passed the Medical Device User Fee and Modernization Act of 2002. The law authorized user fees for premarket reviews to reduce review times and increase device safety and effectiveness.[6]

The agency will face several challenges in the coming years. First, as clinical research continues to skyrocket, the agency must expect an increase in applications for new products. Second, the increased emphasis on national security has put additional pressures on the FDA. Since the FDA regulates products that could be utilized by terrorists, there have been calls for the agency to assume a more prominent role in combating terrorism. Finally, the agency will face several regulatory challenges posed by nanotechnology. Despite these additional pressures and responsibilities, the agency's budget has remained relatively constant.

WHAT IS AT STAKE IN FDA REGULATION OF NANOTECHNOLOGY

FDA regulation of nanomedicine, like FDA regulation of other novel technologies, requires the FDA to engage in a careful balancing act. The agency must attempt to promote timely patient access and foster innovation while also protecting public health by guarding against unsafe technologies. A failure to adequately prepare for reviewing products based on nanotechnology could have significant ramifications for public health, the FDA, and the development of nanotechnology.

First, public health could be jeopardized in two distinct ways if the FDA is not prepared to regulate nanomedical products. If a lack of agency preparation results in hasty approval of dangerous therapies and a failure to effectively monitor clinical research, patients could be exposed to a significant risk of harm during clinical trials.

Alternatively, a lack of agency preparation could take the form of unnecessary delays in patient access. Inadequate resources combined with a growing caseload, an inefficient regulatory structure, a lack of expertise, or FDA reviewers exercising extreme caution could all result in the agency's taking excessive amounts of time to review new technologies. Unnecessary delays could result in patients' being denied access to potentially life-saving and health-enhancing medical devices.

Second, ineffective regulation could have a substantial impact on nanomedical research and development. If a poor regulatory decision results in a publicized casualty, clinical research involving nanotechnology could be brought to a halt. Researchers already fear that it will be difficult to recruit patients for clinical trials involving nanomedical products.[7] If a research subject were to perish as the result of an experimental procedure involving a nanoproduct, it would become nearly impossible to recruit patients willing to engage in human subjects research. Without the volunteers necessary to conduct large-scale trials, the industry would be unable to secure FDA approval for marketing. Start-ups would be severely crippled, and investors could lose confidence in the field.[8]

Other emerging medical technologies as well as industries investing in nanotechnology would also be impacted by a high-profile injury or death. FDA reviewers always fear that approval of a dangerous product will result in an embarrassing interrogation before Congress;[9] the reluctance to approve any new clinical research or product would be significantly magnified in the aftermath of a tragedy. Furthermore, such an incident involving a nanomedical product would be utilized by opponents of nanotechnology to bolster their case for a legislative ban on all nanotechnology research.

If ineffective regulation takes the form of regulatory delays rather than hasty approval, industries investing in nanotechnology will also be crippled. Increased delays in approval for clinical research and marketing result in increased difficulties for start-up companies attempting to secure financing.[10] Increased delays also decrease the likelihood of larger companies devoting resources toward novel research and development.[11]

A failure to prepare for nanotechnology could also impair the efficacy of the FDA. The American public has historically maintained high confidence in the FDA. Indeed, the FDA takes great pride in its "proud tradition" that allows the public to live with "peace of mind."[12] Severe injuries or death during clinical trials and product recalls could reduce the public's confidence in the FDA. At the same time, regulatory delays that result in patient suffering or deaths could also shatter public support. A distrusting public would undermine the effectiveness of the FDA. Recruitment efforts would be hampered and the spirit of managers and employees would be dampened. The agency acknowledges the importance of public support: "[I]n order to keep fulfilling the public's expectations and maintaining its confidence, FDA needs the public's support."[13]

CHALLENGES POSED BY NANOTECHNOLOGY

The FDA will face an explosion of applications for novel therapies in the coming years, and a substantial portion of these new therapies will be based on nanotechnology. Indeed, the FDA itself has identified nanotechnology as a burgeoning arena of science for which the agency must prepare. At a Science Board meeting in November 2000, Elizabeth Jacobson, the Senior Advisor for Science, noted that "[n]anotechnology is no longer science fiction."[14] The Centers have recognized that they will encounter nanotechnology in the near future.[15] In the summer of 2003, Robert Langer declared at an FDA Workshop that nanotechnology "can make a huge impact" and provided several examples of products that will be presented to the FDA in the near future. As nanomedicine comes closer to fruition, it will present complex regulatory issues.

FDA will regulate nanomedical products within the framework provided by current statutes;[16] thus, review will focus on the safety and efficacy of individual products. There are three primary regulatory problems posed by nanomedicine: classification difficulties, product approval dilemmas, and a lack of scientific expertise. Although the FDA has addressed these problems in the context of other emerging technologies, it has not yet taken substantive steps to prepare for these problems in the context of nanotechnology.

The Classification Problem

The first significant regulatory dilemma posed by products based on nanotechnology is that of classification. Although the current classification system has been applied to other emerging technologies, the miniaturization of medical products will compound problems associated with regulating combination products and blur the distinction between the different categories of products to a greater degree than ever before.

The Classification System: Drug, Device, Biologic, or Combination Product

The FDA classifies medical products for regulatory purposes as drugs, devices, biologics, or combination products. The Center for Drug Evaluation and Research (CDER) is responsible for regulating drugs.[17] A drug is defined as:

(A) articles recognized in the official United States Pharmacopoeia, official Homeopathic Pharmacopoeia of the United States, or official National Formulary, or any supplement to any of them; and (B) articles intended for use in the diagnostics, cure, mitigation, treatment, or prevention of disease in man or other animals; and (C) articles (other than food) intended to affect the structure of any function of the body of man or other animals; and (D) articles for use as a component of any article specified in clause (A), (B), or (C)....[18]

In order to manufacture and market a new drug, a manufacturer must first file an Investigational New Device (IND) application to get approval for human subjects research.[19] CDER must approve and monitor the clinical trials. Upon completion of clinical trials that test the product's safety, effectiveness, and dosage, CDER may approve a New Drug Application (NDA) if the benefits of the drug outweigh the risks.[20] The manufacturer must comply with labeling requirements and a set of manufacturing regulations called the current Good Manufacturing Practices (GMPs).[21] CDER can levy a "user fee" on manufacturers for reviewing a new drug application.[22] The revenue generated from user fees must be used to make approval more efficient.

The Center for Devices and Radiological Health (CDRH) is responsible for regulating medical devices.[23] A device is defined as:

an instrument, apparatus, implement, machine, contrivance, implant, in vitro reagent, or other similar or related article, including any component, part, or accessory, which is

1. recognized in the official National Formulary, or the United States Pharmacopoeia, or any supplement to them,

2. intended for use in the diagnosis of disease or other conditions, or in the cure, mitigation, treatment, or prevention of diseases, in man or other animals, or

3. intended to affect the structure or any function of the body of man or other, and which does not achieve its primary intended purposes through chemical action within or on the body of man or other animals and which is not dependent upon being metabolized for the achievement of any of its primary intended purposes.[24]

Devices are classified into three different categories: Class I, Class II, or Class III. Class I devices present the lowest risk and are subject to "general controls."[25] Class II devices are subject to "special controls."[26] Class III devices present the greatest risk and are subject to review for safety and effectiveness. In order to obtain FDA approval for clinical trials, a manufacturer must submit an Investigation Device Exception (IDE). In order to market the device, a manufacturer must submit a Premarket Approval Application (PMA), which imposes strict conditions on the manufacturing and labeling of the device. A new device that is "substantially equivalent" to a device already being marketed is not subject to review as a Class III device if the manufacturer obtains 510(k) approval.[27] CDRH can also levy a "user fee" on manufacturers for reviewing a new device application.

While drugs and devices are regulated under the Food, Drug, and Cosmetic Act, biologics are regulated by Center for Biologics Evaluation and Research (CBER) primarily under the Public Health Service Act.[28] A biologic is defined as:

> a virus, therapeutic serum, toxin, antitoxin, vaccine, blood, blood component or derivative, allergenic product, or analogous product, or arsphenamine or derivative of arsphenamine (or any other trivalent organic arsenic compound) applicable to the prevention, treatment, or cure of a disease or condition of human beings.[29]

CBER is responsible for regulating a wide variety of "biologics": blood and blood components, devices, allergenic extracts, vaccines, tissues, somatic cell and gene therapies, biotech derived therapeutics, and xenotransplantation.[30] CBER's responsibilities in regulating biologics are similar to CDER's responsibilities in regulating drugs. Approval must be granted for clinical testing of new biological products. In order to obtain a license to market, the agency must determine that a biological product is "safe, pure, potent, and manufactured accordingly."[31]

When a product is designated a "combination," the product's "primary mode of action" determines which center has primary jurisdiction over the product.[32] A combination product is defined as:

> 1. A product comprised of two or more regulated components, i.e., drug/device, biologic/device, drug/biologic, or drug/device/biologic, that are physically, chemically, or otherwise combined or mixed and produced as a single entity;
>
> 2. Two or more separate products packaged together in a single package or as a unit and comprised of drug and device products, device and biological products, or biological and drug products;
>
> 3. A drug, device, or biological product packaged separately that according to its investigational plan or proposed labeling is intended for use only with an approved individually specified drug, device, or biological product where both are required to achieve the intended use, indication, or effect and where upon approval of the proposed product the labeling of the approved product would need to be changed, e.g., to reflect a change in intended use, dosage form, strength, route of administration, or significant change in dose; or
>
> 4. Any investigational drug, device, or biological product packaged separately that according to its proposed labeling is for use only with another individually specified investigational drug, device, or biological product where both are required to achieve the intended use, indication, or effect.[33]

The Center chosen to regulate a combination product must apply the appropriate regulatory requirements to each part of the product. For example, if a product incorporates a biologic and a drug, and the primary mode of action is the biologic, CBER would regulate the product, in consultation with CDER, using applicable biologic and drug regulations.

A manufacturer can submit a request to have the product characterized as a drug, biologic, device, or combination product,[34] and the intent of the manufacturer is often evaluated as evidence of how the product should be classified.[35] Ultimately, the FDA is accorded substantial deference in making this determination.[36]

A manufacturer may prefer that the product be characterized in a particular way for two reasons. First, the FDAMA aimed to make the regulatory requirements for biologics and drugs similar,[37] but there are significant differences between the approval process for devices and the approval process for drugs and biologics. There are statutory differences in approval times,[38] and there may be a greater likelihood of securing approval for a product if it is designated as a device.[39] Second, in the case of a combination product, a manufacturer may prefer that a particular Center have primary jurisdiction over the product for several reasons. A manufacturer may be more familiar with a particular Center or a manufacturer may want to target a particular Center for its tendency to evaluate certain types of evidence.[40]

FDA Reform To Address Classification Issues

The original classification system, which designated products as either drugs, devices, or biologics, was adequate as long as products clearly fell into a particular category. Advancing medical technologies that appeared to combine drugs, devices, and biologics led Congress to create the fourth category for combination products in 1990. In 1991, agreements were formed between CDER and CBER,[41] CDRH and CDER,[42] and CDRH and CBER[43] establishing guidelines for determining which Center has primary jurisdiction over a combination product.

Throughout the 1990s, the agency and manufacturers were generally able to determine if a product was a drug, biologic, device, or combination product. However, there have been two regulatory problems associated with combination products. First, there have been disputes over which Center should have primary jurisdiction as determined by the primary mode of action of the product. Not only have manufacturers quarreled with the FDA, but there have been arguments between the Centers. Even with the standards set forth in the inter-Center agreements, the appropriate jurisdictional designation can be "difficult and time-consuming to determine."[44] Second, even when a combination product is efficiently directed to a particular Center, the Center does not always apply the appropriate regulations to all components of the product. For example, under the 1991 agreement between CDRH and

CBER, the latter has been responsible for regulating the medical devices associated with blood collection and processing as well as cellular therapies. Although CBER maintains that it regulates devices according to "the appropriate medical device laws and regulations," the standards used to evaluate device components are more like CBER's licensing requirements than the standards employed by CDRH.[45]

The FDA has initiated reforms to address these problems. First, to eliminate jurisdictional confusions associated with determining which Center has primary authority over products based on tissue engineering and cell therapy, CBER and CDRH established a Tissue Reference Group in 1998.[46] Composed of three representatives from each Center, the group determines which Center should maintain jurisdiction over particular products, clarifies regulations, and writes guidance documents. Second, to make CBER's review of the device component of a combination product more consistent with how review would take place under CDRH, CBER launched the Device Action Plan in 1999.[47] A Device Management Team was established to supervise the regulation of devices at CBER and enhance cooperation between CBER and CDRH. Finally, in 2003, the agency launched the Office of Combination Products to assign combination products to one of three Centers for primary regulatory responsibility. The Office will coordinate any reviews that need the input of more than one Center and resolve any relevant disputes. It is hoped that the new office will increase the speed and effectiveness of premarket review and ensure consistent postmarket regulation.

The Classification Problems That Nanotechnology Creates

As medical products become smaller, classification will become increasingly difficult and confusing for two reasons. First, the ability to operate at the nano level will increasingly enable manufacturers to combine different types of components in producing a single therapy. Second, in the long run, nanotechnology will blur the distinction between "mechanical," "chemical," and "biological" modes of action and make it difficult to determine if a product is a drug, device, biologic, or combination product.

The miniaturization of medical products will result in an increase in combination products. Because it will be difficult to characterize the primary mode of action of these products, there will be jurisdictional confusion and disputes. For example, it is unclear how the novel drug delivery devices, such as polymers or nanoparticles that deliver drugs to cancer cells or nanoshell composites that periodically dispense drugs, should be regulated. The 1991 Intercenter Agreement between CDRH and CDER, which is the primary source of guidance, cannot be unequivocally applied to novel drug delivery systems. The Agreement states that a device "with the primary purpose of delivering or aiding in the delivery of a drug and distributed containing a drug," such as a prefilled syringe, is a combination product with CDER maintaining primary jurisdiction.[48] However, it later states that a device used "concomitantly with a drug to directly activate or to augment drug effectiveness," such as laser activation of oxsoralen for psoriasis, is regulated by CDRH as a separate entity.[49] Assuming the polymers or nanoparticles are activated by infrared light, it is unclear exactly how the delivery system would be regulated. FDA officials responsible for classification have acknowledged the "many shades of gray" involved in classifying novel drug delivery products.[50]

Thus, without updated guidelines governing novel combination products such as drug delivery systems and drugs combined with monoclonal antibodies, there could be an increasing number of jurisdictional disputes (see Figure 6.1). The time-intensive process associated with determining primary jurisdiction will result in increased regulatory delay for nanomedical products.

In the long term, nanotechnology could produce a whole new class of products that will defy easy classification as drugs, devices, biologics, or combination products. The current distinctions between "chemical," "mechanical," and "biological" activity will be rendered useless. First, at the atomic and molecular level, there is no distinction between drugs and biologics, as currently defined. Since biological organisms are comprised of chemical elements, primarily carbon, oxygen, hydrogen, and nitrogen, biological interactions can be characterized as chemical interactions. Second, at the atomic level, the distinction between drugs and devices is worn away. At the macro level, a "mechanical interaction," which conjures images of machinery or tools, can be conceptualized as a change in force and matter but not a change in the chemical composition of the substance. However, when the focus is on the rearrangement of atoms, it makes no sense to distinguish between chemical and physical forces. Thus, the distinction between drugs, device, biologics, and combination products is only tenable to the extent that future nanomedical products are arbitrarily assigned to a particular category. Without guidelines specifically identifying and categorizing different nanomedical products, these products could be characterized as "mechanical," "chemical," or "biological" depending on the framing devices used to depict the product.

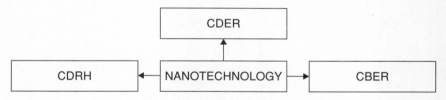

Figure 6.1 Review of nanotechnology at the FDA.

As explained in Chapter 1, complex nanorobot technology is decades away from having, or may never have, any potential commercial impact. Nevertheless, attempting to categorize nanorobots is a useful illustration of the classification issues raised by nanotechnology in the long run. "Microbivores" are conceptual nanorobots that would enter the body, destroy pathogens, and exit the body intact. The microbivore destroys the pathogens by using genetically engineered enzymes. It could be argued that the microbivore functions primarily through mechanical, chemical, or biological means. First, it can be argued that the microbivore, comprised of "ports," "chambers," and "sensors," mechanically destroys the pathogen. Unlike a drug that is metabolized by the body, microbivores exit the body without being fundamentally altered. Indeed, Robert Freitas has described the product as an "artificial mechanical phagocyte," a "device," and a "machine."[51] However, it is arguable that the microbivore engages in a chemical interaction by using enzymes to chemically alter the pathogens. In this respect, it is like an antibiotic or any other drug. Although it is not metabolized like a typical drug, metabolization can be understood as the incorporation of the therapy into the body's bloodstream and the therapy's use of the body's energy as a source of fuel. Finally, since the enzymes used to destroy the pathogens are genetically altered proteins, a careful reading of the Intercenter Agreement between CBER and CDER would appear to support classifying at least part of the product as a biologic.[52]

Ultimately, any determination that the so-called microbivore functions primarily through chemical, mechanical, or biological means would be somewhat arbitrary. A scientific breakdown of how an enzyme operates reveals that both mechanical and chemical methods are used to produce molecular changes.[53] Mechanical forces that involve proton configuration are responsible for the enzyme engaging in chemical reactions involving the production and breakdown of adenosine triphosphate (ATP). Biochemists and molecular biologists have been unable to classify an enzyme's activity as either chemical or mechanical.

Difficulties associated with classifying "pharmacy-in-a-cell" further illustrate the categorization problems that major advances in nanomedicine create. This nanorobot enters a cell, grabs proteins produced by the cell that will not be used, and stores them until they are later needed by the patient. The nanorobot consists of a nickel drum attached to a biological motor. The drum is coated with antibodies that pick up molecules, and an electric field pulls the molecules to a storage chamber and holds them in place. The motor would be powered by ATP. Classification turns on the characterization of the process by which the proteins in the cell are picked up, placed in storage, and then released as needed; it could be argued that this process results in a chemical change or simply the physical movement of matter. The classification is further complicated in two ways. First, antibodies, which are regulated as biologics, play a major role in moving the molecules. Second, the motor which propels the pharmacy-in-a-cell is based on an enzyme, which uses both mechanical and chemical methods to effect molecular changes.

The Problem of Product Approval

In addition to creating confusing issues related to which Center has primary jurisdiction, the miniaturization trend will present complex questions involving data thresholds for product approval. As we argue throughout the book, nanotechnology will improve existing products by making them smaller. In medicine, new therapies based on nanotechnology will be more advanced versions of existing therapies. In many instances, the products being augmented will have already been approved by the FDA. The pressing issue confronting industry and the FDA will be what level of data is required to show that the improved therapy is safe and effective.

Obtaining Regulatory Approval for New Versions of Existing Products

As explained earlier, in order to market a new product, the manufacturer must obtain approval from one of the three centers. Although the specific requirements differ for each center, in general, obtaining approval requires the sponsor to perform exhaustive pharmacology and toxicological studies as well as clinical trials.

The agency has established different approval routes for a sponsor seeking to market a new version of an existing product. A sponsor seeking approval of a newer version of an existing drug has two potential tools to obtain approval without engaging in the rigorous process of obtaining a full NDA pursuant to section 505(b)(1). First, a sponsor can attempt to obtain an abbreviated new drug application (ANDA) by proving that the new product is a duplicate of the previously approved drug product.[54] In order to qualify for an ANDA, the sponsor must scientifically demonstrate that the product is the "bioequivalent" of the approved drug.[55] If a sponsor cannot prove that the drug it seeks to market is the bioequivalent of the approved drug, it can file a 505(b)(2) application. The first type of 505(b)(2) application is a traditional "paper NDA" application supported by published literature.[56] For example, the pharmacology and toxicology data are generated through a literature search while the clinical studies are done by the sponsor. The second version of the 505(b)(2) application relies on prior FDA approval of a drug. The applicant must provide any additional data necessary to demonstrate the safety and effectiveness of differences between the original drug and the 505(b)(2) drug. Thus, while a sponsor does not have to reinvent the wheel, the agency can require specific studies to demonstrate the relevance and applicability of prior findings.[57]

If a device manufacturer makes a change that could affect the safety or effectiveness of the device for which the applicant has an approved PMA, the applicant must submit a PMA supplement for review and approval.[58] The FDA might require additional studies to demonstrate that no adverse effect results from the change. CBER has a similar process for manufacturers wishing to change products marketed under a BLA.

FDA Regulation of New Versions of Other Technologies

Technological advances in medicine have ushered in a variety of new improvements to traditional drugs, devices, and biologics. In some cases, the task of acquiring the data necessary to obtain approval for an improvement technology is relatively simple. For example, the FDA did not have "too much heart burn" in approving the use of antibiotics in bone cement.[59] In other cases, negotiating data requirements with the agency and garnering the necessary data can be a regulatory nightmare. The data required for safety and efficacy of a new version of a product is relatively subjective. When there are no guidance documents, FDA reviewers have substantial discretion. Often, in erring on the side of caution, reviewers tend to request additional clinical data when there is sound scientific basis for approving the newer version without additional information. Even when a new technology slightly modifies the functionality of an approved technology, "all hell [can] break[] loose" in securing regulatory approval.[60] A company improving an existing technology, with ample evidence of safety and efficacy, might find itself conducting a whole new round of clinical trials.

FDA has long dealt with the issue of newer versions of existing technologies and has utilized different tools to smooth the path to regulatory approval. In the 1980s, manufacturers obtained PMAs to market magnetic resonance (MR) scanners. Manufacturers made frequent alterations to MR systems. In each case, they were forced to file PMA supplements, which sometimes involved additional clinical data. Because the process of filing supplement PMAs was costly and the improvements did not impact the safety or efficacy of the technology, FDA approved reclassification of MR technology as a Class II device.[61] As such, newer versions of MR no longer required supplemental PMAs and the possibility of additional clinical trials. Rather, they could be marketed through premarket notification pursuant to section 510(k).

In the 1990s, many biotechnology companies faced a complicated regulatory pathway when improving manufacturing processes of biological products. Often, manufacturers would change the manufacturing process after the clinical trials, but before FDA approval. Traditionally, FDA refused to approve applications unless clinical trials were conducted with the specific product to be licensed because "any change in the product during the clinical trial could be problematic because clinical data obtained with variants of the product...might not be acceptable to support licensure[.]"[62] To alleviate the regulatory burden placed on manufacturers in improving their manufacturing processes, the FDA in 1996 released a guidance document to clarify when additional clinical trials would be needed when improving the manufacturing process.[63] The Guidance Document stated that the FDA could approve biological products based on clinical data from a "precursor product" if it is comparable to the product seeking approval. The Document outlined specific tests that could be used to demonstrate that the later product is comparable to the prior composition.

The Regulatory Approval Issues That Nanotechnology Presents

Nanotechnology will present a host of complicated issues related to regulatory approval. Nanomaterials and devices will primarily be used to develop more advanced versions of a number of existing products. There will be many uncertainties and questions associated with what amounts of additional data will be required to obtain regulatory approval. For example, the novel drug delivery techniques resulting from nanotechnology will present challenging data issues. As discussed in Chapter 1, examples of drug delivery mechanisms include the use of nanostructures that release traditional antibiotics only when near an infection, coatings that prevent digestive enzymes from breaking down the drug in the stomach, and implantable devices that can periodically dispense medicines, such as insulin or morphine. When a new drug delivery system is used in conjunction with a drug that has NDA approval, the FDA will be forced to make difficult decisions about the risks of the new therapy and the data that will be required to market it. Lawyer Jonathan Kahan summarized the regulatory dilemma at an FDA workshop in 2003:

> [T]he question is when you modify the drug formulation to optimize delivery with the device,...are you now about to reinvent the wheel and have to start over.... That is not something that most device companies want to do. They do not want to reinvent the drug wheel. And so the question is: Is a new NDA required for the drug if you have a different mechanism that the mechanism that was described in the NDA-approved label.... And query whether a device company using the 505(b)(2) process can—with a different, let's say, route of administration and a clear drug product—can they rely on 505(b)(2) without a drug manufacturer even on the horizon to get their product through? A real tough issue. I don't have the answer. It is something that a lot of companies are looking at...

Complicated data issues could arise in conjunction with an array of different medical products that could be enhanced by nanotechnology. A number of companies are developing tissue engineering and cell therapies.[64] Some of these companies will receive FDA approval to market their products in the near future. If nanomaterials are later developed that can be integrated into such therapies to increase their effectiveness and reduce immune reaction risks, there will be complex questions associated with what levels of new data are required to continue marketing the improved therapies. Similarly, when approved diagnostic devices are made less invasive by reducing some of their dimensions to the nanoscale, manufacturers will have to negotiate what additional data are needed to demonstrate safety and efficacy.

The uncertainties and costs associated with obtaining regulatory approval for integrating nanotechnology into existing products is magnified by the substantial scientific uncertainties surrounding the insertion of nanomaterials in the human body. As discussed in Chapter 4, there are a variety of unknown health risks associated with nanomaterials. Dr. Vicki Colvin maintains, "Nanomaterials are valuable in many technologies because they interact quite differently with the body than larger materials."[65] Further, many start-ups in the field may overlook the immunological complications associated with different types of nanomaterials. James Baker, a leader in the nanomedical field, notes: "Most of the people proposing this stuff are not biologists and they think they can stick anything in the body if it's small enough."[66]

The ad hoc process of determining the amount of data needed to market a nanotech product could thwart development of the field. Even though the integration of nanotechnology may only slightly alter an existing product and should have no adverse effects, a company might be required to "reinvent the wheel" and start full-blown clinical trials all over again. If a company is not prepared for these stringent data requirements, it might be forced to abandon the entire project. Ultimately, the enormous uncertainties could dissuade industry from embarking on nanomedical research and development efforts.

In the very long run, the introduction of nanorobots could present tremendous regulatory uncertainties. Reviewers will evaluate data for the safety and efficacy of a whole new type of therapy. The agency will have to consider the risks associated with "old nanorobots" being left in the body if they fail, *in vivo* replication, and untested interactions between different nanorobots or nanorobots and drugs. There will also be complicated manufacturing issues. For example, the FDA must ensure that the quality assurance within the manufacturing process is adequate to reduce the possibility of dysfunctional nanorobots as well as the environmental risks associated with nanorobots. As one scientist warns, "A true glitch will come from some direction that nobody anticipated."[67]

The Problem of Scientific Expertise

The regulatory issues outlined above point to a need to obtain expertise in nanoscience at the FDA. Effective regulation of nanomedical products will require the agency to become well versed in nanotechnology. Although the agency has taken steps to acquire the technical abilities necessary for effective regulation of other emerging technologies, the FDA will face unique problems in obtaining the aptitude to effectively regulate nanomedical products.

Scientific Expertise Is Critical to Effective Regulation

Effective regulation requires that the FDA maintain expertise in cutting edge technology and scientific advances. The FDA has recognized the importance of a strong science base in its 2001 performance plan: "The pace of technology innovation in this country and around the world requires the Center's cadre of scientists to keep up with the latest technology and scientific advances, in both the development of medical technology and scientific methodologies."[68] Jane Henney, the former FDA Commissioner, explained the need for the FDA to obtain the scientific expertise to regulate nanomedical products:

> Products on the near horizon will no doubt meld all three: nanorobots that can enter the circulatory system, delivering just the right amount of drug or gene product to the right place. Those who make decisions at the FDA about such traditional or complex and high-tech products must be scientifically equal to the intellectual cognitive development that has invented these advanced technologies as we judge which products are ready for the marketplace. If we are not scientifically strong, our decision-making will become risk-averse or, what is worse, simply wrong.[69]

FDA Preparation For Acquiring Expertise in Emerging Technologies

The FDA has done an adequate job of preparing for novel technologies in the past. The FDA's experience in regulating products based on early biotechnology, artificial intelligence, and advanced biotechnology products demonstrate that the agency is able to acquire the scientific expertise necessary for effective regulation.

The FDA was able to equip itself for effective regulation of early biotechnology products. The biotechnology revolution was launched in 1976 when a human protein was expressed from recombinant DNA in *E. coli*. Recombinant DNA technologies resulted in products such as synthetic insulin to treat diabetes and interferon to treat leukemia, and the biotechnology industry began to take flight in the 1980s. The FDA responded to the emerging industry in several ways. First, the FDA decided not to create a new Center for biotechnology, but to incorporate biotechnology products into the current regulatory structure.[70] Each product was regulated on a case-by-case basis for safety and efficacy. The Office of Biologics Research and Review (OBRR) became the FDA's "expert" in biotechnology review. OBRR hired specialists in molecular biology, protein chemistry, and immunology; almost all biotechnology products, including drugs and devices, were sent to OBRR for review.[71] The agency also began to draft documents called "Points to Consider" in the early 1980s.[72] Although not regulations or guidelines, they were intended to facilitate dialogue and understanding between the FDA and the emerging industry. The FDA has continued to promulgate Points to Consider as the industry has advanced,[73] and they are now considered "dogma in the field of biotechnology."[74] Biotechnology advancements also led the agency to establish the Office of Biotechnology in 1990. The purpose of the Office was to "to enable [the] FDA to meet the new challenges presented by advances in the area of biotechnology."[75] The Office advised the commissioner and other FDA officials on biotechnology science and policy, directed agency research and training, attempted to recruit and retain scientists with needed expertise, and represented the FDA on biotech matters to other agencies, industry, academia, and Congress.[76] The establishment of the Office was lauded as effective in putting the FDA "at the forefront of recent advances in the industry."[77] Having served its purpose of equipping the agency with the ability to effectively regulate biotechnology, the Office of Biotechnology was abolished in 1994.[78]

The FDA also took steps to enhance its science base in preparation for technology based on artificial intelligence. Scientists at CDRH began studying artificial intelligence and preparing for review long before they were presented with any applications. Neural networks, which use biological systems to process information, are now being used to create "smart" devices such as automatic Pap smear readers to do repetitive pattern recognition analysis. As one FDA official explains, "Our scientists saw that the use of artificial intelligence in medical devices was on the horizon and that we needed to have expertise in the area. As a result of our investment in this area, when the first application came in the door, we were ready for it."[79]

In the late 1990s and early 2000s, the FDA was faced with a wave of advanced biotechnology products. Breakthroughs in genomics, proteomics, gene therapy, and tissue engineering resulted in a significant increase in applications for clinical testing of novel technologies. In attempting to keep pace with the explosion of new technologies, the FDA has initiated four policies to improve the agency's effectiveness.

1. Former FDA Commissioner Jane Henney decreed several initiatives to improve the quality of the FDA workforce. She contracted with an outside group to work with the scientific staff and the office of human resources to determine the necessary composition of the scientific workforce in the near future.[80] The contractor was also directed to investigate ways to improve recruitment and retention at the agency.
2. The FDA has made efforts to improve internal training. For example, the Centers put on monthly training sessions, the agency has established an alumni program to keep former staff involved in consulting

and training efforts, and there have been efforts to cross-train staff through a scientific exchange program.[81]
3. The FDA has pursued an aggressive leveraging program involving collaboration with outside experts. There have been numerous "joint training" sessions with industry, where FDA staff tour manufacturing sites to learn about cutting edge research.[82] There are also various Cooperative Research and Development Agreements (CRADAs) between the FDA and different companies,[83] and CBER and CDRH have established "vendor days" to allow manufacturers to provide information about their products and research to FDA staff. The FDA has also pursued partnerships with universities[84] and fostered its relationships with other public health service agencies,[85] existing advisory panels and consultants, professional societies,[86] and domestic and international standards organizations.[87]
4. The agency has taken steps to improve its regulatory science. Research activities allow the FDA to obtain independent laboratory information in reviewing applications, set standards for regulatory assessment, establish test methods, monitor products, and study emerging risks.[88] The FDA highlights the success of its Tissue Proteomics Program as evidence of its ability to engage in cutting edge regulatory science.[89] The research, which involves collaboration with the National Institutes of Health (NIH), focuses on developing proteomic tools for the early detection of cancer and other diseases.[90] The FDA can also boast of cutting edge laboratory research in other areas.[91]

These reforms and initiatives have been relatively successful in enabling the FDA to regulate advanced biotechnology products. The agency has been able to spend some time developing a regulatory framework for genetic testing,[92] tissue engineering,[93] gene therapy,[94] and other novel technologies. Not only has the FDA worked diligently to establish regulations, notices, and guidelines regarding testing and manufacturing procedures, but there is evidence that it has been able to more efficiently review applications and better monitor clinical trials and manufacturing.[95] As senior scientist Donald Marlowe explains, the "Centers can rely on their advisory panels and an array of outside experts to obtain expertise in any area."[96]

However, despite its best efforts to keep pace with advancing medical technologies, there is evidence to suggest that the FDA may begin to experience difficulties in maintaining expertise. In January 2004, a study from the Tufts Center for the Study of Drug Development warned that the FDA will face increased pressure to maintain staff expertise as experienced agency personnel retire or change jobs.[97] Indeed, maintaining technical expertise in the coming years will be "a difficult task in the face of rapid technological change, staff turnover, and the broader context of high employment and movement of knowledge workers."[98]

The FDA Will Face Unique Problems in Attempting to Acquire Scientific Expertise in Nanotechnology

While the FDA has taken steps to acquire the scientific expertise necessary for effective regulation of other emerging technologies, the agency has only begun to prepare for the advent of nanotechnology. As of this writing, there have been few conferences, forums, working groups, leveraging activities, or regulatory science projects aimed at increasing agency expertise in nanotechnology. The agency has conceded that, in the context of nanotechnology, there are "now serious gaps between what the agency needs to do and what it can do."[99]

The FDA will be confronted with complex scientific issues in regulating nanotechnology that are at least as complicated as those raised by the most sophisticated applications of biotechnology. As explained earlier, the agency must make difficult decisions regarding what additional data will be required to approve a therapy that is improved by nanotechnology.

There are several reasons why it may be more difficult for the FDA to maintain scientific expertise in nanotechnology than past and other emerging technologies. First, nanotechnology is unique in that it will touch virtually every aspect of modern medicine. As one scholar put it, "[T]he difference between nanotechnologists and biotechnologists is that the former do not restrict themselves to biological limitations like the latter, and they are much more ambitious about the kinds of accomplishments that they want to achieve."[100] Unlike other past and emerging trends in medical products, nanomedical products will be evaluated by every Center at the FDA; often different Centers will be forced to review similar products. For example, CBER will be primarily responsible for evaluating the efficacy and safety of dendrimers in gene therapy while CDER and CDRH will review dendrimers as drug delivery vehicles. This is different from other emerging technologies where a particular center could establish expertise in a particular area of research. Because CBER was handed the responsibility of regulating nearly all biotechnology products, it was able to develop expertise in the area and develop a working relationship with the biotechnology industry. Staff became intimately familiar with products, ongoing research, and industry players while manufacturers became acquainted with the reviewers, procedures, and requirements of CBER. Division of responsibility to enhance expertise may not be possible with nanomedicine, where every Center will be faced with review of nanoproducts.

Second, it may be more difficult for the FDA to acquire staff with an expertise in nanomedicine than other past and emerging technologies. There are still relatively few experts in this burgeoning field. And as the field begins to take flight in the near future, the most qualified scientists will be lured away by the higher salaries and stock options offered by industry.

Finally, the FDA will be forced to address the scientific issues generated by nanotechnology in the midst of other technological changes and a stagnant budget. Attempting to keep pace with the rapid rate of technological change has already stretched the agency's resources and capabilities. Furthermore, the FDA must assume a more prominent role in national security. From drafting guidance documents to hiring appropriate personnel to acquiring the equipment and facilities needed to analyze nanomaterials, adequately preparing for nanomedicine will require a great deal of focus and substantial monetary investment. Nevertheless, the last several years have witnessed the budget shortfalls in the FDA.

RECOMMENDATIONS

The health care revolution brought about by nanotechnology could dwarf all other trends in the history of medical technology. Nanomedicine will pose unique challenges to the FDA in terms of classification, regulatory approval, and maintaining scientific expertise. The agency must begin to prepare now for this coming revolution in medicine.

First, the agency should contemplate what measures to take in preparation for nanomedicine. It should sponsor conferences and workshops focused on identifying and fleshing out the issues associated with nanomedicine. The fruit of these efforts should be the promulgation of "Points to Consider" documents that initiate a dialogue between the agency and the emerging nanomedical industry. Indeed, the agency appears to be moving in this direction. By the time this book goes to press, the FDA will have put nanotechnology in the spotlight at the 2004 Science Forum.

Second, the FDA should consider establishing an Office of Nanotechnology. Like the former Office of Biotechnology, the Office would advise the commissioner and other FDA officials on nanotechnology science and policy, represent the FDA on nanotechnology matters to other agencies, industry, academia, and Congress, direct agency research and training, and attempt to recruit and retain scientists with needed expertise.

Third, in addressing the categorization problems posed by nanotechnology, the FDA should attempt to identify in advance what Centers will have primary jurisdiction over such products. For example, the Office of Combination Products should draft clear guidelines for drug delivery products based on nanotechnology.

Similarly, in addressing the problem of determining the requisite level of data needed to approve nanotechnology-enhanced products, FDA should consider promulgating guidances. For example, in the case of specific drug delivery mechanisms based on nanotechnology, the agency could clearly define how much additional clinical data are needed when a company submits a 505(b)(2) application. In the short term, because there is so much uncertainty associated with the use of nanomaterials in the human body, the data requirements should probably be relatively stringent. As evidence on the health implications of engineered nanostructures becomes clearer, the data requirements can be loosened. In any case, establishing clear guidelines could reduce the uncertainty for industry.

Fifth, in addressing the problem of maintaining scientific expertise, the agency might need to make efforts to acquire personnel with expertise in nanotechnology. But even if the FDA could employ a sufficient number of qualified scientists, it is impractical to expect that the FDA staff will be able to keep abreast of the rapid changes in this dynamic field. As it has done with biotechnology, the agency should consider utilizing other knowledge bases to increase its expertise. Internal training efforts and continued collaboration with industry and academia to enhance nanotechnology expertise will be critical. It might also be worthwhile to pursue collaboration with the NIH, a major player in cutting edge nanomedical research. The agency may also want to engage in laboratory research involving nanotechnology. The initial focus of research efforts should be on increasing the agency's understanding of immunological complications associated with placing nanomaterials in the human body.

Finally, the most important component of the FDA's strategy for preparing for nanotechnology should be securing additional resources. The agency appears to be aware of the need to prepare for nanotechnology, and it has proven that it harbors the capability to keep pace with emerging technologies in the past. Thus, the primary impediment to the agency's efforts to prepare will be insufficient resources. The agency should actively seek increased funding to regulate the coming revolution in medicine.

CHAPTER 7

National Security and Export Controls

Nanotechnology will clearly be crucial to the nation's security.[1]
—James Murday, director of the National
Nanotechnology Coordination Office

As we saw in Chapter 2, nanoscience could have a major impact on the defense industry. From battlesuits worn by soldiers to materials used to build fighter jets to medicines used to treat chemical and biological attacks, nanotechnology offers substantial improvements in virtually every arena of military defense. Daniel Ratner and Mark Ratner devote an entire book, *Nanotechnology and Homeland Security: New Weapons For New Wars*, to the military applications of nanotechnology.[2] Perhaps the most powerful evidence of the role that nanotechnology will play in security in coming years is the investments being made by the Department of Defense. In FY2005, the Department of Defense intends to pour $276 million into nanotechnology.

While government officials are excited about the potential of nanotechnology to improve American defenses, they must also be weary of its potential to threaten American security interests. Indeed, one of the greatest policy challenges posed by the miniaturization trend will be cultivating nanotechnology while preventing potential aggressors from using it for offensive purposes. In this chapter, we examine how the American government will fare in applying its export control regime to nanotechnology. We first review the state of American export control policy, engage in a detailed summary of each of the separate bodies of laws and regulations, and analyze how they will be applied to nanotechnology. We then discuss the importance of effective export controls. Finally, we offer several broad recommendations to American policymakers confronted with the task of regulating nanotechnology.

THE STATE OF EXPORT CONTROL LAWS

The export control regime was largely fashioned during the Cold War. The persistent threat of a massive confrontation between East and West mandated strict controls on technological transfers. With the collapse of the bipolar world, governments significantly relaxed regulations.[3] American companies convincingly argued that they needed fewer restrictions to effectively compete in the global marketplace. In the aftermath of the 2001 collapse of the World Trade Center Towers, however, a new security frenzy has swept Washington. With the declaration of the war on terrorism and the establishment of the Office of Homeland Security, there is a new emphasis on preventing dissemination of dangerous technologies.[4] In coming years, export control laws could be applied with greater scrutiny than ever before.

In crafting and enforcing export control laws in the twenty-first century, the federal government must carefully balance American economic concerns with national security interests. When new technologies pop up on the radar screen, the government must attempt to understand them and apply appropriate regulations. Generally, agencies responsible for administering export control laws are slow to adjust to technological change. One commentator explains, "The cumbersome, conflict-ridden and dilatory regulatory process often obstructs timely adjustments to the control level in order to keep up with advances in technology."[5] In the coming years, the federal government will be confronted with the task of applying export controls and regulations to nanotechnology.

THE REGULATORY LANDSCAPE

Navigating through the various export control laws can be a confusing and frustrating endeavor. Even the most "expert government personnel are often hard pressed to interpret the complex, often ambiguous, and frequently changing export control regulations."[6] In one case, the defendant was an expert who "gave lectures and published newsletters on the subject" of export control.[7] He claimed innocence by arguing that the laws were simply too complicated to comprehend. In reviewing the case, the Eighth Circuit noted that even courts can be confused "by these massive legislative and bureaucratic artifacts."[8] Notwithstanding established doctrine that "ignorance of the law" is no excuse, many courts have overturned convictions because defendants did not have knowledge of and specific intent to violate the law.[9] Indeed, virtually "every major commentator and expert who in recent years has reviewed the rules that must be examined and complied with by American business has found them to be unwieldy, complex, convoluted, and extremely confusing."[10]

Part of the reason for the complexity is that there are a number of different bodies of law administered by several agencies. The departments of State,

Commerce, Customs, Defense, Energy, and the Treasury all have some jurisdiction over export controls. In light of their different missions, agencies often squabble over particular policies and cases. According to the House Government Operations Committee, there is "incessant bickering" that will "continue as long as multiple agencies are routinely involved."[11] Further, within each agency, firms might encounter several different offices.

We spent several weeks gazing at hundreds of pages of convoluted regulations. This is our best attempt to summarize laws relevant to firms involved in nanotechnology. We discuss laws and regulations administered by (1) the Department of Commerce, (2) the Department of State, and (3) other agencies.

Department of Commerce

Since the terrorist attacks of September 11, 2001, the Department of Commerce has become more stringent in enforcing export control laws.[12] The Bureau of Industry and Security (BIS) is responsible for administering the Export Administration Regulations (EAR).[13] We first summarize the provisions of the EAR and then describe how they apply to nanotechnology.

Summary of EAR

EAR regulations apply to companies exporting "dual-use" technologies to countries that are not subject to economic sanctions.[14] "Dual-use" technologies are technologies that can be used for military purposes, but are primarily intended for commercial applications. A relatively small percentage of total American exports and reexports require a license from BIS. License requirements are dependent upon an item's technical characteristics, the destination, the end user, and the end use. Determining whether a company must apply for a commerce export license before engaging in a particular transaction involves several steps.

First, a company must determine if the technology is "subject to the EAR."[15] Items subject to the EAR include:

1. All items in the United States, including in a U.S. Foreign Trade Zone or moving in transit through the United States from one foreign country to another
2. All U.S. origin items wherever located
3. U.S. origin parts, components, materials or other commodities incorporated abroad into foreign-made products, U.S. origin software commingled with foreign software, and U.S. origin technology commingled with foreign technology, in quantities exceeding *de minimis* levels
4. Certain foreign-made direct products of U.S. origin technology or software

5. Certain commodities produced by any plant or major component of a plant located outside the United States that is a direct product of U.S. origin technology or software.[16]

Items not subject to the EAR include:

1. Items that are exclusively controlled for export or reexport by certain agencies
2. Publicly available technology.[17]

If a technology is subject to the EAR, the exporter must then determine if the technology is classified on the Commerce Control List (CCL).[18] A company must painstakingly review the CCL to determine the technology's Export Control Classification Number (ECCN). The CCL is divided into nine categories:

1. Materials, Chemicals, Microorganisms, and Toxins
2. Materials Processing
3. Electronics Design, Development and Production
4. Computers
5. Telecommunications
6. Sensors and Lasers
7. Navigation and Avionics
8. Marine
9. Propulsion Systems, Space Vehicles and Related Equipment

If a technology is not listed on the CCL, it is designated as "EAR99."[19] It is unlikely that a license is needed if the technology falls in this category.[20] Technology "required" for the "development," "production," or "use" of technology on the CCL is controlled according to the specific CCL technology category that covers the technology at issue.[21]

If the technology is listed on the CCL, the company must then determine if its activities fall under one of the General Prohibitions.[22] General Prohibition One restricts export and reexport of controlled items to listed countries. In addition to direct shipment or transmission of items, "export" can include visual inspection of the technology by foreigners, oral exchanges of the information, and application to situations abroad of technical expertise acquired in the United States.[23] Under the "deemed export" rule, regulations also apply when foreign nationals are hired and given access to controlled technology.[24] Foreign nationals are defined as temporary immigrants, including those holding H-1 and H-2 visas. Thus, the hiring of permanent residents to work with controlled technologies does not constitute an export. Even when a foreign national is not directly involved in R&D efforts, if he or she has access to the firm's computer system, a "deemed transfer" might have taken place.

General Prohibition Two restricts reexport and export from abroad of foreign-made items incorporating controlled American technology. A license is necessary when the technology is made with more than the *de minimis* amount of controlled American content. The value of U.S.-origin technology incorporated into foreign-made technology is determined by dividing the total value of the controlled American parts, components, and materials incorporated into the foreign-made item by the sale price of the foreign-made item.[25] The *de minimus* thresholds are different for different technologies and countries.

General Prohibition Three restricts reexport or export from abroad of the foreign-produced direct product of American technology and software. Foreign-made items are subject if they are manufactured using certain types of American technology subject to the EAR, and the foreign-produced items are subject to national security controls as designated on the proper ECCN of the Commerce Control List.

General Prohibitions Four through Ten generally target intentionally "bad behavior," such as:

- engaging in actions prohibited by denial order
- exporting with knowledge that the technology will be used in defined nuclear, missile, chemical and biological activities, nuclear maritime uses, and certain aircraft and vessels
- exporting to embargoed destinations
- supporting proliferation activities
- shipping through certain countries
- violating any order, terms, and conditions
- proceeding with knowledge that a violation has occurred or is about to occur

Firms then use the information contained in the "License Requirements" section of the ECCN in combination with a country chart to determine whether exportation is prohibited.[26] Companies must also analyze supplements of the EAR to find out if the buyer is on a list of denied countries, institutions, or individuals.

Even if technology is found on the CCL and the particular transaction is prohibited, a company must sift through a number of possible exemptions.[27] Examples of exemptions include certain goods to certain countries, shipments of limited value, temporary exports, and the servicing and replacement of parts and equipment.

If a company is required to file a license application with BIS, the process of obtaining a license can be arduous and lengthy. BIS must consult the Departments of State, Defense, and Energy, as well as the Arms Control and Disarmament Agency, on licensing decisions.[28] If the reviewing agency does not take some action within thirty days, BIS can presume there is no objection. If the reviewing agency does raise any questions or objections, the process can be substantially delayed.[29] An Operating Committee resolves disputes and disagreements between the agencies on licensing decisions.

If a company violates the EAR, both the company and the individuals responsible for the violation are subject to criminal and civil / administrative penalties.[30] At the time of this writing, BIS is increasing enforcement efforts.[31]

Application of EAR to Nanotechnology

Due to the broad scope of the EAR and the far-reaching potential of nanoscience, most nanotechnologies are subject to the EAR. It is unknown at this time whether firms will be required to seek licenses from BIS for particular technologies.

First, it is unclear whether different types of nanotechnologies are included on the CCL. We identified a number of different categories that might encompass different materials, tools and processes, and devices and systems based on nanotechnology. (See Table 7.1.) In many instances, the items on the CCL are defined in such a way that it is clear they encompass materials, devices, and systems at the nanoscale. For example, the list of items controlled under magnetic metals clearly specifies certain types of nanocrystalline materials.[32] Similarly, category 1C011 includes "metals in particle sizes of less than 60 [micrometers] whether spherical, atomized, spheroidal, flaked, or ground..."[33]

In other places, however, the language does not clearly establish that nanoscale forms of the listed items are also subject to regulation. For example, the list of items controlled in category 1C210 includes certain types of carbon "fibrous or filamentary materials." Filament is defined as the "smallest increment of fiber, usually several [micrometers] in diameter." The wording of the section makes it unclear how bundles of carbon nanotubes should be treated. A literal reading of the text might lead to the conclusion that bundles of carbon nanotubes are not included. A more contextual interpretation could result in a different conclusion. Items in this category are subject to controls, because their properties make them desirable building materials for certain military applications. Nanotube-based materials would likely have enhanced structural properties rather than "fibrous or filamentary materials." As such, it is arguable that the regulations should be understood to encompass carbon materials that are smaller than fibers and filaments.

In addition to uncertainties regarding the scope of items on the CCL, it is unclear how certain EAR regulations apply to nanoscale devices. For example, Part 744 specifically restricts export or reexport of "'microprocessor microcircuits,' 'microcomputer microcircuits,' and microcontroller microcircuits having a 'composite theoretical performance' (CTP) of 6,500 million theoretical operations per second (MTOPS) or more and an arithmetic logic unit with an access width of 32 bits or more, without a license if, at the time of the export or reexport, you know, have reason to know, or are informed by BIS that the item will be or is intended to be used for a 'military end-

TABLE 7.1 Examples of ECCN Categories That May Be Applicable to Different Areas of Nanotechnology

Materials	Tools & Processes	Devices & Systems
1A002—"Composite" structures of laminates…	1B002—Equipment for producing metal alloys, metal alloy powder, or alloyed materials…	2B007—"Robots" having any of the following characteristics…
1C002—Metal alloys, metal alloy powder and alloyed materials	1B001—Equipment for the production of fibers, prepregs, performs, or "composites"…	3A001(a)—General purpose integrated circuits…
1C003—Magnetic metals, of all types and of whatever forms…	1B102—Metal powder "production equipment"	3A228—Switching devices…
1C005—"Superconductive" "composite" conductors…	2B227—Vacuum or other controlled atmosphere metallurgical melting, casting furnaces, and related equipment	3A233—Mass spectrometers…
1C010—"Fibrous or filamentary materials" which may be used in organic "matrix," metallic "matrix," or carbon "matrix"	2E003(f)—"Technology" for the application of inorganic overlay coatings or inorganic surface modification coatings to non-electronic substrates…	5A001—Telecommunications systems, equipment, and components
1C210—"Fibrous or filamentary materials" or prepregs…	3B001—Equipment for the manufacturing of semiconductor devices or materials…	6A002—Optical Sensors
3C001—Hetero-epitaxial materials consisting of a "substrate" having stacked epitaxially-grown multiple layers…	3D003—Computer-aided design (CAD) "software"…	6A005—"Lasers," components, and optical equipment

use'...."[34] As mentioned in Chapter 1, efforts are underway to commercialize circuits and processors based on nanotechnology. The use of the word "micro" to describe the technology might lead one to decide that the regulation does not apply to nanoelectronic devices. However, since the processing capabilities of nanoelectronics are likely to dwarf microelectronics, the regulations should be interpreted to apply to nanocircuits and the processors based on them.

Finally, the EAR does not address molecular manufacturing. Although this technology could be decades away (if it is even possible) from having any impact, it could be used for a number of different military applications. The CCL does identify certain types of "robots" as items subject to controls,[35] but the descriptions of these "robots" are very different from the self-replicating nanorobots envisioned by nanotechnologists.

State Department

While the Commerce Department reviews commercial technologies that could be used for military applications, the State Department regulates technologies developed primarily for defensive purposes. The State Department is generally more aggressive in applying export controls than the Commerce Department. The Office of Defense Trade Controls (DTC) at the State Department administers the International Traffic in Arms Regulations (ITAR).[36] We first summarize ITAR and then analyze how it applies to nanotechnology.

Summary of ITAR

Technologies subject to ITAR, which are primarily related to the military, are described on the U.S. Munitions List.[37] When the issue arises as to whether an item is primarily "military" or "commercial," DTC has statutory authority to determine which agency has jurisdiction.[38] The Munitions List is divided into 20 categories.[39]

If a company is in the business of either manufacturing or exporting defense articles, it must register with the Office of Defense Trade Controls.[40] In order to export a defense article, a company must obtain the approval of the Office of Defense Trade Controls, unless an exemption is available.[41] It is more difficult to secure a license from DTC than BIS. Not only is DTC more reluctant to approve exports for security reasons, but it also takes more time to obtain a license. The State Department often imposes severe civil and criminal penalties on violators.[42]

Application of ITAR to Nanotechnology

Nanotechnology will be used in a number of different military technologies. Nanotechnologies could conceivably enhance products in nearly every category of the Munitions List. For example, nanoparticles are already being used to enhance the effectiveness of rocket fuel. Future battle suits might be made of carbon nanotube armor and have nanosensors that detect chemical weapons. Nanoelectronic devices might even be used to develop "smart" aircraft, tanks, and even robotic soldiers. Because defensive applications of nanotechnology generally fit into the categories on the Munitions List, applying ITAR to nanotechnology will be a relatively simple task.

Other Laws and Regulations Administered by Other Agencies

In addition to EAR and ITAR, there are a number of other laws and regulations administered by different agencies that might be relevant to firms investing in nanotechnology. The Office of Foreign Assets Control (OFAC), a subdivision of the United States Treasury Department, enforces sanctions programs targeted against specific states such as Cuba, Iran, Iraq, Libya, North Korea, and Sudan. Because different sanctions result from different legislative and executive actions, the sanctions regimes can be very distinct. In general, OFAC prohibits "U.S. persons" from being involved in activities that benefit sanctioned countries.[43] OFAC regulations cover a broad range of activities, including the transfer of technology, regardless of origin, as well as financial dealings. Regulations also prohibit transactions with "Specially Designated Nationals"—entities or individuals determined by OFAC to be acting on behalf of sanctioned countries.[44] OFAC restrictions are unclear and imprecise.[45]

There are also two different bodies of laws restricting investment in or acquisition of American companies. The National Industrial Security Program (NISP), which is administered by the Defense Security Service (DSS) of the Defense Department, regulates foreign investment in companies with classified government contracts.[46] When a company with a classified government contract is acquired by a foreign entity, it must notify DSS, and DSS will impose restrictions on disclosure of sensitive information.[47] In order to continue to perform under a classified contract, the acquired company must implement special security measures that restrict involvement of the foreign parent in management of the company.

The Committee on Foreign Investment (CFIUS), an interagency committee chaired by the Treasury Department, can block any transaction that places an American corporation under foreign control if the transaction threatens American national security.[48] Parties to a transaction have the option of filing notice with CFIUS to receive clearance for the transaction.[49] If the parties do not file a voluntary notice and obtain approval, the transaction can be investigated, and potentially divested, *ex post facto*. Although CFIUS has traditionally focused on transactions involving defense suppliers, in recent years, it has extended its investigations to Internet and telecommunications companies.[50]

Finally, the Energy Department regulates technology related to the production of "special" nuclear materials such as plutonium and enriched uranium,[51] and the Nuclear Regulatory Commission controls exports of "peaceful" nuclear technology such as nuclear power plants, nuclear reactor vessels, equipment, and components for reactors.[52]

THE IMPORTANCE OF EFFECTIVE EXPORT CONTROL LAWS IN NANOTECHNOLOGY

A failure by agencies to clearly address how novel nanoscale, materials, devices, and systems are to be regulated could have serious repercussions for American industry, the development of nanotechnology, and national security.

First, there is a risk of overregulation of nanotechnology. The export control regime places on industry the burden of understanding the regulations and determining when it is necessary to apply for a license. If firms are unsure how regulations apply to their technologies, they might seek regulatory clearance when it is unnecessary to do so. Navigating through the bureaucratic channels necessary to obtain an export license can be a lengthy and frustrating process. American companies that get caught in regulatory webs can lose out to competitors in forming strategic partnerships and capturing global markets. As one expert remarks, "unclear export control policy and bureaucratic maneuvering provide no assurances for buyers of U.S. high-tech goods which forces them to seek other suppliers."[53] The government's experience in addressing encryption technology illustrates the impact of ineffective regulation on technology and industry. Throughout the late 1990s, different export control policies relating to encryption technology were promulgated. These policies were criticized for being unclear, unworkable, and hindering American competitiveness.[54] According to one export control lawyer, "[T]he government's attempts to restrict transfers of encryption software were a major hindrance not only for technology companies, but also for mere users of encryption."[55]

In other cases, uncertainty regarding the scope and application of regulations could undermine national security. If companies find loopholes in the regulations, nanoscale materials, devices, and systems that could be used for military purposes might wind up in the wrong hands. The risks posed by underregulation of export controls are illustrated by tracing the origin of Iraq's military buildup. In the 1990s, American exporters shipped advanced computing systems to Iraq that were used in Saddam Hussein's nuclear weapons program.[56] Similarly, fiber optic technology that American companies sold to China was later installed by Chinese engineers in Iraq to improve Iraq's air defense system.[57]

RECOMMENDATIONS

Several commentators have proposed broad, fundamental reforms of the export control regime such as placing export authority under a single federal agency. Notwithstanding the imminent need for a lone regulatory body, fundamental restructuring is unlikely to take place in the near future. The task of rewriting thousands of pages of complicated and overlapping regulations is monumental. One scholar writes that the export provisions are "infinitely more complex than the criminal and tax codes."[58] Further, federal agencies are unlikely to support efforts to strip them of their powers under export control laws.[59]

We make two suggestions to federal agencies in confronting the challenge of applying export control laws to nanotechnology. First, some of the regulations should be revised to clarify if they encompass nanoscale versions of materials, devices, and systems. For example, the CCL should be revised to clarify whether items described as carbon "fibrous or filamentary materials" include materials incorporating carbon nanotubes, and Part 774 should clarify whether regulations on microprocessors extend to nano-sized processing technology. Second, at some point in the future, regulations should directly address molecular manufacturing technology. Although molecular manufacturing is unlikely to have any impact for decades or even centuries, the government will undoubtedly have an interest in controlling the spread of early breakthroughs.

CHAPTER 8

Federal Funding

In the public sphere technological optimism leads government officials to respond to a serious national problem by throwing technology at it. Even when science provides no basis for believing that a "technical fix" is feasible, scientists and engineers are called on to invent not only the appropriate technical solution but the fundamental new knowledge that will make it possible. And usually, amid controversy about its feasibility, important figures in the scientific community rally to the challenge and predict their own eventual success. Often, but not always, they are right.[1]

—Economists Linda R. Cohen (University of California at Irvine) and Roger G. Noll (Stanford University)

Perhaps the gravest and most timely issue confronting policymakers is the amount of funding that should be channeled toward research and development of nanotechnology. Throughout history, the fates of different technologies have been influenced by the changing tides of public policy. As early as 1836, Congress appropriated $30,000 to subsidize Samuel Morse's first telegraph. In the twentieth century, the Eisenhower administration planted the seeds for the semiconductor industry, the Kennedy administration gave birth to the space program, and the Nixon, Ford, and Carter administrations allocated billions of federal dollars to the pursuit of alternative energy technologies. At the dawn of the twenty-first century, it appears that public officials are rallying behind nanotechnology. The unveiling of the National Nanotechnology Initiative (NNI) by President Clinton in 2000 has ignited a global race to develop this technology. Billions of dollars in public funds are being invested in the field. While the massive public investment bodes well for the future of nanotechnology, simply throwing money at nanoscience does not ensure that the field will reach its full potential. The size of the nano wave will largely depend on how public funds are managed.

This chapter reviews the federal commitment to nanotechnology. First, we provide an overview of the current financing landscape in nanotechnology by examining funding provided by domestic and foreign governments, R&D expenditures made by large corporations, and private equity investment. We show that while governments and large corporations are pumping large amounts of cash into research and development, private investors are still reluctant to make substantial investments at this early stage. Second, we address whether government should be involved in financing nanotechnology in the first place. We argue that there are several persuasive justifications for the NNI. Finally, we analyze whether federal expenditures are being managed in a manner that increases the likelihood of developments in the field. We review the implementation of the National Nanotechnology Initiative, summarize the details of the 21st Century Nanotechnology Research and Development Act, and provide some suggestions for future funding priorities.

THE FINANCING LANDSCAPE

Billions of dollars were spent worldwide in nanotech research and development in 2003. The majority of that funding came from governments and large corporations. Despite the excitement among some venture capitalists, venture capital investment only comprises a small percentage of the total financing. In the following pages, we provide a summary of:

- Federal and state funding in the United States
- Public spending in other countries
- Private investment

American Funding

Federal Funding

In July 2000, President Clinton launched a $422 million National Nanotechnology Initiative (NNI) to galvanize research and development in the field. Specifically, the initiative intended to increase funding for fundamental research, establish nanotechnology research centers, instill a research infrastructure, support workforce education and training, and study the ethical, legal, and societal implications of nanotechnology. In FY2004, the Bush administration proposed a total budget of $849 million for the NNI, up from $774 million in the previous year. Congress codified the existing administrative framework by passing the 21st Century Nanotechnology Research and Development Act in the fall of 2003. The legislation authorizes $3.7 billion in nanoscale science and engineering between FY2005 and FY2008. The details of the NNI and the new legislation will be fleshed out later in the chapter. Figure 8.1 shows American funding for nanotechnology since 1997.

Federal Funding

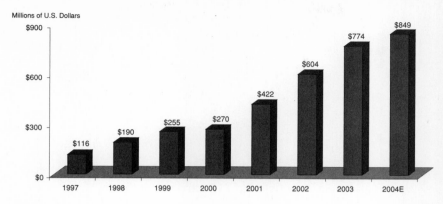

Figure 8.1 American funding for nanotechnology.

State Funding

States have also begun to channel money toward nanotechnology investment. California has lead the way by contributing $25 million to establish the California Nanosystems Institute (CNSI).[2] Jointly funded by private industry, CNSI is situated on the campuses of UC Santa Barbara and UC Los Angeles. Other states, envious of the economic rewards enjoyed by Silicon Valley, have also made substantial investments. For example, the Pennsylvania Technology Investment Authority wrote a check for $10.8 million to establish a Nanotechnology Institute intended to "transform the Delaware Valley into Nanotech Valley."[3] Similarly, the Texas Nanotechnology Initiative was launched to establish a consortium of industry, university, government, and venture capitalists "to establish Texas as a world leader" in nanotechnology.[4] The regional race to capture nanotechnology has only begun. In 2002, the Nanotechnology Business Alliance launched a "Nanotech Hubs Initiative" to aid states and regions in establishing "Nanotech Valleys" and "clusters." Mark Modzelewski explained to Congress the regional frenzy to cultivate nanotechnology: "Though we have launched efforts in six regions—as well as affiliates in the EU and Canada—we have been inundated with calls from states and 11 countries to help develop this capacity. These states and regions are already looking to nanotechnology to ignite economic development."[5]

Foreign Public Spending

As shown in Figure 8.2, international competition in the nanotechnology arena is heating up. Japan and the European Community can boast of programs in nanotechnology comparable to that of the United States. According to Japanese government data, Japan's budget for nanotechnology

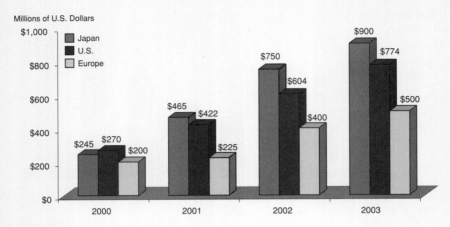

Figure 8.2 The international race.

was approximately $900 million in 2003, and the budget request for 2004 seeks to increase funding by 20 percent.[6] A report from the *Journal of Japanese Trade and Industry* reveals that Japan believes developing nanotechnology is necessary for the "restoration of the Japanese economy."[7] The EU has displayed similar euphoria by earmarking $1.2 billion for nanotechnology research and development in 2003 and 2004.[8] Canada has shelled out $140 million to develop a National Institute For Nanotechnology,[9] and South Korea, Taiwan, China, and Russia have started making significant investments in nanotechnology.

Private Investment

Corporations

A substantial portion of the money channeled to nanotech R&D in 2003 came from large corporations. Indeed, sifting though nanotechnology-related patents reveals that many of the patents on upstream research are held by large corporations like Hewlett-Packard, IBM, and NEC.

Start-ups

There are dozens of start-up companies in this field. Start-ups can obtain private financing from several different sources such as friends and family, angel investors, strategic alliances with large corporations, and venture capitalists. We discuss each of these financing mechanisms in detail in Part III of the book. Specifically, we provide a great deal of data on venture capital investment in different sectors of nanotechnology.

SHOULD GOVERNMENT FUND NANOTECHNOLOGY?

Before analyzing how the National Nanotechnology Initiative is being implemented, one should consider whether government funding of nanotechnology is justified in the first place. In this section, we identify the different rationales for government funding of R&D and explore what rationales can be used to support the existence of the NNI.

Public funding of research and development can be motivated by three different goals. First, government funding can be intended to support "basic" scientific research.[10] This goal is rooted in the assumption that the private sector is prone to underinvest in basic research because the returns to the investment may be appropriated by others, and the risks are difficult to measure.[11] For example, government should support scientists such as Richard Feynman, a physicist who made major strides in quantum mechanics, because private companies are unlikely to engage in such research. Publicly funded basic research is usually conducted by government agencies or universities. This goal is generally accepted as a legitimate basis for government financing of R&D.

Second, public funding can be intended to advance technology in the production and performance of goods and services for which government is the dominant customer.[12] For example, federal expenditures on national defense are generally intended to yield products that will be purchased by the Department of Defense. This goal is also relatively uncontroversial.

The third goal—a more contentious one—is that government financing can be intended to aid the private sector in developing commercial products. According to this rationale, certain industries are particularly susceptible to the problems of insufficient appropriability and market imperfections that preclude optimal levels of private investment. These "strategic" industries are characterized by rapid technological progress and close linkages with other industries.[13] Scholars have criticized this goal on the basis that the government does a particularly poor job of deciding what types of commercial R&D should be funded.[14]

To some extent, government funding of nanotechnology is motivated by all three of the preceding goals. Public funds are intended to support basic research taking place at universities and government labs. Additionally, a great deal of money is used by government agencies to facilitate the production of goods that will be purchased by the government. Finally, government funding is intended to support private commercialization of products. This goal only provides a rationale if nanotechnology is "strategic" and is susceptible to market failures in private R&D decisions. Many areas of nanotechnology are simply not yet ripe for private investment. While nanoscience is still in its infancy, the field is experiencing rapid advances which could impact a variety of different industries. Large government expenditures are not only justified, but critical to the survival and flourishing of nanotechnology.

MANAGING FEDERAL FUNDING OF NANOTECHNOLOGY

General Problems in Managing R&D Programs

Even if the goals of government funding of nanotechnology are justified, it is possible that the program is not being implemented in a manner that achieves those goals. Before reviewing implementation of the NNI and the 21st Century Nanotechnology Research and Development Act, we provide a brief summary of general problems encountered in managing public R&D programs. Critics of publicly funded research and development highlight two general problems associated with managing R&D programs.

First, decision makers are prone to fund projects that would develop and thrive absent public expenditures.[15] In an attempt to develop a track record for "choosing winners," bureaucrats are inclined to fund projects that are certain to succeed. However, these are the projects that are least worthy of funding since they are likely to take place with or without government involvement. The most deserving projects are ones that are too risky to be funded by the private sector but could yield revolutionary breakthroughs. Second, decision makers are susceptible to political pressures in making funding determinations. Critics highlight several egregious failures that were arguably supported for political and not technical reasons. As examples, consider the following:

- The Clinch River Breeder Reactor project cost taxpayers over $5 billion without yielding any benefit
- $920 million was poured into the original American Supersonic Transport without achieving any meaningful result
- $20 billion was spent on the Synfuels Program, which produced a single technology (Cool Water) valued at about $263 million.[16]

According to critics, the underlying cause of these failures was classic "pork barrel" politics.[17] Government funding results in the formation of special interest groups claiming that they are worthy of support. Such interest groups provide biased information and have a major influence on the decision-making process.

Without engaging in a rigorous empirical or theoretical analysis, we generally proceed under the assumption that these problems do not automatically render R&D programs ineffective. Rather, they can be accounted for by a well-designed program. As economists Roger Noll and Linda Cohen conclude, "the factors influencing the success of a project are not only predictable but in part controllable, in the sense that the details of the way the program is set up can affect its expected performance."[18] The remainder of this chapter attempts to identify weaknesses in the implementation of the NNI and provide a framework to combat the problems associated with implementation.

The National Nanotechnology Initiative

Structure of the National Nanotechnology Initiative

The National Nanotechnology Initiative (NNI) is managed by the National Science and Technology Council (NSTC). The National Science and Technology Council was established by President Clinton's Executive Order on November 23, 1993.[19] This Cabinet-level council serves as a "virtual agency" by enabling the president to coordinate science, space, and technology policies across federal agencies. The NSTC's Subcommittee on Nanoscale Science, Engineering, and Technology (NSET), through the National Nanotechnology Coordination Office (NNCO), coordinates planning, budgeting, and implementation of the NNI between federal agencies.[20] NSET directs interactions among program officers from the participating agencies, periodic management meetings and program reviews, and joint scientific and engineering workshops. NSET works with the Office of Science and Technology Policy (OSTP) and individual agencies to establish NNI priorities, budgets, and metrics for evaluating various research activities.

Current participants in the NNI include the Departments of Defense (DOD), Energy (DOE), Justice (DOJ), Transportation (DOT), Agriculture (USDA), State, and Treasury; the Environmental Protection Agency (EPA); the National Aeronautics and Space Administration (NASA); the National Institutes of Health (NIH); the National Institute of Standards and Technology (NIST); the National Science Foundation (NSF); the Nuclear Regulatory Commission (USNRC); the Central Intelligence Agency (CIA); and White House offices (NEC, OMB, and OSTP).[21] Under the NNI implementation plan, each

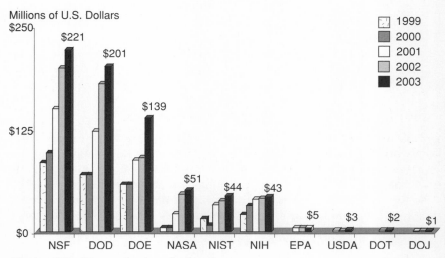

Figure 8.3 Nanotechnology funding by agency.

agency has discretion to invest NNI funds in projects that support its own mission.[22] Figure 8.3 shows how much funding each agency has devoted to nanotechnology research activities under the NNI over the last five years.[23]

The NNI implementation plan includes five primary themes.[24] They are:

1. Long-term fundamental nanoscience and engineering research that will build upon a fundamental understanding and synthesis of nanometer-sized building blocks with potential breakthroughs in areas such as materials and manufacturing, nanoelectronics, medicine and healthcare, environment and energy, chemical and pharmaceutical industries, biotechnology and agriculture, computation and information technology, and national security. This investment will provide sustained support to individual investigators and small groups doing fundamental, innovative research and will promote university-industry-federal laboratory and interagency partnerships.
2. Grand Challenges.
    - Shrinking the entire contents of the Library of Congress into a device the size of a sugar cube through the expansion of mass storage electronics to multi-terabit memory capacity that will increase the memory storage per unit surface a thousandfold
    - Making materials and products from the bottom up, that is, by building them up from atoms and molecules. Bottom-up manufacturing should require less material and pollute less
    - Developing materials that are 10 times stronger than steel (but a fraction of the weight) for making all kinds of land, sea, air, and space vehicles lighter and more fuel-efficient
    - Improving the computer speed and efficiency of minuscule transistors and memory chips of mullions making today's Pentium IIIs seem slow
    - Using gene and drug delivery to detect cancerous cells by nanoengineered MRI contrast agents or target organs in the human body
    - Removing the finest contaminants from water and air to promote a cleaner environment and potable water
    - Doubling the energy efficiency of solar cells
3. Centers and Networks of Excellence that will encourage research networking and shared academic users' facilities. These nanotechnology research centers will play an important role in the development and utilization of specific tools and in promoting partnerships in the coming years.
4. Research infrastructure will be funded for metrology, instrumentation, modeling and simulation, and user facilities. The goal is to develop a flexible enabling infrastructure so that new discoveries and innovations can be rapidly commercialized by American industry.
5. Ethical, legal, societal implications and workforce education and training efforts will be undertaken to promote a new generation of skilled

workers in the multidisciplinary perspectives necessary for rapid progress in nanotechnology. The impact that nanotechnology has on society from legal, ethical, social, economic, and workforce preparation perspectives will be studied. The research will help us identify potential problems and teach us how to intervene efficiently in the future on measures that may need to be taken.

Reviewing the NNI

In 2002, the NNI was reviewed by the National Research Council.[25] Specifically, the Committee sought to answer the following questions:

- Does the NNI research portfolio address the skills and knowledge that will allow the United States to fully benefit from the new technology?
- Is the balance of the research portfolio appropriate?
- Are the available American resources (human, infrastructure, and funding) being applied appropriately within the portfolio?
- Are the correct seed investments being made now to provide needed infrastructure for future years (2002 to 2005 and beyond)?
- Are partnerships (government-industry-university, international) being used appropriately to leverage the public investment in this area?
- Is the portfolio of programs being coordinated in such a way as to maximize the effectiveness of the investment?
- Does NNI give sufficient consideration to the societal impact of advances in nanotechnology?
- Are the processes for evaluating the effectiveness of the NNI (determination of metrics, milestones, and so on) appropriate and meaningful?
- How should the program be evaluated in light of the long-term (10- to 20-year) nature of many of its research goals?
- What are some important areas for future investment in nanotechnology?

The Report reviewed five different aspects of the NNI:

1. The balance of the research portfolio
2. Program management and evaluation
3. NNI partnerships
4. Relevant skills and knowledge
5. Social science research

We summarize the key findings. First, in reviewing the research portfolio, the Report emphasized that more must be done to facilitate interdisciplinary research.[26] Specifically, universities do not have incentives to fund interdisciplinary research, and such research is not valued by tenure and promotion committees. The Report argued that the greatest achievements in nanoscale research will be the result of high-risk endeavors. NSF awards modest (up to $100,000), one-year Nanoscale Exploratory Research grants for proof-of-concept for early-stage ideas. However, the number of these grants is relatively limited. Early-career researchers often say "that they cannot submit proposals for funding until they have already conducted enough experiments to have all but proved the expected result of the proposed investigation—which, of course, they do not have the funding to do."[27] Finally, the Report argued that establishing an interdisciplinary culture in science and engineering and supporting high-risk research requires long-term funding commitments.

Second, in reviewing management of the NNI, the Report highlighted the lack of interagency partnerships and coordination.[28] The Report also identified a need to establish a system for measuring the progress of the NNI as a whole.[29] Finally, the Report criticized the NNI "strategic plans" as redundant and called for the development of a "crisp, compelling, overarching strategic plan."[30]

The three most important components reviewed in the category of "NNI Partnerships" were university-industry partnerships, federal-industry partnerships, and international partnerships. In reviewing university-industry partnerships, the Report praised the increasing number of university-industry collaborations in nanoscale science and technology.[31] For example, the NSF has established six large university-based nanoscale research and engineering centers, which will receive $65 million over 5 years. Each of the six centers is required to have industrial partners collaborating in its research.

The Report also recognized the various mechanisms that have been employed to support private research and development.[32] The DOD, DOE, NASA, and NSF have utilized small business innovative research (SBIR) and small business technology transfer (STTR) federal grants to support nanoscience and technology research at small firms. Agencies can also use cooperative research and development agreements (CRADAs) to participate in research with a private partner. Finally, the Advanced Technology Program (ATP) can be used to match private funds in high-risk research.

The Report identified international collaboration as playing an important role in the development of nanotechnology. Substantial international cooperation exists among individual scientists.[33] The Report highlighted the need for American participation in international research centers devoted to nanoscale topics.

In reviewing whether the labor force is being adequately trained to work in nanotechnology, the Report evaluated education and training at universities as well as K-12 education. NSF has devoted a great deal of attention to education and training programs.[34] For example, the NSF utilized its combined research and curriculum development (CRCD) and its research experiences for undergraduates (REU) programs to strengthen undergraduate and graduate education in nanoscale science and technology. NSF has also leveraged its research groups, centers, and networks for educational purposes. The agency has even made efforts to address more immediate technology workforce education needs through the Nanotechnology Research and Teaching Facility at the University of Texas at Arlington and the Regional Center for Nanofabrication Manufacturing Education at Pennsylvania State University.

Several reports reviewed by the committee highlighted the importance of improving K-12 science and math programs to ensure a labor supply capable of filling the demand created by nanotechnology. Some efforts have been made under the NNI to improve K-12 education. For example, the NSF has used its nanoscience research centers and programs as vehicles for nontraditional outreach efforts to K-12 students and teachers.[35] NSF has also sponsored a number of outreach efforts specifically targeted at the general public.[36] Despite these efforts, the Report concluded that a more coherent outreach and education program is needed.

The Report criticized the lack of social science research being conducted in nanotechnology. Other than the establishment of a center for environmental research at Rice University, no social science projects were funded in 2001. The committee was disappointed by the lack of attention devoted to social science research.

Based on the analysis, the NRC made the following ten recommendations:[37]

1. The Office of Science and Technology Policy should establish an independent standing nanoscience and nanotechnology advisory board (NNAB) to provide advice to NSET members on research investment policy, strategy, program goals, and management processes. Such an advisory board is necessary to identify and recommend research projects that do not fit within any single agency's agenda. The NNAB would be composed of leaders from industry and academia with scientific, technical, social science, or research management credentials.
2. NSET should develop a "crisp, compelling, overarching strategic plan." The plan would describe short (1 to 5 years), medium (6 to 10 years), and long-range (beyond 10 years) goals and objectives. Specifically, it should focus on long-term commercial applications.
3. NSET should support long-term funding in nanoscale science and technology.
4. NSET should increase multiagency investments in research at the intersection between nanoscale technology and biology.
5. NSET should create programs for the invention and development of new instruments for nanoscience.
6. A special fund should be established to support interagency research programs, particularly among the National Institutes of Health, the Department of Energy, and the National Science Foundation.
7. NSET should focus on facilitating the development of an interdisciplinary culture for nanoscale science and technology within the NNI.

8. Industrial partnerships should be stimulated and nurtured, both domestically and internationally, to help accelerate the commercialization of NNI developments. NSET should create support mechanisms for coordinating and leveraging state initiatives to organize regional competitive clusters for the development of nanoscale science and technology.
9. NSET should develop a new funding strategy to ensure that the societal implications of nanoscale science and technology become an integral and vital component of the NNI.
10. NSET should develop performance metrics to assess the effectiveness of the NNI in meeting its objectives and goals.

The 21st Century Nanotechnology Research and Development Act

Based on the NRC recommendations, Congress passed the 21st Century Nanotechnology Research and Development Act.[38] We summarize the most important parts of the legislation.[39]

First, the legislation authorizes appropriations for nanotechnology R&D programs at the NSF, DOE, NASA, NIST, and EPA, as described in Table 8.1. Because the legislation only authorizes the increased funding, there is no guarantee that this money will be spent on nanotech R&D. A number of different appropriations subcommittees must elect to channel the funds to nanotechnology in annual spending bills.

Second, the legislation established the National Nanotechnology Program ("Program") to fund sustained and interdisciplinary nanotechnology R&D through grants to researchers and through the establishment of research centers and advanced technology user facilities, encourage greater coordination between federal agencies, and support research focused on the societal, ethical, educational, legal, and workforce implications of nanotechnology. The National Science and Technology Council (NSTC) was charged with the task of managing the Program. Specifically, NSTC must:

TABLE 8.1 Funding Authorizations of 21st Century Nanotechnology Research and Development Act ($ Millions)

Agency	FY 2005	FY 2006	FY 2007	FY 2008
NSF	385.0	424.0	449.0	476.0
DOE	317.0	347.0	380.0	415.0
NIST	68.2	75.0	80.0	84.0
NASA	34.1	37.5	40.0	42.3
EPA	5.5	6.1	6.4	6.8
Total	809.8	889.6	955.4	1024.1

- Establish goals and priorities
- Establish program component areas to implement those goals and priorities
- Develop a strategic plan to be updated triennially
- Develop a plan to encourage commercialization through programs such as SBIR and STTR
- Consult stakeholders
- Offer a Program budget to OMB

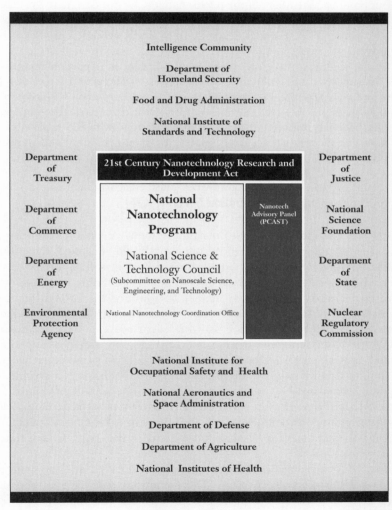

Figure 8.4 Structure of the NNI after the 21st Century Nanotechnology Research and Development Act.

The Council is also required to submit an annual report outlining federal nanotechnology budgets for the current and next fiscal years. The report must also include an assessment of progress made toward achieving goals and priorities as well as implementing the recommendations of the advisory panel. The legislation also provides more resources to the National Nanotechnology Coordination Office and directs it to provide technical and administrative support to the Council and Advisory Panel.

Third, the legislation established an advisory panel to review the Program on a biennial basis and provide recommendations to improve it. The president must appoint the members of the panel. As such, he can designate the existing President's Council of Advisors on Science and Technology (PCAST) as the advisory panel. The National Academy of Sciences will also conduct a trienniel review.

Finally, the legislation provides specific direction to particular agencies. NIST is supposed to conduct basic research on issues related to the development and manufacture of nanotechnology, the Secretary of Commerce is required to establish a clearinghouse of information related to commercialization of nanotechnology, and the DOE is provided with $25 million to create consortia to conduct research related to systems biology and molecular imaging.

RECOMMENDATIONS

Efforts taken by policymakers to install a framework for implementing and managing federal funding of nanotechnology should be lauded. By authorizing $3.7 billion in funding and establishing an administrative framework, the bill sends a signal that the federal government is committed to feeding and nurturing nanotechnology from infancy to adulthood. The legislation also addresses many of the issues raised by the National Research Council. Specifically, it encourages greater coordination between federal agencies, increases the focus on high-risk, long-term, and interdisciplinary research, enhances support for societal, ethical, and workforce issues, and establishes an advisory panel. At the time of this writing, we are hopeful that the National Nanotechnology Program, as managed by the National Science and Technology Council, will give rise to a tidal wave of new and useful technologies based on nanoscience. Notwithstanding our confidence, we offer three broad themes to policymakers and bureaucrats as they embark on the task of managing federal funding in the coming years.

First, policymakers should carefully monitor the role that the nanotechnology advisory committee plays. The committee will have an enormous influence over what types of projects receive federal funding. In the months prior to passage of the bill, a political debate unfolded regarding the structure and composition of the advisory panel. Some legislators supported an independent advisory panel while the White House insisted that the president be allowed to designate PCAST as the advisory panel.

Theoretically, the advisory panel could combat the traditional problems associated with managing public R&D funds. The panel is supposed to provide an objective review of the research agenda, supervising and correcting mismanagement of funds. The members would not be seeking to enhance their reputations as bureaucrats by "choosing winners," nor would they be susceptible to lobbying by interest groups. We fear that, by relying on PCAST, some of the value in establishing an advisory panel may be lost.

First, PCAST may not be technically equipped to serve as an effective advisory panel for federal funding of nanotechnology. PCAST's mission is to provide "advice to the President from the private sector and academic community on technology, scientific research priorities, and math and science education."[40] PCAST evaluates an array of different issues including the role of science and technology in homeland security, aggregate spending on research and development, broadband technologies, and energy efficiency.[41] Placing nanotechnology on PCAST's crowded platter could severely tax the resources and energy of the committee. The complexity and scope of nanoscience combined with the procedural difficulties of funding interdisciplinary research demand an advisory board solely focused on nanotechnology. Further, many PCAST members do not have backgrounds in nanoscience or related disciplines. Members of a panel who recommend how billions of dollars be spent should have some minimal level of expertise in the subject matter.

Second, PCAST may be less objective in its review than a truly independent nanotechnology advisory panel. An objective advisory panel would be comprised of a group of people with strong backgrounds in science and who represent a range of different interests. For example, there could be requirements that members have distinguished scientific service or that only a certain number of members come from industry. Additionally, there might be requirements that a cross-section of views be represented by the panel and the president give due consideration to recommendations from Congress and the scientific community in making appointments.

There are no constraints on the president's selection of PCAST members. Because the president can appoint anyone he pleases to PCAST, there is a greater risk of interest group politics subverting the objectivity of the review. For example, the president might stack PCAST with members from industry. Indeed, while several members of the current PCAST are distinguished scientists at universities, most are CEOs and investors. These members might recommend that increased funding should be channeled toward applied research or that a particular field of research deserves greater support. Even if the panel's recommendations are not biased by the backgrounds of the members, the appearance of impropriety might undermine the credibility of the review.

Second, policymakers should continually monitor the total amount of public money spent on nanotechnology. Because the new legislation is an authorization bill, it does not guarantee that any funds will be allocated to nanotechnology research and development. The amount of money actually spent will be determined by thirteen different appropriations subcommittees in annual spending bills. Even if the entire amount authorized is channeled toward nanotech R&D, policymakers should continue to analyze whether greater total expenditures would be desirable. Based on our interviews with researchers and government officials, agencies will only be able to fund a small fraction of the numerous nanotech proposals submitted in the coming years. Even when projects do receive funding, the amount provided could be insufficient. In 2002, NSF's average grant was $115,000 per year and lasted for 2.9 years.[42] Tom Kalil, who played an instrumental role in introducing the NNI in 2000, argues that current funding levels are "totally inadequate for supporting the kind of research groups required to make progress in nanoscale science and engineering, which almost always involve multiple graduate students and postdocs." The need for policymakers to keep a watchful eye on total funding levels in coming years is heightened by the international race. Other countries around the world are seeking to gain competitive advantage by spending as much or more than the United States. For example, the Japanese government spent approximately $900 million on nanotech R&D in 2003. If the United States wishes to maintain its technological and economic leadership, it must keep pace with other countries in funding nanotechnology research and development.

Third, government officials should consider taking additional steps to promote commercialization of nanotechnology. Breakthroughs in nanotechnology could have ripple effects throughout the economy. Just as the Internet radically changed businesses ranging from law firms to bookstores, nanotechnology will impact a number of different industries. Thus, commercialization of nanotechnology is worthy of greater public support than commercialization of other technologies. One way in which government might assist commercialization of nanotechnology is by increasing SBIR funding. As pioneering breakthroughs continue to be made at universities and government labs, there will be a proliferation of start-up companies. Most of these companies will seek SBIR grants, and the number of nanotechnology solicitations could dwarf available resources. Additionally, the traditional amounts of funding provided by SBIR might not be sufficient for nanotech start-ups. In general, Phase I SBIR grants are made for $100,000, and Phase II awards are for up to $750,000. As explained elsewhere, the costs of nanotech R&D are exorbitant. Policymakers should carefully evaluate whether total SBIR resources available for nanotechnology should be increased and whether nanotech grants should be larger and of longer duration. The NIH appears to be taking the lead in increasing the size of SBIR grants for nanotech companies.

CHAPTER 9

Conclusions

We began this tale of law, business, policy, and science by describing the hopes and ambitions of scientists, government officials, and entrepreneurs. As of this writing, it is uncertain if or when these hopes and ambitions will come to pass. Will Dr. Smalley live to see buckyballs result in a cure for cancer? Can organic molecules lead to the next generation of computing power that allows machines to have intellectual capabilities comparable to those of human beings? Will humans ever witness the birth of self-replicating nanomachines? In 20 years, when the world looks back upon the dawn of the twenty-first century, will it compare the first steps on the journey to the Nano Age with other great technological revolutions such as the internal combustion engine or the computer? Or will it chastise those leading the charge to the Nano Age for planting the seeds of the shredding of humanity? Or will nanotechnology simply be a forgotten field described in dusty books and journals that some may have heard of, but about which no one really cares?

No one knows for sure. The answers to these questions depend on a variety of circumstances and forces. Perhaps the future has more to do with mere chance than anything else. More likely, the answers to these questions depend on the aptitude, work ethic, and wisdom of scientists and engineers as well as those who know little about science, but much about government, law, and economics. As we wrote in Chapter 1, the flame of nanotechnology relies as much on the oxygen supplied by government officials as it does on the wood provided by scientists.

While scientists are racing ahead on the journey to the Nano Age, their brethren in government are lagging behind. As far as we are aware, this book is the first attempt to provide a comprehensive analysis of the legal and policy issues that loom on the horizon of the nano world. Four central themes can be derived from our exploration of these issues.

1. Policymakers Should Distinguish Hype from Reality.

In light of the frenzy that nanotechnology has generated in the press and financial communities, it is difficult for policymakers to know what is actually taking place in the field. Some reports claim that nanotechnology has already matured while other reports caution that it is decades away. Of course, as discussed in Chapters 1 and 2, the truth is somewhere in between—it depends on how the term "nanotechnology" is defined. We divide nanotechnology applications into three categories: "simple" nanotechnology, "building small" nanotechnology, and "building large" nanotechnology. "Simple nanotechnology," which primarily describes the use of nanoparticles in different materials, is already making a market impact. For example, stain-resistant jeans and transparent suntan lotion incorporate nanoparticles. "Building small" nano-technology, which will improve a range of different products from computers to drug delivery techniques to solar cells, are not likely to come to market for three to fifteen years. And "building large" nanotechnology, which refers to self-replicating nanorobots, will not be a realistic possibility for decades, if ever.

2. There Should Be a Well-Informed Policy Debate in the Immediate Future About the Desirability of Nanotechnology.

As argued in Chapter 3, the speed at which nanotechnology is developing creates an urgent need for well-informed policy analysis and debate. If the arguments advanced by Bill Joy and other environmentalists are compelling, the next two years provide a window of opportunity to stop the technology in its tracks. Once the technology takes off and becomes an engine of economic growth, it will be virtually impossible to contain. If, however, the case against nanotechnology is feeble, policy action is still required in the immediate future. The arguments promulgated by nanotechnology opponents must be shattered before they gain momentum and threaten the entire industry. A cursory review of the press reveals that grassroots opposition is already emerging, and the recent release of the book (and soon to be movie) *Prey* promises to fuel the public fears and misperceptions of nanotechnology. The stories of nuclear power and biotechnology illustrate the harms of allowing technologies to proceed without serious policy discussion. Public backlashes against these technologies impeded their potential.

We argue that this policy debate should be resolved in favor of continuing the journey to the Nano Age. Although we have reservations about a world filled with "spiritual machines" and self-replicating nanorobots, our rejection of the call to relinquish nanotechnology is based on several persuasive arguments. First, it is likely that the benefits of nanotechnology outweigh the costs. Technological progress has always been a double-edged sword—the power to make substantial improvements in the health, welfare, and capabilities of humans is inevitably accompanied by the potential for destruction. But time and again, humans have done a decent job of exploiting the benefits of technology while minimizing the harms. We are optimistic that this will hold true for nanotechnology. Second, and more compelling, is that a policy to end the journey is unlikely to succeed. Not only would an international treaty be impossible to establish, supervise and enforce, but restricting research might even be unconstitutional. To the extent such a policy cannot prevent development of nanotechnology, it will increase the likelihood of an accidental catastrophe or intentional misuse of the technology. Using the advents of the atomic energy and biotechnology industries as models, a more prudent course of action is to cultivate nanotechnology while promulgating regulations to prevent the harms.

3. Nanotechnology Will Present Challenges to the Fundamental Organization of the Regulatory State.

Policymakers construct categories to define the world for regulatory purposes. For purposes of describing and regulating the macro and even micro worlds, these categories are adequate. As industry enters the nano realm, however, traditional categories will become less useful. The regulatory challenges facing the Environmental Protection Agency, the Patent and Trademark Office, the Food and Drug Administration, and the various agencies administering export control laws illustrate this theme.

Environmental Protection Agency: Chapter 4 explored the regulatory challenges that the Environmental Protection Agency (EPA) will face in regulating nanotechnology. Environmental laws require regulation of new chemicals. Since nanomaterials are just smaller versions of existing chemicals, these substances are entering the environment without any real regulatory review. Because nano-sized particles have different properties than bulk materials of the same substance, some environmentalists point out that they could present a host of new health and environmental hazards. Some environmental groups are calling for a ban on the commercial production of nanostructures until they are proven safe.

After analyzing the available data on the toxicity risks associated with nanomaterials and weighing the risks against the benefits, we conclude that, while more data is needed, the current data dictate that the EPA should not subject nanomaterials to stringent regulations. Nevertheless, the EPA should garner more data and subject nanostructures to additional review procedures.

Patent and Trademark Office: Chapter 5 revealed that the Patent and Trademark Office (PTO) has encountered similar categorization problems in reviewing nanotechnology patents. Since there is no center dedicated to nanotechnology or group of reviewers specializing in nanotechnology, patent applications in the field have been directed to different centers for review. The disjointed and substandard review combined with an explosion of patent filings in recent years have produced several intellectual property hurdles in this field. The PTO is issuing some broad and overlapping claims on upstream inventions. The resulting patent disputes could could stifle downstream innovation. Further, the sheer number of overlapping patents that have issued and will continue to issue make it difficult for firms to commercialize products.

We offer several recommendations to alleviate some of the problems on the intellectual property front. First, the PTO should adequately equip itself with the knowledge and resources to effectively review applications in nanotechnology. Second, courts should be willing to narrow the scope of overly broad patents when they are wielded as monopolistic weapons. Finally, government should explore innovative ways to encourage the formation of patent pools on the building blocks of nanotechnology.

Food and Drug Administration: Chapter 6 turned to the issue of regulation of nanomedicine. The Food and Drug Administration (FDA) classifies products as either drugs, devices, or biologics, and regulates each category differently. Since nanomedical products are often a combination of drugs, devices, and biologics, classification will become increasingly difficult and confusing. Further, because nanotechnology will primarily improve existing products, the agency will encounter complicated issues associated with clinical data necessary for product approval. Finally, the agency has done little to acquire the expertise necessary for effective review of the safety and efficacy of nanomedical products. A failure to adequately prepare for reviewing products based on nanotechnology could have significant ramifications for public health, the FDA, and the emerging nanotechnology industry.

The FDA should consider implementing several reforms to ensure that it is adequately prepared to regulate nanomedicine. The agency should continue its dialogue with industry. It should also consider establishing an Office of Nanotechnology, promulgating clearer guidelines governing which Center has jurisdictional authority over different types of nanoproducts, and acquiring more personnel with expertise in nanotechnology.

Export Control Laws: In Chapter 7, we reviewed challenges posed by nanotechnology to the export control regime. There are a variety of convoluted laws and regulations administered by several different agencies. Sifting through hundreds of pages reveals that they were not drafted with nanoscale materials, devices, and systems in mind. While most nanotechnology companies will encounter these laws, it is unclear in some cases how these laws should be applied. We urge agencies responsible for administering such laws to clearly address how they apply to nanotechnology.

4. Government Support Should Be Substantial and Sustained. Effective Management Is Key.

After concluding that nanotechnology should be embraced and identifying ways for agencies to smooth the regulatory bumps, we turned to what are perhaps the most important questions. Should government devote resources to cultivating nanotechnology? If so, how should these resources be allocated and managed? The answer to the first question is a resounding "yes." Government financing is critical to sustain basic research, fund the production of goods for which the government will be the primary customer, and support a "strategic" industry. The answer to the second question is not as obvious. Historically, public R&D programs have been plagued by implementation and management problems. Just months before this book went to press, Congress enacted legislation codifying the current structure of the NNI and authorizing additional funding for nanotechnology. While we are hopeful that the National Nanotechnology Program is an effective framework for public funding of nanotechnology, we advise policymakers to monitor the role that the nanotechnology advisory committee plays, scrutinize the total amount of money being spent on nanotechnology, and take additional steps to promote commercialization of nanotechnology.

We are very passionate about the potential of nanotechnology to change the world. But we are also very concerned that social, political, and regulatory issues could stifle development of the field. In the end, we are optimistic that scientists and policymakers can rise to the challenge and effectively prepare industry and society for the coming revolution in nanotechnology. It is our hope that this book will assist those leading this noble journey to the Nano Age.

PART III

Nanotechnology Business

Part I of this book defined nanotechnology, described its potential applications, and proposed a model for understanding the industrial structure giving rise to its development. Part II identified broad policy and regulatory issues that confront legislators and regulators. In Part III we turn to issues that face individual nanotechnology businesses. We identify and analyze opportunities for and challenges to nanotechnology development in order to assist the scientists, entrepreneurs, lawyers, and investors who are responsible for building this new industry.

CHAPTER 10

Starting a Nanotech Company

Business inside the university is very different from business outside the university. Successful companies that spin out of universities are ones where the professors quickly realize that they cannot run the company. Scientific founders would be wise to tell the business folks to "take this baby" and let it go.[1]

—Dr. Max Lagally (founder, nPoint)

Many scientists and managers launching companies in the exciting new field of nanotechnology have dreams of creating the next Intel or Hewlett-Packard. While some companies may turn out to be successful and get acquired or make initial public offerings, others will meet less auspicious fates and they will fail. Often, success depends on the legal and business decisions made in the earliest days of the company. This chapter surveys the issues faced by newly born nanotech companies and makes recommendations to entrepreneurs preparing to enter the industry.

We begin by addressing the most important consideration for any new nanotech company: intellectual property (IP). In order to launch a company, founders must obtain rights to the basic technologies that the company seeks to commercialize. We discuss the different IP challenges for start-ups spinning out of universities, government labs, and other corporations. We then analyze the issues faced by entrepreneurs in organizing the business. Specifically, we discuss choosing a name for the company, selecting an organizational form for the company, structuring a board of directors and management team, the issuance of stock to founders and employees, and deciding where to locate the business.

INITIAL INTELLECTUAL PROPERTY CONSIDERATIONS

Nanotechnology companies are formed around technology and the intellectual property protection that surrounds it. Nanotech companies' intellectual property portfolios usually consist of both IP developed internally and IP licensed from other parties. Chapter 13 is devoted to explaining how companies can protect IP generated internally. Because nanotech R&D is so expensive, almost all nanotech companies at conception must license at least some IP from external sources. Indeed, most nanotech startups thus far are spinouts of research projects at larger organizations. For many entrepreneurs, the decision to start the company may depend on whether they can obtain licenses to IP rights held by these organizations.

Depending on the circumstances giving rise to the birth of a new company, it may face different challenges obtaining and protecting its IP. There are three common ways a new nanotech company may be born. First, and most typical of this field, a company can spin out of research taking place at a university. Second, the company can be based on a pioneering breakthrough in a government lab. Third, a group of researchers at a large company might elect to leave to start their own business. There are unique considerations for entrepreneurs in each of these circumstances; we will discuss the licensing issues and IP preparations that must take place.

The University Spin-Off

A large portion of funds allocated under the National Nanotechnology Initiative (NNI) are channeled to research universities. As shown in Chapter 2, most nanotech start-ups are spinoffs from pioneering breakthroughs at universities. Table 10.1 provides a list of universities that have established nanotechnology centers, institutes, and networks. We expect research at these universities to generate a large number of patentable inventions in the near future.

When an invention is made at a university, the inventing professor works with the university's technology transfer and licensing office (OTT or OTL) to file a patent on the invention. While the patent is pending, the OTL office researches the value of the technology and evaluates the potential licensees. Although most discoveries at academic institutions result from federally funded research, the 1980 Bayh-Dole Act granted universities ownership rights to inventions developed with federal monies.[2]

If the inventing professor wishes to launch a company to commercialize the invention, he must convince the university to exclusively license the intellectual property to the future company. In general, university technology offices are sophisticated. The officers we have worked with possess, or have access to, expertise in nanoscience and are experienced in licensing and business matters. License agreements between universities and companies can have several dif-

TABLE 10.1 Nanotechnology Research at Universities

University	Center / Institute / Network
University of California Los Angeles	Institute For Cell Mimetic Space Exploration Center for Scalable and Integrated Nano-Manufacturing
California Institute of Technology	Kavli Nanoscience Institute
University of California Berkeley	The Molecular Foundry
Texas A&M University	Institute For Intelligent Bio-Nanomaterials and Structures For Aerospace Vehicles
Princeton University	Bio-Inspection, Design and Processing of Multifunctional Nanocomposites
Purdue University	Institute for Nanoelectronics and Computing Network For Computational Nanotechnology
Cornell University	Center For Nanoscale Systems Nanoscale Science and Engineering Center Nanobiotechnology, Science, and Technology Center Cornell Nanoscale Science and Fabrication Facility (National Nanofabrication Infrastructure Network)
Rice University	Center For Biological and Environmental Nanotechnology
Northwestern University	Integrated Nanopatterning and Detection Center for Nanofabrication and Molecular Self-Assembly Center for Transportation Nanotechnology Network For Computational Nanotechnology
Columbia University	Center For Electronic Transport In Molecular Nanostructures
Harvard University	Nanoscale Systems and Their Device Applications National Nanofabrication Infrastructure Network
Rensselaer Polytechnic Institute	Directed Assembly of Nanostructures
Massachusetts Institute of Technology	Institute for Soldier Nanotechnologies
University of California Santa Barbara	Center for Nanoscience Innovation for Defense Nanotech Fabrication Facility
University of Minnesota	Center for Nano-Energetics Research National Nanofabrication Infrastructure Network

TABLE 10.1 (continued)

University	Center / Institute / Network
University of Delaware	Center for Nano-Energetics Research
University of South Dakota	Center for Nano-Energetics Research
University of Oklahoma	Center for Nano-Energetics Research Center For Semiconductor Physics In Nanostructures
University of Arkansas	Center For Semiconductor Physics In Nanostructures
Northeastern University	Nanomanufacturing Research Institute
University of North Carolina	North Carolina Center For Nanoscale Materials
North Carolina State University	North Carolina Center For Nanoscale Materials National Nanofabrication Infrastructure Network
University of North Carolina Chapel Hill	Nanoscale Science Research Group
Howard University	Keck Center for the Design of Nanoscale Materials for Molecular Recognition (National Nanofabrication Infrastructure Network)
Pennsylvania State University	Nanofabrication Facility (National Nanofabrication Infrastructure Network)
Stanford University	Stanford Nanofabrication Facility Network For Computational Nanotechnology IBM-Stanford Spintronic Science and Applications Center
Georgia Institute of Technology	National Nanofabrication Infrastructure Network
University of Michigan	National Nanofabrication Infrastructure Network
University of New Mexico	National Nanofabrication Infrastructure Network
University of Texas at Austin	National Nanofabrication Infrastructure Network
University of Washington	National Nanofabrication Infrastructure Network
University of Illinois	Network For Computational Nanotechnology
University of Florida	Network For Computational Nanotechnology
University of Texas, El Paso	Network For Computational Nanotechnology
Morgan State University	Network For Computational Nanotechnology
University at Albany – SUNY	Albany Nanotech

ferent types of conditions. First, an OTL office can grant a start-up a nonexclusive license, which enables the university to license the technology to others. Second, it can grant a start-up an exclusive license to all rights to the invention. Exclusivity enables the licensee to prevent any other entity from using the technology for any purpose. Third, it can grant a "field of use" exclusive license to develop the technology in a particular manner. Finally, it can grant a license based on predefined territorial rights, for example a license to develop a product for only a particular country or region.

Technology licensing offices struggle with whether they should grant exclusive licenses or nonexclusive licenses.[3] On the one hand, entrepreneurs need an exclusive agreement in order to attract investment to develop a particular technology. On the other hand, technology transfer officers are weary of granting exclusive rights to processes and tools that will be widely needed. The ability of a company to secure an exclusive licensing agreement with a university depends on numerous factors including the technology itself, what other companies are interested in licensing the technology, and the policies of the particular university that developed the technology.

In biotechnology, universities often grant nonexclusive licenses for research tools. For example, Stanford University's patent on recombinant DNA techniques was licensed on a nonexclusive basis. Similarly, the University of Wisconsin intends to grant nonexclusive licenses for its embryonic stem cell patents.[4] In the case of nanotechnology, university technology transfer offices are still wrestling with the questions of which kind of licenses to grant. In general, it appears they are choosing to grant exclusive licenses to broad patents that cover fundamental processes. Some universities are being cautious about the breadth of the licenses they grant. For example, Stanford University only provided Molecular Nanosystems with a license to use the chemical vapor deposition process for synthesizing nanotubes in a few fields.[5] In contrast, Rice University granted Carbon Nanotechnologies a broad license to use patents governing the laser ablation process to develop products in any field. Some OTL offices also insist on mandatory sublicensing provisions in the license agreements.[6] Thus, if a technology becomes widely needed, the licensee must sublicense it to entities that are not direct competitors.

Assuming the university determines that an exclusive license should be granted, the ability of a company to obtain the license may turn on who is interested in licensing the technology. If no other companies are interested in obtaining an exclusive license, the company will secure the license as long as it can persuade the technology office that it possesses the resources and expertise to develop the technology and successfully market it. If there is demand for the license, the licensing decision could turn on who is competing for the technology. If a start-up is competing with a large company to obtain the license, the university may prefer to license to the start-up company. In order to encourage small business innovation, the Bayh-Dole Act encourages academic institutions to favor small companies over larger companies when making licensing decisions.[7] If there is competition between an existing small company and a start-up company, the licensing decision becomes more difficult. Quite often the decision on which company receives the license may depend on the relationship of the company's founders to the university. Some universities encourage professors to pursue entrepreneurial ventures and will be likely to support these ventures. Other universities frown on such activities. As Max Lagally, a University of Wisconsin professor who started nPoint, a company to develop tools for positioning nanostructures, put it: "At Stanford and MIT, starting a company provides a badge of legitimacy. At other places, you don't want to mention it."[8] Additionally, some universities expressly prefer to give licenses to companies founded by professors at the university while others do not. However, even when there is no express partiality, established professors generally have a strong relationship with the technology transfer office, and are therefore more likely to obtain the license.

Once a type of license is agreed to, the price of the license must be negotiated. When one licensing associate we interviewed was asked about pricing, she chuckled and jokingly responded, "Why don't you ask a difficult question?"[9] A range of different variables is used to determine the appropriate pricing scheme, but all technology licensing officers with whom we spoke agreed that it can be an arbitrary and haphazard process. Licensing officers attempt to estimate the value of the technology by surveying the market. The price will depend on a host of variables including the type of license granted, the demand for the technology in the market, the validity of the patent, other patents in the field, royalty rates for similar products, the expected cost of developing and bringing the product to market, and the cost of the discovery. Most university OTL offices charge an up-front fee and royalty on the income generated by the patented invention. Some universities prefer to receive the up-front fee in cash while others take equity in the company.[10] Up-front fees can be anywhere from a few thousand to several hundred thousand dollars. Royalty rates generally range from 1 to 10 percent.

A variety of different terms are included in the license agreement. Some universities require licensees to reimburse patent application fees, and others call for a minimum annual payment. The agreement may include terms allowing the university to reacquire the technology if the company fails to meet certain development milestones. Universities might also require access to different types of information during the licensing period. The license will probably also contain language entitling the university to use inventions resulting from the licensed technology. The agreement should clarify a number of other rights and responsibilities of both parties, including provisions for enforcement of the patent, sublicensing, indemnification of the patent holder, and procedures for dispute resolution and termination of the agreement. Table 10.2 summarizes common provisions found in license agreements.

TABLE 10.2 Standard Provisions in a University License Agreement

Terms	Explanations
Definitions	Technology
	Field of use / territory
	Net sales
	Licensee and affiliates
Grant of License	Exclusive versus nonexclusive
	Field of use (for example, carbon nanotubes for field emission)
	Territory (worldwide versus United States only)
	Sublicensing rights (can licensee license the technology)
	University and government can use the technology
	Grant back rights (does university have rights to future inventions resulting from technology?)
Consideration	License fee
	Equity
	Costs of patent prosecution
	Royalty on net sales
	Percentage of sublicense income
	Annual maintenance fees
	Milestone payments
	Assignment rules and fees
Reporting	Companies must provide quarterly or annual reporting and payment
Milestone Terms	To maintain license, company must keep certain diligence milestones such as technological development, securing funding, and so on
Sublicense Provisions	Sublicense agreements must contain certain provisions such as disclaimer of warranties, maintenance of university rights
Infringement	Who enforces the patent
	Who pays the expenses
	Distribution of damages between licensee and university
Representation and Warranties	Licensee assumes all risks associated with licensed technology
Limitation of Liability Indemnification	University will not make any warranties as to the validity of the patent(s) or merchantability of the technology
	Licensee will indemnify university against any claims
	Licensee is required to obtain certain amounts of product liability insurance prior to commercial sale of a product
Term and Termination	Duration of license
	Causes for termination
	Notice requirements for termination
	Dispute resolution
Other Provisions	Patent numbers on products
	Prohibition on using university's name without consent
	Licensee will comply with all applicable laws and regulations

The Government Laboratory Spinoff

Another place where nanotech companies are born is the federal laboratory system. The more than seven hundred federal laboratories in the United States perform more nanotech research than any other group of laboratories in the world, public or private.[11] These laboratories receive substantial funding from the National Nanotechnology Initiative (NNI), and their research extends to areas that are not "...well suited to university or private sector research facilities because of its scope, infrastructure, or multidisciplinary nature, but for which there is a strong public and national purpose."[12]

Federal research laboratories are required by law to transfer their technologies to private industry for commercialization purposes, making them a treasure chest for entrepreneurs interested in obtaining access to nanotech inventions. In the Stevenson-Wydler Technology Innovation Act of 1980, Congress mandated that a fixed percentage of all laboratory budgets be allocated toward "the rapid movement of federal laboratory research...into the mainstream of the U.S. economy."[13] Accordingly, the federal laboratories fund numerous groups whose sole mission is to help the private sector navigate through the maze of constantly changing research areas, priorities, and programs of the federal laboratories. These groups promote either cooperative research arrangements or the licensing of their technology. The most important group charged with overseeing the dissemination of technology from federal laboratories to the private sector is the Federal Laboratory Consortium for Technology Transfer (FLC).[14]

With 60 percent of all federal research funds dedicated to the DOD, DOE, and NASA, the vast majority of nanotechnology research is conducted at these labs.[15] We summarize the relevant administrative offices involved in commercialization and licensing activities at each agency.

The Department of Defense (DOD), which has 86 laboratories, is heavily engaged in nanotechnology research. For example, the Naval Research Laboratory operates multiple programs for the development of nanotechnologies for military use, and the "Institute for Soldier Nanotechnologies" is a research collaboration with MIT to create sensors, tracking devices, and uniforms that better protect soldiers.[16] In order to help private industry navigate through its many projects, the DOD maintains numerous groups charged with technology dissemination to private industry (see Table 10.3).

The Department of Energy (DOE) operates 24 laboratories with four main focuses: energy, national security, the environment, and science. DOE laboratories are actively pursuing the development of nanotech through the Nanoscale Science Research Centers (NSRC).[21] The NSRCs, which support research in the synthesis, processing, and fabrication of nanoscale materials, include the Molecular Foundry at Lawrence Berkeley, the Center for Functional Nanomaterials at Brookhaven, the Center for Nanophase Materials at Oak Ridge, the Center for Integrated Nanotechnologies jointly at Sandia/

TABLE 10.3 DOD Groups Charged with Technology Dissemination

DOD Group	Responsibility
Office of Technology Transition (OTT)	The OTT operates a database of current projects and coordinates between the DOD's many laboratories and those at the DOE and NASA.[17]
Office of Research and Technology Applications (ORTA)	ORTA serves as a liaison between DOD facilities and private industry and has has demonstrated "nontraditional" working relationships.[18]
Defense University Research Initiative on NanoTechnology (DURINT)	DURINT is a public/private technology cooperative "to enhance universities' capabilities to perform basic science and engineering research and related education in nanotechnology critical to national defense."[19]
Defense Advanced Research Projects Agency (DARPA)	DARPA provides direct funding for start-up nanotech companies whose research meets the requirement of developing technologies for military use.[20]

Los Alamos, and the Center for Nanoscale Materials at Argonne. Generally DOE laboratories are run by independent operators such as Batelle or by universities such as Iowa State University, and these groups have some licensing discretion. The DOE also maintains its own groups charged with dissemination of technology to the private sector (see Table 10.4).

Nanotech research at the National Aeronautics and Space Administration (NASA) is conducted through ten installation centers known as Field Instal-

TABLE 10.4 DOE Groups Charged with Technology Dissemination

DOE Group	Responsibility
Office of Industrial Technologies (OIT) now called Industrial Technologies Program (ITP)	ITP focuses on deploying technologies to the private sector that improve energy efficiency and are beneficial to the environment. ITP manages collaborative R&D projects, offers financing, and connects companies with appropriate laboratory groups within the DOE.[22]
The Laboratory Coordinating Council (LLC)	LLC promotes joint laboratory research with private industry by renting laboratory facilities for private research and disseminating intellectual property "to the best interest of all concerned."[23]

lation Centers. Each center is assigned a unique role for leadership in a given area in order to prevent overlap in research. The leading centers for nanotech research are the Ames Center for Nanotechnology, the Langley Research Center, and the Goddard Center.[24] Their focuses are computational, material, and applied nanotechnology research, respectively. NASA's organization-wide nanotech goals include creating new sensors, new materials and advanced micro-electronics.[25] The NASA Commercial Technology Division (NCTD) coordinates research across facilities while each NASA center operates its own commercialization and technology transfer office in order to tailor services to the industries with which they are most closely associated.[26] (See Table 10.5.)

TABLE 10.5 NASA Groups Charged with Technology Dissemination

NASA Group	Responsibility
The NASA Commercial Technology Division (NCTD)	NCTD oversees collaboration efforts and offers information resources such as NASA's publication of partnership opportunities, *Spinoff*, which features successfully commercialized NASA technologies. NCTD also operates a web portal (http://nctn.hq.nasa.gov/) with links to the different NASA websites and news of new funding opportunities for private companies.[27]
Other Programs Examples: • NASA Commercialization Center (NCC) • Center for Technology Commercialization (CTC)	While NCTD has an oversight role, each NASA center operates its own commercialization and technology transfer office in order to tailor services to the industries with which they are most closely associated.[28] Each program is designed to foster commitment to commercialization of new technology. NCC is an incubator for start-ups commercializing JPL technology. The goal of the NCC is to provide private companies with "the necessary physical, management, and product development infrastructure to successfully commercialize NASA technologies."[29] CTC is NASA's Northeast Regional Technology Transfer Center and provides an array of services to industry partners including "technology assessments, technical research small business turn-arounds, echnology commercialization, patent research, and strategic planning."[30]

Entrepreneurs seeking to start a company based on laboratory research must enter in to a licensing agreement. Like universities, the laboratory staff we spoke with had extensive technology backgrounds and experience collaborating with industry. Nanotech companies seeking to negotiate licenses with federal labs are likely to encounter issues similar to negotiating licenses with universities. Like universities, government officials might be reluctant to grant an exclusive license due to fears of preventing widespread adoption of the technology. Standard government license agreements include many of the same types of terms and conditions found in university licensing agreements. In contrast to universities, federal laboratories may impose more extensive restrictions on the use of the technology.

The Corporate Spinoff

A third means by which nanotech companies may be started involves employees, usually researchers, who leave a large corporation. There is an old tale of Silicon Valley that "once upon a time everyone working there worked for IBM."[31] At the time of this writing, few nanotech start-ups have spun out of large companies. Yet, as the field develops, some brave entrepreneurs are likely to take the leap.

The intellectual property issues associated with a group of researchers leaving a large company to start a nanotechnology company are more complex than those associated with a university or laboratory spinoff. In some instances, circumstances will be ripe for a peaceful divorce, and the parent company will support the launch of the start-up. The large corporation may choose to invest in the people leaving rather than litigate against them. A parent company is more likely to support the spinoff under two circumstances: (1) when the start-up's technology will complement, rather than compete with, the parent's technology; and (2) when the invention still requires a great deal of research and development for commercialization.[32] If the parent company is cooperative, the startup could seek to license technology and/or obtain their financial assistance. Large corporations are interested in these kinds of arm's-length relationships, where they can maintain a financial interest in a related technology but not be responsible for fully funding or managing the company. For example, corporate-backed venture capital funds are becoming increasingly popular, because they can be used to maintain influence over new companies and pave the way for possible acquisitions or alliances later.

When a parent company does not support the birth of the start-up, complicated and costly intellectual property battles can ensue. Most scientists and engineers have agreements with their employers that any innovations made by the employees at the workplace belong to the company. Thus, employees who leave to start their own company face an undesirable prospect: being sued by their former employer. A carefully timed lawsuit can cripple a start-up before it can raise capital and begin to take flight.

Suits launched by the parent company against its former employees are likely to be based on either trade secret law or contract law. First, the parent company might argue that, because the concept for the invention originated from research at the parent company, it owns the rights to the technology. In theory, the ideas generated in the workplace are subject to trade secret protection. In practice, however, courts generally side with former employees when the issue presented is timing of the invention.[33] As one legal scholar writes, "Ex-employees usually receive the benefit of the doubt when a case presents a close question of timing, i.e., when the employer suspects (but cannot prove) that an ex-employee actually came up with the idea for an invention while still employed. Hence it is in many cases quite feasible to leave a firm after one arrives at the general notion of an invention, but before any of the provable milestones of invention arrive."[34] As long as the employees take nothing tangible from their former employer, they are likely to prevail.[35]

Even if the parent company cannot argue that it owns the invention, it can invoke the doctrine of inevitable disclosure to prevent former employees from starting a competing company. The doctrine creates an evidentiary presumption that an employee cannot help but rely on her knowledge of the parent company's trade secrets in her new job. Some courts have used the doctrine to enjoin employees from working for competitors for a period of time or limiting the scope of the employee's duties.[36] However, most courts will only apply the doctrine when the former employee acted in bad faith.[37] For example, in the leading case, the court emphasized the former employee's lack of candor and "out-and-out lies" to his former employer.[38] Generally, courts do not invoke the doctrine when the former employer simply fears that the departing employee may inadvertently or unconsciously use or disclose his or her knowledge of trade secrets.[39]

Parent companies might also assert two different arguments against former employees under contract law. First, many employers include "trailer" or "holdover" clauses in employment contracts, which state that inventions made after the employee leaves the company are still owned by the employer. Courts construe these provisions narrowly. Some courts have completely voided the provision when the period of time is unreasonable[40] while other courts have upheld trailer clauses that only apply to inventions made using the parent company's trade secrets.[41] As one legal scholar concludes, trailer clauses "have limited effect[.]"[42]

A second legal attack might rely on post-employment covenants not to compete. A non-compete clause prevents an employee from working for a competing enterprise after leaving the company. Different states have different rules regarding the enforceability of non-compete clauses. California law voids any "contract by which anyone is restrained from engaging in a lawful profession, trade, or business of any kind..."[43] While many states do enforce non-compete clauses, they are construed very narrowly. Generally, to be enforceable, a non-compete covenant must be reasonable "in view of the totality of the circumstances, including the scope of geographical, temporal, and competitive activity restrictions."[44]

Notwithstanding the limited utility of trade secret law and anti-compete covenants in court, the parent firm can wield a lawsuit to strangle the spinoff. Indeed, the uncertainty of the law creates "incentives for frivolous litigation designed to harass competitors rather than to obtain relief for trade secret misappropriation."[45] When starting the company, entrepreneurs should take several steps to minimize the risk and costs associated with litigation involving the former employer. First, and most important, entrepreneurs should make efforts to establish positive relations with the former employer to minimize the risk of a suit. Second, to increase the likelihood of a successful and rapid conclusion, the start-up should take clear and definite measures to prevent the former employer's trade secrets from having any impact on research and development. Finally, the start-up should maintain copious development records, such as inventor's notebooks and invention logs. Comprehensive records are powerful evidence to rebut claims of trade secret misappropriation.

NAMING THE COMPANY

After the founders of the company have identified the necessary IP and decided to start a company, a second major task is choosing a name for their business. They will need to decide whether to follow the growing number of companies incorporating "nano" in their names. The reason for the growing use of the term is clear: "nano" draws attention from investors. As one venture capitalist put it, some investors are putting money in companies because they "simply have nano in their name."[46] Just as the addition of ".com" attracted money in the 1990s, many entrepreneurs and investors are betting that "nano" will be a magnet for capital. At the time of this writing, stock prices of companies with the word "nano" in their name are soaring. As shown in Table 10.6, publicly traded "nano" companies have large market capitalizations relative to their small revenue streams. Nanometrics, whose stock ticker is NANO, witnessed its stock rise from $11.98 on November 2 to $16.06 per share on December 3 for no apparent reason. According to one press account, even Nanometrics was surprised by the attention because the "company's been around for 25 years, well before nanotechnology was even thought of[.]"[47] The run-up prompted a popular press analyst to include the surge in "nano" stocks in his 2003 list of "The Five Dumbest Things on Wall Street" stating, "...the nanotech bandwagon has proved unstoppable, gaining momentum despite its occasional alarming resemblance to the dot-com bubble."[48]

Even companies that originated under different names have changed their names to incorporate the word "nano." For example, Nanomix was formerly Covalent Materials. Companies not directly engaged in nanoscale research have also changed their names to include "nano." NanoPierce, which was

TABLE 10.6 Publicly Traded Companies with "Nano" in Their Name

(Millions of U.S. Dollars)

Company	Market Cap (12/31/03)	2002 Revenues
Altair Nanotech: ALTI	$128.2	$0.3
NanoProprietary: NNPP	263.1	1.4
NanoPierce: NPCT	25.2	0.2
NanoScience Tech: NANS	86.8	—
U.S. Global Nanospace: USGA	97.2	—
Nanophase: NANX	140.5	5.4
Nanogen: NGEN	230.9	17.2
Nanobac Pharmaceuticals: NNBP	67.0	—
Nanosignal: NNOS	18.1	0.1
Nanometrics: NANO	180.3	34.7
Total	**$1,237.4**	**$59.3**
Implied P/S ratio	20.9x	

originally named "Sunlight Systems," develops electrical connections for microelectronic devices. Similarly, Nanometrics manufactures metrology systems for measuring thin films. Companies are also now using the word to describe their products. Nano-tex has trademarked "nano-care" and "nano-pel" for stain and wrinkle-resistant materials.

While companies may want to take advantage of the "nano" exuberance, several other considerations should influence the decision of a company name. First, the founders may wish to choose a name that is likely to receive trademark protection. In Chapter 13, we provide a detailed description of names that are eligible for trademark protection. Second, the founders should seek a company name whose web domain name, or close match, they can acquire. Many domain names with the word "nano" in them have already been purchased. Finally, companies should consider potential disadvantages to having the word "nano" in their names. If nanotechnology does not live up to market expectations, the name could cause investors to shy away in the future. Further, the word "nano" might raise red flags and trigger additional scrutiny at regulatory agencies enforcing export control laws, the Environmental Protection Agency, or the Food and Drug Administration.

ORGANIZATIONAL STRUCTURE AND CORPORATE GOVERNANCE

A third major task for founders is incorporating the company and establishing its governance structure. In this section, we compare organizational forms, identify critical governance and hiring issues, and explore employee stock option plans for nanotech companies.

Choosing an Organizational Form

Founders must register their business with the government and can elect between three organizational forms: a limited liability company (LLC), an S Corp or a C Corp. S Corps and C Corps are owned by shareholders and operated by directors and officers. Corps have an established structure and must follow certain corporate formalities. By contrast, LLCs are owned by their members, who have flexibility in structuring the administrative and managerial functions of the business. In general, although LLCs and S Corps offer tax advantages over C Corps, the latter are more prevalent for nanotechnology companies for several reasons.

Tax Advantages of LLCs and S Corps

Some nanotechnology companies are organized as S Corps and LLCs[49] due to their tax advantages. Both the LLC and S Corp are "flow through" entities for tax purposes—there is no tax at the entity level; income passes through and is taxed at the individual owner level. As such, the losses of an early stage nanotech enterprise organized as an LLC or S Corp can be utilized as deductions by investors. In contrast, income generated by C Corps is taxed at corporate rates at the entity level and then taxed again at individual rates when profits are distributed. The early losses of a nanotech start-up organized as a C Corp are unlikely to ever be utilized for a later tax reduction. A C Corp can carry forward losses to deduct in the 20 following years.[50] Unsuccessful nanotech companies will be unable to utilize future deductions, because they will probably never generate any profits to be taxed on. Successful nanotech companies will likely undergo a change of control at some point in their development. They may accept additional venture financing, be acquired, or go public. In these cases, the "change of ownership" rules for C Corps substantially diminish the amount of tax deductions allowed.[51]

Advantages of C Corps

Notwithstanding certain tax benefits of LLCs and S Corps, there are several reasons why nanotech start-ups usually prefer the C Corp structure. As we discuss later, C Corps must maintain corporate records such as articles of incorporation, financial records, and stock ledgers. These documents increase the perceived accountability of management and make finding investors and acquirers easier. Second, many venture funds will not finance LLCs or S Corps, because they find them too risky. In addition, many institutions that invest in venture funds are already tax-exempt (for example, pension funds). Therefore, they do not benefit from the special tax treatments that occur as a result of these organizational forms.[52] Third, in order to go public, a company organized as an LLC would be required to convert to a C Corp, so many companies with this goal prefer to begin immediately down that track. Fourth, LLCs are not allowed to compensate employees through stock options. As we discuss later, stock options can be a valuable management tool for start-ups; they can lessen cash burdens and attract and motivate employees. Fifth, a nanotech start-up may not want to be organized as an S Corp because there are limitations placed on who can own them. For example, there can only be one class of outstanding shares and there can be no more than 75 shareholders.[53] As such, an S Corp. is unlikely to be able to sell preferred stock to venture capitalists.

Choosing Between LLC, S Corp, and C Corp

Entrepreneurs involved in nanotechnology should consider several factors in deciding between a C Corp, S Corp, and LLC. First, the decision turns on the initial investors. If the early investors can utilize the initial losses, they may have more interest in LLC status. Second, the company must determine when and if it will require venture capital financing. As noted above, most venture capitalists are reluctant to invest in LLCs and cannot invest in S Corps. If the company can delay venture financing for a substantial period of time, then it may be advantageous to start the business as an LLC and convert later to a C Corp in order to seek venture financing. Finally, the decision depends on the amount of time that it will take for the company to generate income. If the company will have income within a short period of time, the carry forward losses can be utilized and the C Corp option becomes more attractive. As one nanotech CEO summarizes, "... timing is everything."[54]

When entrepreneurs are uncertain, they should err on the side of the C Corp. We conclude this section with the advice from Terry Lowe, CEO of Metallicum: "In retrospect, I'd say that it is best to start as a C Corp unless the stakeholders have a real good reason to go the LLC route."[55]

Governance

Because most nanotech companies are organized as C Corps, the remainder of this section discusses the governance structure of C Corps in detail. In addition to establishing a charter, the three crucial elements of nanotech companies organized as C Corps are the board of directors, scientific advisory board, and management.

Articles of Incorporation

In order to create a C Corp, founders must draft the articles of incorporation, also known as the charter. The charter describes the purpose, place of business, and other details of the corporation. The charter also includes bylaws, which are the official rules and regulations governing the corporation's management.

Board of Directors

Once companies are incorporated, shareholders elect a board of directors to be the governing body of the corporation. The board's primary job is to ensure that management is acting in the interests of the owners. The members of the board of directors are paid in cash and/or stock, meet several times each year, and vote on many major company decisions. When a company is first formed, the board may be composed of only the scientific founders and the CEO. As the company grows, it may seek board members with industry experience and financial contacts in order to take advantage of their management experience and access to capital sources. Over the company's lifetime, new board members will be added and others will resign, reflecting changing needs of the business and changes in ownership of the company that result from new rounds of fundraising.

Scientific Advisory Board

While it is not a requirement of the C Corp structure, nanotech companies often establish scientific advisory boards (SABs). The SAB does not have the kinds of formal powers held by the board of directors. Instead, the SAB serves a consulting role for management and is a source of technical expertise and additional industry contacts. The advisory board should consist of professors or industry officials that are familiar with the technology being developed at the company. Often, the SAB may initially consist of two or three members and then expand as the company matures.

We believe that establishing an SAB can be extremely important in several respects. First, members of the advisory board should be active participants in technology development at the company. Members should periodically meet with the company's management to provide information regarding the most recent developments in research and development. In addition to giving scientific advice, board members from industry can be valuable in discussing business strategy. For example, one company we interviewed narrowed the scope of its applications after speaking with SAB members whose many years of experience revealed flaws with the initial plan.[56] Further, the advisory board can provide industry recognition and serve as a lure for the company to attract other key employees.

Finally, and most important, a credible scientific advisory board may be critical to many nanotech companies seeking funding. As discussed in Chapter 12, except for a few firms, many venture capitalists do not currently have a deep understanding of nanoscience. They look for strong scientific advisory boards that comprise renowned university professors in order to feel confident in the technological merits of the company. For example, Nanosys, which has been extremely successful in attracting venture financing, can boast of a "Dream Team" of scientific advisors. The illustrious founding team includes famous scientists Drs. Paul Alivisatos, Moungi Bawendi, Louis Brus, James Heath, Charlie Lieber, and Peidong Yang. The backing of these experts gives Nanosys greater credibility with technologists and non-technologists alike.

Management

The management team for a nanotech company must include people with a wide mix of business, law, finance, technology, sales, and operational backgrounds. This means that the initial founders of the company must transition many duties to new members of the team. Some scientific founders play an active role in managing the company while others limit their activities to duties on the scientific advisory board. For example, James Baker, founder of NanoBio Corporation, maintains his faculty position at the University of Michigan while serving as CEO and CFO of the company. In contrast, Chad Mirkin, founder of NanoInk and Nanosphere, is a director and scientific advisor but does not have formal management positions in his companies.

While a founding technologist must provide scientific guidance to the company, it may be undesirable for him or her to be involved in day-to-day management of the business. First, the time requirements can turn a professor's 60-hour week into a 100-hour working week. Most of the founding professors we interviewed warned that attempting to divide time between the academic office and the corporate office is too ambitious. Both the university and the company are likely to suffer.

Second, founding professors may not have the business skills required to effectively run a company. In some sense, being a professor is like managing a business. The professor must solicit funding from outside sources, supervise graduate students, maintain a lab, and publish papers. However, the legal and business issues confronting entrepreneurs are very different from a professor's concerns. Intellectual property disputes, crafting creative financing arrangements, negotiating corporate partnerships, and commercializing products are areas in which business professionals have more direct experience. One founding professor concluded, "Business inside the university is just very different from business outside the university."[57]

Third, it is important to note that investors are wary of academics' running companies. Even if they can do it, their lack of direct business experience reduces the company's credibility and chances of securing financing. Possibly the most important factor in obtaining financing is the credibility of a company's management. Nanosys CEO Larry Bock explains the investor's position: "There is a venture capital dictum that VCs only invest in people that have successfully done it once before with someone else's money."[58] Indeed, Nanosys is a good example of a case where the company complemented its scientists with strong managers. All of the executives had been involved in other major platform technology companies before Nanosys.

Issuance of Stock

From the very beginning, stock ownership in the company must be explicitly spelled out to prevent disagreements later. In a C Corp, maintaining a stock ledger is a legal requirement. When properly done, the initial issuance of stock to founders and employees should not have any significant tax consequences[59] and should be exempt from registration under securities laws.[60] A complete treatment of the tax consequences of stock issuance is beyond the scope of this discussion.

Employee Stock Options

With limited capital and a need to compete for a relatively small supply of suitably-skilled scientists and engineers, the company may elect to develop a stock option plan ("Plan") to attract employees. The Plan must be approved by the board of directors and should be drafted to achieve several goals. First, the Plan should empower the board's compensation committee with flexibility in implementing the Plan. This is necessary to address changing market conditions but cannot be a "blank check" in which option agreements are tailored for every executive. Non-standardized stock option agreements can create problems when the company is looking to sell or go public. Second, the Plan should be developed to comply with securities laws and to maximize tax benefits for employees and the company. Finally, the Plan should be detailed enough to prevent ambiguities from giving rise to future disputes, but concise enough for negotiating purposes. A number of complicated issues must be addressed in crafting a Plan. We survey several of the most important.

First, the Plan should identify what types of options will be issued. The company can issue "nonqualified options" (NQOs) or "incentive stock options" (ISOs). There are different rules associated with using each type of option.[61] Generally, employees prefer ISOs for their tax advantages.[62] To maximize flexibility, the company should adopt a Plan that enables it to grant both ISOs and NQOs.

Second, the Plan should establish the size of the employee option pool. Due to the relatively small size of the labor supply, nanotech start-ups may require larger employee option pools than start-ups in other fields. Employee option pools range in size, but we have seen some companies with allocations for employees reaching greater than 20 percent of the total shares outstanding.

Third, the Plan must set a vesting schedule that states the length of time required for employees to be eligible for their stock options. Vesting schedules can vary depending on the ways in which the company seeks to attract, motivate, reward, or retain employees. For example, shorter vesting schedules may attract employees, but longer vesting schedules retain employees. The most common schedule vests 25 percent of options every year for four years, but three- and five-year schedules are not uncommon.[63]

Setting up the Plan requires a number of other decisions such as acceleration of vesting, repurchase rights at employment termination, and rights of first refusal on transfer.[64] Finally, developing the Plan involves wading through complex securities laws.[65]

LOCATION

There are many considerations that a nanotech entrepreneur should evaluate when determining where to locate the business. First, as has been described in detail in this chapter, nanotech companies are springing up around certain leading research universities and federal laboratories. Locating nearby may enable companies to stay close to the original researchers, have a voice in their ongoing research, stay up-to-date on industry developments, and benefit from the network of experts, funding sources, and industry representatives drawn to these institutions. Second, many cities and states have established aggressive programs to lure nanotech companies. They offer tax breaks, direct subsidies, and access to top-tier research facilities at little or no cost. Third, with a dearth of qualified scientists and engineers, nanotech companies must locate in one of the centers of technical expertise. Put another way, nanotechnology is not an industry where employees will initially follow jobs, but where companies will grow as a result of the infrastructure of technology expertise that is present locally. Finally, locating nearby potential customers can provide a significant competitive advantage. For example, Inframat Corporation attributes some of its success in marketing nanostructured industrial coatings to the proximity of its customers. We have identified five hubs for nanotech development and characterized some of their key advantages.

California—Silicon Valley

California's Silicon Valley remains the center of technology development in the country. We estimate that approximately 50 percent of all US nanotech companies are based in California, and 65 percent of these are located in the northern half of the state. The Silicon Valley pioneered the process of university and industry collaboration, and it created a unique culture very conducive to developing new technologies. The history of companies such as Hewlett-Packard and Apple starting out of Palo Alto garages or companies like Yahoo! and Google spinning out of Stanford class projects has created an unparalleled entrepreneurial culture.[66] Stanford and UC Berkeley are leading research universities and the Silicon Valley offers new companies a strong infrastructure consisting of a highly educated populace, federal research facilities, and venture capital firms. The Silicon Valley was the birthplace of the last boom in technology innovation and early signs indicate that it will play the leading role in fostering new nanotech companies.

California—Southern

Another promising nanotech region is Southern California, including the greater Los Angeles area. Los Angeles has long been the home to numerous aerospace/defense firms and federal research facilities. Including its suburbs, Los Angeles is also the largest metropolis in the country, and the intellectual capital base is large. Caltech is a leading research university actively pursuing nanotech development. The state-funded California Nanosystems Institute (CNSI) on the campuses of UC Santa Barbara and UC Los Angeles promotes university and industry collaboration for nanotech R&D. The non-profit group Larta also conducts conferences and attempts to foster university and corporate technology transfer.[67]

Northern Illinois

Some of the most exciting research in nanotechnology is occurring in the northern half of Illinois. The University of Illinois at Urbana-Champaign, Northwestern University and the University of Chicago are conducting world-class research. Argonne National Laboratory also attracts federal funding and scientific talent. The area is home to a highly educated populace, a relatively low-cost of living and has a strong business and financial community. Northern Illinois has been slower to attract nanotech business investment than California, and it is unclear how its research strengths will translate in to commercial development. AtomWorks is an Illinois-based consortium that brings together business, government, and university leaders to bridge the gap between research and business development in the state.

Massachusetts

Massachusetts is another area that is a center of nanotech development. Route 128 is home to many high-technology companies. Medical diagnostics and biotech are industries with a strong presence in Massachusetts. MIT and Harvard are leading centers of technology development and MIT is currently a strong leader in nanotech research. The state is aggressively trying to maintain a competitive position and keep technologies in state through incentive-based programs. The Massachusetts Technology Collaborative is attempting to expand the state's "critical mass of activity in nanotech"[68] and is fostering awareness and support for the industry in the state. The state also benefits generally by a highly educated populace and numerous other excellent research universities in the New England area.

New York

The New York area offers many of the same advantages for prospective nanotech companies as Southern California. New York possesses a large, educated populace, established industry, and an existing federal research infrastructure. The NanoTech Resources Inc. is a group responsible for coordinating nanotech research efforts among the SUNY schools. To date, however, this area lags far behind California, Illinois, and Massachusetts in creating a hub for nanotech development.

Texas

Texas was an early leader in nanotech but has been less successful in attracting new investment than other places. Most activity is located around Dallas and Austin/Houston. The Strategic Partnership for Research in Nanotechnology (SPRING) is an information and resource-sharing cooperative between Rice and the University of Texas (at the Dallas, Arlington, and Austin schools). The Texas Nanotechnology Initiative is another group attempting to advance nanotech development in the state. At this point, however, it is unclear to what extent a nanotech industry will develop there.[69]

Washington D.C

Stretching from Arlington, Virginia, to Washington, D.C., and on to Bethesda Maryland, this area is in close proximity to many research universities, highly educated populaces, and numerous federal facilities and programs. However, except for money spent at federal research facilities, the area has not witnessed much activity in nanotech. Given the importance of government funds to nanotech development, it will be interesting to see how much nanotech activity grows in this region.

CHAPTER 11

Business Plans and Strategy

The biggest red flag in nanotech business plans is people's estimation of how easy it is to penetrate a market.... We see too high of a hockey stick in those early years. Unless they've already been around a number of years building experience, naïve expectations in a business plan are probably the first sign of weakness.[1]

—Peter Grubstein (managing partner, NGEN Partners)

The investment environment that nanotech entrepreneurs encounter when seeking capital will be very different from the world that Internet entrepreneurs encountered in the late 1990s. Investors have learned bitter lessons about the risks of funding companies prematurely. As a result, they will not be pouring in large amounts of capital to companies that possess little more than "three scientists and a dog."[2] In order to fund companies, nanotech investors will require tested technology, defendable patents, growing target markets, a clear shot at profitability, and a strong management team whose members possess diverse skill sets. This new funding environment means that nanotech entrepreneurs will need to conduct more rigorous self-assessment of their strengths and weaknesses compared to their "high tech" predecessors of the Internet era. The company business plan is the place where this self-assessment takes form, and it is the place where entrepreneurs make their best case that the company is a good investment. This chapter describes the fundamentals of business plan writing for the nanotech investment community. It makes recommendations to entrepreneurs to help them write plans and prepare sound strategies in order to maximize their likelihood of obtaining capital under the best terms.[3]

The chapter begins by describing different types of business plans and the need to tailor them to different target audiences. The discussion then turns to issues facing nanotech companies. First, an effective executive summary requires that entrepreneurs succinctly explain complicated technology to investors that may not have a science background. Second, nanotech companies will need to prove to investors that their market projections are justified and not "naïve." We discuss the key issues involved in choosing target markets. Once entrepreneurs have chosen which market to pursue they should think carefully about how to organize the company to address that particular market. The plan must be realistic and show how the company will reach its intended market. The chapter concludes with a minimally technical discussion of financial modeling and valuation issues for nanotech companies. Although this chapter was tailored for people without finance backgrounds, the discussion is specific enough to nanotech that even experienced financial professionals will benefit.

GENERAL BUSINESS PLAN CONSIDERATIONS

A businesses plan is a document that defines business goals and priorities. It may be used by entrepreneurs to obtain venture funding or by established corporations to communicate the corporate vision to interested parties such as business partners and employees. There may be many versions of the same business plan tailored to different audiences such as employees, investors, and potential strategic partners. What is constant about good business plans is that they clearly describe business objectives and they realistically project how a company intends to obtain and apply the resources necessary to achieve its goals. Instead of differentiating between different versions of business plans, we will describe the archetype of the business plan—the version required by and intended for investors.

Business Plan Uses

There are at least four reasons why a prospective entrepreneur should write a business plan before attempting to launch a nanotech company. First, a business plan is essential for securing outside investment. The 2001 crash in Internet stocks has made investors wary of unproven technologies such as nanotech. Venture capitalists, corporate investors and Wall Street bankers are only interested in well-defined strategies and are looking for entrepreneurs with a firm grasp of financial reality. A good business plan addresses these concerns by both demonstrating the potential of the technology and showing that the entrepreneur has carefully considered the important obstacles the company will face in reaching its goals. Indeed, other things being equal, a well-constructed plan can be the difference between begging for capital and receiving multiple offers for investment.

A second reason to write a business plan is to explain the company vision to customers. It is said that customers are just investors receiving goods and services in place of equity. Consequently, like investors, many customers require assurance and belief in the underlying business model of the company. Large customers, in particular, fear the waste of time, expense, and embarrassment of partnering with a defunct entity. As nanotech venture capitalist Peter Grubstein explains, "[N]othing fast happens in these industries. If Ford had to change the paint and it was wrong that would mean 1.5 million recalls."[4] Thus, large customers are very cautious about changing how they do business; nanotech companies will have to convince bigger clients of their viability as long-term partners.

Third, writing a business plan provides significant help in launching and managing the business. Performing an analytical review of the company's strategy before launch will enable managers to anticipate problems before they arise. Since the company's earliest decisions also will be some of its most critical, predicting obstacles and key variables will significantly increase likelihood for success. Chapter 10 explores some of the issues involved in the nanotech formation, and in this chapter we elaborate on the issues that must be discussed in the business plan.

Fourth, a business plan is a valuable management tool for conveying the company's mission, organizational philosophy, and priorities to employees. Sharing the business plan can bring together employees across the organization—everyone from engineering to sales—and give them a common purpose. Because a good business plan clearly lays out the priorities of management, employees can use the business plan to better understand their own roles in the organization and to navigate through their daily tasks in ways most consistent with the goals of management. Ensuring that everyone is working towards a common vision creates more efficient organizations.

Key Elements

All business plans tell the story of the company—where it comes from, who is involved, what its goals are, and most importantly, why investors should want to be involved with the company. Table 11.1 describes eleven sections commonly included in business plans. This list is not intended to be a fixed prescription, but it is a frame of reference that entrepreneurs can use to consider whether they have covered the issues that investors may consider most relevant to the company's particular situation.[5] We describe these issues in the context of nanotech.

EXECUTIVE SUMMARY

The executive summary is an introduction to the company. For investors who review many business plans and may have limited time, it may be the only part of the plan they read. Therefore, the executive summary is a very impor-

TABLE 11.1 Sections Commonly Included in Business Plans[6]

Name	Description
Executive Summary	A two- to five-page summary of the business mission and strategy. It includes an "elevator pitch," or very short explanation of why the opportunity is attractive.
Market Opportunity	A description of the target customers: who they are, what they want, and how many of them there are. It should project how customers change over time.
Risks	A thorough outline of the factors that could undermine the success of the business. Both issues within and outside the company's control should be explained.
Competition	An analysis of the companies and technologies currently serving the target customers. There should be a projection of how new entrants may try to compete for these customers and how existing businesses will respond to the threat posed by the company.
Technology	A description of the science behind the company's product and an analysis of how it is different from existing technology. The company's intellectual property should be described in this section.
Strategy	An explanation of decisions that better enable the company to meet market needs. For example, choosing to partner may be an important strategy consideration.
Operations	A description of hiring needs, including research and development, manufacturing, and sales and marketing departments.
Management	A list of executives, directors, and scientific advisory board members. The list should describe managers' duties, as well as qualifications for their positions.
Financial Model	A detailed projection of revenues and expenses for the business, including complete income statements, balance sheets, and cash flow statements.
Sources and Uses	An estimate of the cash the company has available, the amount being raised and how it intends to use fundraising proceeds.
Valuation/Return on Investment	Estimates the current and future value of the company. It should project the time it will take for investors to realize a return and the rate of return for investors during that time.

tant part of the business plan and should be concise, typically not more than three pages.[7] What is most important is that it provides a clear mission statement of the company and it explains the technology and potential market to investors. It may also describe other issues such as marketing and sales, the company's capital position or projected returns to investors. The purpose of including these other items is to create additional context for describing the overall mission, vision, and strategy of the company.

In some cases, this may be a simple task for nanotech companies. For example, it may be easy to convey the benefits of a lighter material, a cleaner fuel additive, or a longer-lasting battery. In other cases, explaining nanotech to investors who lack the appropriate technology background may be difficult. It is a major challenge for nanotech business plans to provide investors with essential scientific information and not lose focus of the business plan's primary purpose of persuading investors of the benefits of the company's innovation. Catalytic Solutions, a developer of advanced coating solutions, has raised $72 million to date from a wide range of investors. Based on our discussions with their investors, we believe their success is owed in no small measure to their ability to reach a broad audience. The following excerpt shows that their company summary is persuasive to people of different levels of scientific understanding:

> We invented and own a fundamentally unique and patented material technology for the development of catalysts. Our technology significantly improves catalytic performance, is highly durable and cost-effective. We have developed unique nanostructures that are extremely thermally stable and resistant to sintering. This enables superior catalytic performance over time and at extreme temperatures...
>
> Although our technology has broad applicability, we have focused our commercialization efforts on environmental applications, namely automotive, diesel and energy applications. We are currently manufacturing products for the automotive and diesel markets and our technology has undergone extensive lab and field testing by customers in each of these markets. In addition, we are developing energy-related products in conjunction with strategic partners. This testing and product history have demonstrated the following competitive advantages over existing technology.[8]

Catalytic Solutions' summary clearly identifies the potential advantages of their technology. It also describes the differences between the technology and the current state of the art in a way that nontechnologists can understand. Scientists and business people alike will be interested in learning more about the company after reading this introduction. Finally, it describes the fact that the technology has many different uses, and it details specifically where the company intends to focus. Prioritizing between the different advantages of the technology avoids a key pitfall particularly common to nanotech business plans. Entrepreneurs must focus their plans on the most attractive markets and we will next describe the process of selecting a market on which to focus.

CHOOSING THE MARKET

The most important part of developing a business plan is choosing the market or markets to pursue. We explain the importance of choosing markets and discuss how entrepreneurs should choose target markets.

The Importance of Market Focus

During the Internet boom, many investors were persuaded to invest in companies based on optimistic market estimates from investment banking analyst reports.[9] However, appeals to greed are less acceptable to nanotech investors than they were during the Internet boom. Nanosys has successfully raised money by pursuing a strategy of attacking various markets without any clear focus. However, many investors are wary of companies trying to focus on too many markets.[10] By trying to focus on too many markets, resources for any individual market are diluted, and the company can become unfocused.[11] Companies with more specific focus will be better able to understand their customers and competitors, and they can more effectively target their technology development efforts to creating *products* that meet market needs.

In many cases, tools, materials, systems, and devices based on nanotechnology are broadly enabling. As mentioned, the National Science Foundation predicts that nanotechnology will be a $1 trillion industry by 2015. If a company liberally refers to capturing part of a $1 trillion market, investors are likely to be skeptical. They will want to know who the customers are, what specific products they are buying, and how much of the total value being purchased is directly attributable to the company's particular technology. An effective business plan will argue that the company's technology can be used in a range of different applications, but it has chosen to focus on one or two in the immediate future. For example, NanoOpto's business plan identifies telecommunications applications as its target market, even though the company's nanoimprinting technology can be used in a variety of ways. As NanoOpto Chief Executive Barry Weinbaum put it, "If the company was going after 10 or 12 markets, then it would be like trying to boil the ocean: You can generate a lot of heat, but not a single bubble."[12]

How to Choose the Best Markets

Choosing which markets to focus on involves two steps: (1) identifying target customers through market segmentation and (2) assessing the risk factors that influence the attractiveness of particular markets.

Market Segmentation

The first step in creating credible market projections is to define the market for the technology. This is called market segmentation, and it relates to how suppliers and buyers within a given industry are different.[13] Although it may seem counterintuitive, the goal of good market segmentation is to limit the scope of the technology to its most important markets. There are almost an infinite number of ways to segment a market. As Michael Porter of Harvard Business School explains, "In theory, every individual buyer or product variety in an industry could be a segment.... To segment an industry, each discrete product variety (and potential variety) in an industry should be identified and examined for structural or value chain differences from others."[14] Porter's analysis means that anyone performing market segmentation should take into account all the different types of people who buy products and the many different suppliers in that industry. Understanding differences between each market actor is helpful in getting a better feel for the precise way a new technology will impact the market. While it is usually not possible to explore the preferences of each individual buyer in the industry or to fully investigate every seller in the market, understanding how the new technology will impact different buyers and suppliers clarifies the strategy of the business.

In Chapter 1, we separated nanotech into three categories: "simple," "building small," and "building large" technologies. For example, within the building small category, we identified markets such as electronics, energy, and life sciences that are likely to be impacted by nanotech. For market segmentation purposes, separating out these three categories is a good start. But companies looking for funding must further subdivide their target markets based on the particular product they intend to create. For example, the market for life sciences can be segmented into many categories including diagnostic devices, pharmaceuticals, drug discovery, drug delivery, and tissue engineering. Within diagnostic devices, there also exist more specific subcategories. For example, there are diagnostic devices based on genetic and proteomic markers. Each is suitable for diagnosing different diseases and each has a discrete group of customers. Even within a product area with a discrete group of customers, the company should try to be more specific in their analysis. For example, within each of these diagnostic devices there are different markets for purification, amplification, and identification technologies. As a company gets more specific in defining its product and intended market, it will be in a better position to discuss its target market, financial projections, and business strategy with investors.[15]

Factors in Choosing Markets

Nanotech companies must carefully evaluate risks associated with different markets before deciding where to focus their efforts. Factors that must be considered include technology, competition, partnerships, public misperceptions, and regulatory issues.

Technological Development Speed: First, nanotech companies need to carefully consider the development time frame in deciding which market applications

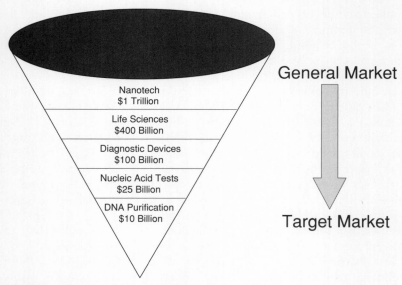

Figure 11.1 Narrowing a market through segmentation.

to pursue. If their core technology allows, they may wish to focus on developing "simple" nanotechnology applications before "building small" nanotechnology, because large-scale deployment of "simple" applications are generally within current technological reach. By and large, start-up companies working with "simple nanotechnology" are likely to develop commercial products much more quickly and with a higher rate of success than companies experimenting with more complex devices and systems. For example, Nanogate develops multifunctional nanomaterials. While their technology might ultimately be used in biotechnology and opto-electronics, the company has focused on developing simpler applications such as coatings and powders. The company has already established a customer base of manufacturers by developing surface coatings for consumer goods, printing, and automobiles.[16]

Likewise, companies that focus on a "building small" application may want to pursue applications that involve a lower scale of integration. These kinds of applications include chemical sensors, field emission devices, thin films for electronics, photovoltaics, and optical detectors. Products that are more densely integrated or, alternatively, have a large degree of complexity—such as memory or processors—will certainly take more time to develop.

The development of products in nanotechnology may also depend on the development of complementary technologies that do not yet exist. For example, in molecular diagnostics, many companies are creating devices that can detect variations in a sequence of DNA. However, in order for these devices to be useful in disease detection, complementary mathematical algorithms that correlate sequence to function—genotype to phenotype—must be developed. Consequently, even if suitable detection devices are developed, the lack of adequate statistical techniques will diminish their value as it still would be impossible to take the crucial step from detection to diagnosis.

Competition: Competition should also influence the decision of what application to target. A commonly used model for understanding competition and the source of competitive advantage is Michael Porter's "Five Forces" model. (See Table 11.2)

Nanotech companies should evaluate these different factors for each of the different markets they are considering. Some companies look for markets where there is no competition. NanoBio Corporation, for example, chose to focus on using its antimicrobial agent to treat toenail infections, because it felt it could establish a monopoly in this field. Although the drug might be used to treat other infections that may represent larger total markets, the toenail market is a new market without competitors. NanoBio CEO Dr. James Baker described the decision, "[t]oenail fungus is an enormous clinical problem in up to 25 percent of adult Americans, and currently there's no topical therapy that's effective.... So we believe that we would be moving into essentially a virgin market of fairly significant size in that area."[18]

Most nanotech companies also focus on intellectual property when evaluating the competitive landscape. As we explain in Chapters 5 and 13, many companies are racing to acquire patents and patent licenses. A dominant intellectual property arsenal can provide a company with sweeping powers over a market. As such, a company might be more likely to choose a market

TABLE 11.2 Five Forces Model[17]

Force	Description
Substitutes	Alternatives to the company's goods/services offering
Rivalry among existing competitors	Extent to which companies will go to undercut their competitors
Barriers of entry	The company's ability to prevent others from competing
Bargaining power of suppliers	The ability of suppliers to dictate price to the firm
Bargaining power of buyers	The ability of the firm to dictate price to buyers

where few patents currently exist and in which the company can establish broad patent protection.

While companies should focus on markets where they can achieve competitive advantages, they should be mindful that few firms ever establish complete control over a market. Even when a firm has monopoly power, alternative technologies are likely to develop, and new companies will compete. Indeed, because nanotech is developing at such a rapid rate, new companies can find innovative ways to circumvent patented technologies. For example, although Konarka was one of the first companies to attempt to create a patent position around nanotechnology in photovoltaic cells, a number of other companies (Nanogram, Nanosolar, Nanosy) have also targeted this market. Generally, as new technologies develop and new companies enter the market, the sources of competitive advantage eventually erode. Thus, business plans should show how firms are positioned to continually adjust to changing competitive market landscapes.

Partnership Considerations: A third consideration for choosing a market relates to the company's prospects for partnering with a larger corporation. In many cases, a large corporate partner will be necessary to enable the company to reach its target market. Start-ups pursuing global markets may require the distribution and marketing capabilities of a company with a multinational sales force. For example, Nanosphere's partnership with TakaraBio Inc., one of Japan's largest biotech companies, enables it to reach Asian markets. Bill Athenson, vice president of business development for Nanosphere stated, "This alliance provides Nanosphere with a premier distribution partner in Asia and access to...the Asian marketplace."[19] In other cases, certain companies control target markets. In the life sciences industry for example, Roche, Abbott, and Bayer control 67 percent of all diagnostic test distribution.[20] A start-up's ability to sell to their target markets may be highly dependent on their ability to strike partnerships with these large players.[21] We devote most of Chapter 14 to issues involved in corporate partnering.

Public Perception: It is not yet known how the public will respond to the many different kinds of nanotechnology applications. As we explained in Chapter 3, the public seems ambivalent towards nanotechnology, and resistance among certain groups is brewing. In Berkeley, California the public health commission recently held a series of meetings to address health and environmental concerns regarding the effects of nanotech research and development on local communities.[22] These meetings made clear that many segments of the public are carefully watching nanotech applications they believe may pose environmental or health risks. Companies may wish to avoid researching and developing nanomaterials that are potentially toxic. As discussed in Chapter 4, the respiratory risks associated with carbon nanotubes warrants caution in how they are used and further study to determine their actual toxicity. Similarly in Chapter 3, we identified books such as Michael Crichton's *Prey* that have left many in the public concerned of the possibility of uncontrollable self-replicating nanorobots. Entrepreneurs may want to avoid products based on related technologies or business plan language that might leave room for confusion.[23]

Regulation: Companies should also consider regulatory risks in deciding which markets to focus on. Part II of the book engaged in a detailed discussion of environmental regulation, FDA regulation, and export control laws. As explained in Chapter 4, the EPA has not yet put forth strict regulations on production of nanomaterials. Nevertheless, there is always a risk that new evidence on the toxicity of certain nanomaterials could cause the agency to pass strict regulations. Thus, companies may want to avoid developing technologies that carry health and safety risks.

Chapter 6 discussed FDA regulation. Performing the necessary clinical trials to obtain FDA approval can be an expensive and time-consuming process. Further, there is always the risk that trials will not be rigorous enough to satisfy the FDA. FDA requirements can be a "moving target" in which it is not always clear what level of scrutiny a product will encounter. To minimize regulatory risk, start-ups may elect to focus on market applications that do not require precise engineering of drugs or drug delivery devices. Many start-up companies in this sphere were launched to develop novel drugs and drug delivery devices. However, the complexities associated with developing such "smart" drugs have caused these companies to rely on more "simple" forms of their technology for their initial revenue. For example, the founders of C Sixty assembled the company to develop fullerene-based drugs to treat HIV and amyotrophic lateral sclerosis (Lou Gehrig's disease). However, the company's first successful products are likely to be fullerene antioxidants for dermatological and cosmetic applications relating to the effects on skin of aging and exposure to the sun.[24]

Finally, as explained in Chapter 7, the export control laws are convoluted and burdensome. To avoid getting caught in this complicated regulatory web, companies may want to steer clear of technologies that can be used for military purposes.

RISK DISCLOSURE

While entrepreneurs may not want to think about the many things that can go wrong in starting a new nanotech business, the business plan must disclose all relevant risks. First, experienced investors understand that all new companies face many obstacles, and they will be highly skeptical of a business plan that does not directly address these issues. Second, all private companies raising money in the United States are required by law to fully disclose significant risks known to management.[25] Companies that do not fully disclose risks to investors unnecessarily expose themselves to investor lawsuits and permanent loss of credibility.

OPERATIONS

Once companies have chosen which markets to focus on, they must provide a description of how research and development, manufacturing operations, sales and marketing, customer service, and management will be organized in order to effectively target the markets they have chosen.

Research and Development

Almost all nanotech companies must conduct research and development to bring products to market. Because companies with research facilities must purchase sophisticated tools, nanotech R&D can be extremely expensive. While equipment prices vary, in total they can be prohibitive for poorly capitalized companies. State-of-the-art chemical vapor deposition systems currently cost upwards of one million dollars.[26] Atomic force microscopes currently cost upwards of a hundred thousand dollars. Clean rooms, super clean water, reliable backup power systems and commercial disposal systems are also expensive. The business plan must provide an overall estimate of how much the company must spend on R&D and how long it will take to develop the product.

Manufacturing

Scaling up for mass production in nanotechnology can be extraordinarily challenging. While some production of nanomaterials has already begun, it is uncertain whether many companies can overcome the tremendous engineering hurdles to mass-produce devices and systems based on nanostructures. Further, even if efficient mass-production techniques can be developed, manufacturing facilities will be extremely expensive. As we explain in Chapter 12, most nanotech venture capital rounds are between $5 million and $25 million. While this is a significant amount of money, it is not generally enough to create large scale manufacturing facilities.

Nanotech companies must tackle these issues in the business plan and explain how they will manufacture their products. As we discuss in Chapter 14, many nanotech start-ups should structure their business plan to rely on corporate partnerships for manufacturing capabilities. The plan must also identify other issues related to manufacturing such as EPA regulations (see Chapter 4), FDA regulations (see Chapter 6), and export control laws (see Chapter 7) to ensure that it is operating within the law. Complying with regulations can be an expensive and time-consuming process, and entrepreneurs cannot overlook this fact when writing their business plans.

Sales and Marketing

Nanotech companies must also describe the sales and marketing efforts they will undertake in order to reach their target markets. They may need to spend money on advertising, promotions, public relations plans, direct mail, attendance at trade shows, a direct sales force, and telesales forces.[27] These expenditures are likely to be greater for nanotech companies than other start-ups for two reasons. First, as we explained in Chapter 3, there is a substantial risk that public misperceptions will impede adoption of the technology. Companies must not only be able to convince customers that they have the best product, but they must also assure the public that the technology is safe. Second, as we argue in Chapter 16, nanotechnology hype could have an enormous influence on valuations for nanotech companies. Thus, nanotech companies have a greater incentive to spend more resources on marketing than companies in other industries where awareness does not have such a large payoff. Of the ten public companies with the word "nano" in their names that we profiled in Chapter 10, only three currently have revenues greater than their sales expenditures.[28] The most mature company on this list, Nanometrics, still spends more than 30 percent of revenues on sales efforts.[29]

The large expenditures that nanotech companies must devote to sales and marketing creates a risk that the company will burn more cash than it can afford. One company we interviewed had an excellent technology and high estimates of future earnings. However, it was discredited because it did not include a realistic estimate of the number of salespeople required to go out and achieve the sales they predicted. Thus, companies must realistically assess the costs of marketing and sales in their business plans.

Customer Service

Nanotech companies must also address customer service in their business plans. This is an area that is often overlooked but that is a significant cost and source of risk.[30] This is especially true in nanotech where clients may not have the technical capabilities to solve problems on their own. Technical problems can undermine customer confidence and delay sales of future product versions. Companies that accurately anticipate customer requirement needs can design service plans for customers that charge a fee; in this way, the company can tap into a potentially large source of continual revenue and at the same time ensure customer satisfaction.

Key Management

The business plan should include a description of the background of key managers and board members. As described in Chapter 10, it is highly advantageous for new nanotech companies to gain credibility by hiring well-known and experienced employees. However, hiring high-profile employees is accompanied by the risk that if these employees leave, the company will lose their particular expertise and damage investor confidence. The plan should mention the risks associated with key employees leaving the company.

FINANCIAL MODELING

Nanotech business plans include financial models. A financial model is a forecast of key business variables and their effects on cash flow.[31] This section begins by describing the importance of good methodology, the time period covered in financial projections, and the data that must be included in financial models. Data sources for creating nanotech market estimates are given and a methodology for dealing with uncertainty in nanotech is proposed. Finally, we provide an example of how a nanotech market estimate and projection should be prepared.

The Importance of Good Methodology

Financial projections for start-up technology companies are never fully realized and rarely come close to the original estimates.[32] Experienced investors know this and approach projections with a very critical eye. While they do not expect entrepreneurs to possess a crystal ball, they do expect them to be able to defend projections against many tough questions.[33] Sloppy estimates are one of the first ways investors weed out weak business plans, and poor estimates reflect negatively on management's credibility. Because the cash flows predicted by the financial model are the basis for determining the valuation of the company, creating a good model with strong assumptions is also essential for getting the best possible financing terms.

Time Frame

Traditionally, financial models cover a five-to seven-year period and show only annual results for the company.[34] Projections typically cover the early periods in the establishment of the business when there are no sales and the company is "burning" cash through the maturation stage of the business when it is able to sell its product. Five to seven years may not be enough time to show a nanotech company's full growth potential. Nanotech entrepreneurs may need to extend the time frame for the projections to ten or more years. While this is not unprecedented, investors may look doubtfully upon revenue projections made too far out in to the future. The consensus in the financial community is that, all else being equal, projections in later years are less reliable than projections in earlier years.

Key Components of a Financial Model

A financial model is based on three key components: income statements,[35] balance sheets,[36] and cash flows.[37] A complete description of accounting issues is beyond the scope of this book, but we do identify related issues in creating a financial model for a nanotech company.[38]

Market Estimates

Company projections of income statements, balance sheets, and cash flows are based on the company's estimates of the future state of the market. There are three commonly accepted sources for market estimates in a business plan. First, business plan market estimates can be quoted from third-party sources. Second, market estimates can be derived from inductive analysis of each of the components of the market. This generally involves creating independent estimates by aggregating and analyzing primary data related to each of the underlying components of the market. A third approach is to combine the first two approaches and compare both third-party sources and inductive estimates in order to provide a range of possible outcomes for the total size of the market.

Third Party Sources

Third party sources include research reports, government studies, and estimates from industry insiders. Table 11.3 provides a list of the most valuable sources of primary and secondary data for nanotech market estimates. The key to using these data sources is to perform a cross-comparison of differing market estimates in order to understand the assumptions made in each estimate. As mentioned earlier, investors may approach projections with a critical eye. Entrepreneurs can expect questions and should be able to answer inquiries as to the source of the data, its corroboration from other sources, the specificity of the claims, the legitimacy of the source, and the timeliness of the projection data. Being able to answer such questions provides proof to investors that entrepreneurs are doing the "homework" necessary to make their business idea viable.

Predictions without a full understanding of the underlying issues can turn investors off to an otherwise solid business plan. A very prevalent example of this issue is the National Science Foundation's estimate that the total market for nanotechnology is $1 trillion. Because no one knows where this analysis comes from, or what assumptions such a prediction are based on,[39] the number evokes skepticism and derision on the part of investors looking to address real and immediate markets.

Creating Independent Estimates

Predicting markets in well-established industries is much easier than projecting growth in emerging markets such as nanotech. In well-established industries, detailed information in public filings can be used to make projections. To predict growth in emerging markets, entrepreneurs must identify the needs that exist in the market and the utility provided by the new technology.[40] The intersection of these two factors is the total market opportunity.

Consider the following example. A company is trying to estimate the total market and future growth of the market for a nanobio tool to predict and

TABLE 11.3 Sources of Information for Nanotech Business Plans

Industry Journals	Description	Website
Foresight Institute	One of the founding groups of the nanotech industry. The website provides links to news, careers, events, and public information related to nanotech.	www.foresight.org
Nanotechnology Research Institute (Japan)	Website of the National Institute of Advanced Industrial Science and Technology Japan (AIST). Includes industry and government news in nanotech related to developments in Japanese investments.	http://unit.aist.go.jp/nanotech
Institute of Nanotechnology (UK)	Website committed to tracking and advancing the development of a nanotech industry in the United Kingdom.	http://www.nano.org.uk/
American Association for the Advancement of Science (AAAS)	Includes links to Science Magazine and related archives. Includes the latest information on the most recent developments in the field.	http://www.aaas.org/
Nanotechnology Law & Business Journal	Website provides expert insights and analysis related to nanotech patent prosecution and litigation, venture financing of start-up companies, and legislative and regulatory initiatives.	www.nanolabweb.com
Nanobusiness Alliance	Industry association founded to advance the emerging business of nanotech. Supporters and contributors include some of the biggest names in business and nanotech.	www.nanobusiness.org

Government Sources	Description	Website
National Nanotechnology Initiative (NNI)	Overview of the NNI, including budgets, conferences, R&D funding priorities, links to places to apply for funding, and departments and their research priorities.	www.nano.gov
National Science and Engineering Nanotechnology	National Science Foundation (NSF) website for NNI program. Includes much of the same information as NNI's website, but specific to NSF projects.	www.nsf.gov/home/crssprgm/nano
Advanced Technology Program (ATP)	ATP funds high-risk, high-payoff projects from all technology areas. Website includes priorities, partnerships, products and services, and recent news.	http://www.atp.nist.gov/

Government Sources	Description	Website
National Institute of Standards and Technology (NIST)	NIST works with industry to create technology, measurements and standards and also funds research. The website provides contact information and the latest news.	http://www.nist.gov/
FedBizOpps	FedBizOpps.gov is the single government point-of-entry (GPE) for federal government procurement opportunities over $25,000.	http://www.fedbizopps.gov/
Defense Advanced Research Projects Agency (DARPA)	DARPA is the central research and development organization for the Department of Defense (DOD). It manages and directs selected basic and applied research and development projects and pursues research and technology where risk and payoff are both very high and where success may provide dramatic advances for traditional military roles and missions. Website includes budget, contact, and project specific information.	http://www.darpa.mil
DOD Comptroller	DOD Office of the Comptroller, provides detailed information on the DOD budget, new programs, and requirements for firms interested in procuring government contracts.	http://www.dod.mil/comptroller/
Federal Laboratory Consortium	Network of federal laboratories linking the laboratory mission technologies and expertise with the marketplace.	http://www.federallabs.org
FirstGov	U.S. Government's official web portal. Includes numerous links to various government organizations.	http://www.firstgov.gov/
Small Business Administration (SBA)	SBA includes Small Business Innovation Research (SBIR) and Small Business Technology Transfer Programs (STTR). These organizations provide funding opportunities for small and emerging technology companies, including nanotech companies. Website includes information on funding requirements, contact information and news on companies recently receiving funding.	http://www.sba.gov/sbir/
Computer Retrieval of Information on Scientific Projects (CRISP)	A searchable database of federally funded biomedical research projects and provides information on obtaining federal awards	http://crisp.cit.nih.gov/
U.S. Census Bureau	Provides statistics on historical and projected U.S. population and demographic characteristics.	www.factfinder.census.gov

TABLE 11.3 *(continued)*

Popular Press	Description	Website
Nanodot	Foresight-related website dedicated to latest discussions of issues critical to the development of the nanotech industry, including government programs and contributions from technology leaders.	www.nanodot.org
Nanotechnology Magazine	Website for the *Journal of Nanotech*, a leading print publication. Articles include reader submissions reviewed by leading academic and industry leaders. Website includes archives of issues, daily updates of industry developments and articles on trends in nanotech.	www.nanotechweb.org
Nano Investor News	Portal for media research in MEMS/nanotech with news articles, presentations, research, marketing, and company and investor data targeted at the venture capital and business communities.	www.nanoinvestornews.com
Small Times	The largest publication dedicated exclusively to developments in nanotech. Website includes comprehensive information on companies and venture capital investments as well as news on industry and government investment and regulation.	http://www.smalltimes.com

Private Research	Description	Website
Edgar Online	Edgar Online provides Securities and Exchange Commission (SEC) filings from U.S. public companies. These filings include business, financial, and competitive information on the companies and their industries. Some services require subscription.	www.freeedgar.com
LexisNexis	Information service providing business, legal, and government sources of data. Requires subscription.	www.lexisnexis.com
Thomson Research	Data collection service providing information on corporate finance including company profiles, financial information, and SEC filings. Requires subscription.	www.research.thomsonib.com
Multex	Provides access to investment banking and research reports including forecasts on industries, markets, and companies. Requires subscription.	www.multexnet.com
marketresearch.com	An aggregator of leading sources of market intelligence. A very large collection of pay-per-use research reports on virtually every market services require subscription.	www.marketresearch.com

Business Plans and Strategy

detect diabetes and related glucose levels. A bottom-up approach to identifying this growing market requires identifying the key variables that will influence adoption of the product. It might take the following three-step approach:

1. Find out who might have use for this product.
   - Number of Americans with heart disease
   - Growth in heart disease
2. Determine who might buy this product.
   - Percentage of cases actually diagnosed
   - Annual number of tests for diagnosed patients
3. Examine the impact of the nanotech on the market.
   - Determine a price per unit (hardware and disposables)
   - Estimate percentage likely to purchase (penetration rate by year)
   - Estimate number of disposables purchased per year

The goal of these questions is to identify the key variables that will influence the revenues associated with the product. Researching the answers to these questions enables the company to estimate the market and revenues for this product. (See Table 11.4)

The data in Table 11.4 are based on official government estimates, realized financial results from existing companies, and information provide by doctors and experts in the field. Compiling data from various sources allows a company to effectively defend its projections. Entrepreneurs should not stretch market estimates to support overly optimistic conclusions. Conservative assumptions will enhance entrepreneur credibility and stature in front of investors.

SOURCES AND USES

The business plan should include a description of "sources and uses," the name for the company's sources of funding, and the uses to which it intends to apply those funds.[41] A typical sources and uses chart is shown in Table 11.5.

If the company cannot raise its intended funds, then the uses will change. Similarly, if the company faces unanticipated challenges, such as technology development delays, the uses of the funds it does raise will almost certainly change. As a result, companies must describe the risk that either actual sources or actual uses might be different from what is anticipated in the business plan.

VALUATION

The business plan should include a valuation analysis of the company. The valuation of the company is determined by the profit and cash projections of

TABLE 11.4 Market Projections for a Hypothetical NanoBio Company

(Data in million)

	2004	2005	2006	2007	2008	2009	2010
Hardware							
# of Patients[1]	18.0	19.0	20.0	21.0	22.0	23.0	24.0
% Diagnosed[1]	66.8%	68.3%	69.8%	71.3%	72.8%	74.3%	75.8%
Potential Customers	12.0	13.0	14.0	15.0	16.0	17.1	18.2
Hardware Penetration[2]	0.0%	0.0%	0.0%	1.0%	2.0%	4.0%	5.0%
Cumulative Hardware Units	0.0	0.0	0.0	0.1	0.3	0.7	0.9
New Units	0.0	0.0	0.0	0.1	0.2	0.4	0.2
Revenue/Unit ($)[3]	N/A	N/A	N/A	$66.49	$63.17	$63.17	$63.17
Hardware Revenue	**$0.0**	**$0.0**	**$0.0**	**$9.96**	**$10.78**	**$22.94**	**$14.28**
Disposables							
Implied Number of Patients	0.0	0.0	0.0	0.1	0.3	0.7	0.9
Annual Disposables[2]	4.0	4.0	4.0	4.0	4.0	4.0	4.0
# of Full Compliance Disposables	0.0	0.0	0.0	0.6	1.3	2.7	3.6
% of Patients Fully Complying[4]	0.0%	0.0%	0.0%	26.0%	27.0%	28.0%	30.0%
# of Disposable Units	N/A	N/A	N/A	0.2	0.3	0.8	1.1
Revenue/Unit ($)[3]	N/A	N/A	N/A	$35.63	$33.84	$33.84	$33.84
Disposable Revenue	**N/A**	**N/A**	**N/A**	**$ 5.5**	**$11.7**	**$25.9**	**$ 36.9**
Total Market	**$0.0**	**$0.0**	**$0.0**	**$31.0**	**$45.0**	**$97.7**	**$102.4**

(1) Center for Disease Control, Report on Trends in Diabetes, 2003.
(2) 2002 Annual Report for Inverness Medical Innovations, Inc. (AMEX: IMA). These estimates were created using Inverness' home diabetes test as a proxy. The estimates here are lower than Inverness' actual historical penetration rates in order to reflect greater competition.
(3) Prices based on Lifescan's OneTouch home diabetes testing system. Prices have been reduced below actual OneTouch prices to reflect greater competition in the future.
(4) Estimates based on telephone interviews with doctors who currently treat diabetes patients.

the company that we explained in the section on the financial model.[42] The valuation of "private companies, especially those in the early stage of their life cycle, is a difficult and often subjective process."[43] Studies show that different investors value companies in "startling different ways."[44] As one commenter observes, "What is the true value of [the start-up]? Take your pick: Forty-six cents a share, figures one venture capital firm, or $4.42 per share, according to another."[45]

There are several reasons why it is particularly difficult to ascribe value to nanotech start-ups. First, the technology is still in its infancy. With so many technical uncertainties, it is unclear if and when nanostructures will make their way into different commercial products. Second, the primary value of these companies is their intellectual assets—patents, scientists, and trade secrets.[46] It is particularly difficult to attach a value to these kinds of intangible assets.[47] Finally, it is still unclear whether investors will have an appetite for nanotech public offerings. As this book goes to press, the proposed IPO of Nanosys could provide the first direct test of public market interest in these young companies. The likelihood of a public exit is an important factor in determining the value of a private market company. The combined effect of these subjective factors can produce orders-of-magnitude differences in the valuation of new nanotech companies.

Despite the difficulties inherent in nanotech valuation, there are several tools available for valuing companies,[48] and many creative methodologies can be applied when trying to value such early stage growth companies. Most valuation techniques are derivations of two approaches.[49] The first approach is "Comparable Company Analysis," in which the value of the company is determined by evaluating other similar companies whose value is known (for example, public companies and recently acquired companies). The second method is the "Discounted Cash Flow" or "DCF" calculation.[50] The DCF is a mathematical approach to valuing a company that is determined by estimating the value of the firm's current and future earnings available to investors (profits and losses). Nanotech entrepreneurs should understand

TABLE 11.5 Sources and Uses

Sources		Uses	
New Equity Contribution	$361,368	Cash at Closing	$100,000
New Senior Debt	350,000	Purchase of Outstanding Stock	16,087
Assume Capital Leases	33,269	Payoff Senior Debt	389,089
		Payoff Subdebt	206,192
		Assumed Capital Leases	33,269
Total	**$744,637**	**Total**	**$744,637**

both approaches in order to make the strongest case for their company valuation. In this section, we describe these two approaches.

Comparable Company Analysis

One of the easiest and most reliable ways to value a company is to compare its value to the value of other similar companies. This is called "comparable company analysis" and it involves valuing a company based on a *relative* comparison with other companies in the same industry and subject to the same economic trends in order to evaluate valuation metrics. The idea is that if two companies are similar, then their valuation metrics should be similar. In order to conduct a comparable company analysis one must take two key steps. First, a group of similar peer companies must be identified whose valuations are known. Second, the financial metrics (known as "valuation multiples") upon which those valuations are based must also be identified.

Identifying Similar Companies

"Similar" or "comparable" in this case means that companies being evaluated are in the same industry and are similar in size, maturity, profitability, and growth rate.[51] Since no two companies are identical, a good comparable company analysis should include as many comparable companies as possible so as to make the average or mean from the calculation less dependent on differences in value from any single other comparable company.

For early-stage technology companies, such as those in nanotech, the main problem with this valuation technique is that there are not many comparable companies whose value is known. In most industries, a company's valuation is determined by evaluating similar public companies whose value is known. The few public companies that describe themselves as nanotech companies are almost entirely unrelated to the vast majority of research and development currently going on in nanotech. In addition, these companies are in such early stages and their values are so driven by hype that we do not believe a comparable company approach is appropriate. For example, companies might use Nanogen as a starting point for arguing that they should command similar valuations. However, as we discussed in the Chapter 10, the company's valuation tripled due to a well-timed press release regarding a patent. We do not believe the tripling of Nanogen's valuation had any underlying economic cause, since the company's earnings, revenue, and announced growth outlook did not change.[52] As we mentioned, in most industries, it is relatively easy to test aberrant valuations against industry norms with a large enough group of comparables. But there are no industry norms in nanotech and few other comparables against which to check. Second, even if one assumes that Nanogen's change in price was a rational reflection of the value of their new patent, Nanogen's technology is too specific to the company to be considered comparable to most nanotech companies. Nanotech companies dealing in electronics, energy, or any one of a number of other industries would have little in common with Nanogen. Thus, with a dearth of comparable companies and questionable valuation trends relating to existing companies, it may be difficult for nanotech companies to find enough relevant comparables to make strong arguments for their value.

Valuation Multiples

In the case where a set of comparable companies can be found, a company must determine the proper valuation multiples. A valuation multiple is "an expression of market value relative to a key statistic which is assumed to relate to that value."[53] There are many "key statistics" that can be used as a basis for multiples including revenues, net income, dividends, assets or growth in sales.[54] For purposes of a nanotech discussion, the important factor in multiples to consider is that the underlying metric used for multiples depends on the industry and the company involved. The most common multiple in established industries is the Price-to-Earnings (PE) multiple. The PE multiple is the total value of the firm divided by its earnings for the most recent year or trailing twelve months of operations. The PE multiple for General Electric as of December 31, 2003, was approximately 22.1 times earnings from the last twelve months.[55] Companies that are considered comparable to GE might also expect that their value can be ascertained by multiplying their profits for the last twelve months by 22.1. A company that is comparing itself to GE might argue that since it is a higher-growth company, it deserves a higher valuation multiple than GE. Those on the other side of the bargaining table—that is, investors—would argue that since GE is a stronger and more viable company, the valuation multiple should be lower.

In nanotech, as in other new sectors, it is not possible to value companies based on a PE ratio. This is because most companies in the industry do not yet have earnings. Based on valuation metrics applied to other early-stage and high-growth companies, we expect the key metrics for valuation to be revenue (price divided by revenue) and revenue growth (price divided by revenue growth).[56] However, as we have explained, we expect the issue of determining proxy multiples and appropriate comparables to be a very challenging issue for nanotech companies for the foreseeable future.

Discounted Cash Flow

In Discounted Cash Flow (DCF) analysis, there is no need to evaluate peer group companies.[57] Consequently, we expect DCF to be the method of choice for nanotech companies. In DCF, the value of the company is simply the present value of its future earnings for investors. This present value (PV) of the company's earnings for any year is equal to the earnings available to investors in that year (CF) divided by 1 plus the discount rate (R) raised to the time period elapsed since the present (T):

$$PV = CF/(1 + R)^T$$

The data for this calculation will be based on the financial model we discussed earlier. The earnings available to investors (CF) number will be the company's income number post tax and post all investment expenditures needed to support future operations of the business. The discount portion of the methodology (R) is based on the theory of time value of money, which holds that all money is more valuable today than it is in the future. Hence the term *discount* relates to the reduction or discount placed on future revenues in order to determine the present value. The discount rate that is applied to these future earnings is one of the areas of highest subjectivity in DCF calculations and will be particularly subjective for nanotech companies.

As shown in the Figure 11.2, typically a company loses money at the beginning of operations; if successful, financial returns will be realized at a later time. It takes more dollars at the later period to make up for the dollars spent today in supporting the business. For nanotech companies, expenses will be very high in early years, and revenues and earnings will take longer than in most other industries. Consequently, in DCF calculations, nanotech valuations will be particularly sensitive to the value of the discount rate. Most of the earnings that count towards the value of the company will be further in the future and thus more heavily discounted than shorter-term earnings.

Risk Factors and Discount Rates

The Discount Rate is "the investor required return or cost of capital."[58] The return that investors require depends on the risk they perceive in the company.

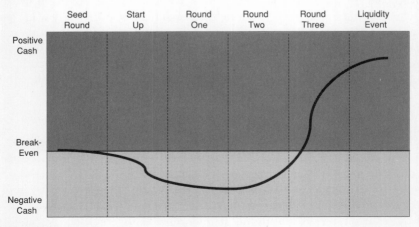

Figure 11.2 Typical cash for a start-up nanotech company.

The greater the perceived risk of the company, the greater the required return will be for investors to put money in to the company.[59] Thus for early-stage start-up companies such as those in nanotech, the discount rate is significantly higher than for established companies. For established companies such as Wal-Mart or Microsoft, an appropriate Discount Rate might be 12 to 18 percent. For new technology companies, such as nanotech companies, the discount rate may approach 100 percent depending on the specific company and technology involved. For first-round nanotech companies where the technology is not yet developed, we have seen discount rates between 75 and 100 percent. In the second round, 45 to 75 percent is typical. Finally, in a third round, depending on the maturity of the company and their closeness to profitability, a discount rate might be 35 to 45 percent.[60] While these are common deal terms that are helpful to know, the final discount rate is dependent on negotiations between investor and entrepreneur. In this sense, the many different issues regarding commercialization obstacles that we discuss in this book should be helpful considerations in negotiating the final rate.

Terminal Values

A "terminal" or "residual" value is "the value at the end of the discrete projection period in a discounted future earnings model."[61] For example, if a company creates a financial model that projects financials for seven years, the terminal value captures the value of all earnings beyond the seven-year horizon. Since many nanotech companies will realize most of their financial rewards after the period forecasted in the financial model, we expect terminal values to also be a very important issue in nanotech valuation.[62] The terminal value is estimated using a DCF approach for future earnings; a constant growth rate and a discount rate are applied to a base earnings number for all periods beyond those covered in the model itself.[63] The growth rate that is used should reflect the company's prospects for long-term growth. So, a higher growth company will have a higher terminal growth rate used in the calculation than a company that is in a lower-growth area. Conversely, higher levels of risk for earlier-stage growth companies will require the use of higher discount rates, creating a steeper discounting of future earnings and reducing the final value of the terminal value estimate.

Exogenous Variables that Determine Valuation

While it is important to reduce the perception of risk through solid projections and thorough evaluation of variables such as competition, other factors that are not in the company's control will be critical to determining the actual discount rate that is used. First and foremost is investor sentiment. No matter how much work is done preparing a business plan or financial model, investor sentiment may be the most important factor determining funding and valuation for nanotech companies. It is difficult to predict the moods of

TABLE 11.6 Sources and Fees of Business Plan Outsourcing

Type of Companies Firm	Pricing Structure	Nanotech Capabilities	Notable in Nanotech
Big 5 Consulting Firms	Flat fee based on scope of project and research required	Little	*Boston Consulting Group:* Strong capabilities in emerging technologies including nanotech and deep connections in industry *Mercer Management Consultants:* Excellent pool of technology talent and qualification in market sizing *McKinsey & Co.:* Lack of nanotech focus countered by strong health care practice and deep industry contacts
Bulge Bracket Investment Banks	Require engagement for strategic transaction	Very Little	*Merrill Lynch:* Leading nanotech investment bank with strong technology practice and excellent research on future nanotech trends *UBS Warburg:* Very strong biotech and group technology and a very strong Silicon Valley presence
Small Consulting Practices	Flat or hourly fee $50–$250/hr.	Varies	*Technolytics:* Strong background evaluating emerging technologies with a focus on nanotech *Sygertech:* An early leader helping nanotech ventures write business plans and evaluate markets
Niche Investment Banks	May require engagement for strategic transaction	None	No one of note with a strong nanotech practice

the capital markets. Other factors beyond the control of the business or its management that influence success include macroeconomic growth, globalization, trade liberalization, and the rate of technology development and adoption in other nations. Each of these exogenous variables will have a role in determining the final valuation of the company.

While many of these factors are out of the control of the entrepreneur, he or she should not ignore them. Because investors want to know that management is aware of all of the factors that will determine return on investment, it is essential to be able to discuss all of the risk factors of the business. Also, by separating controllable factors from uncontrollable factors, one can better focus available resources and develop more appropriate strategies.

BUSINESS PLAN OUTSOURCING

There are a number of professionals who offer help for business plan writing. Their contributions include conducting research, providing analytical support, preparing financial and valuation support, or writing the business plan entirely. Investment bankers, consultants, and others offer these services. Entrepreneurs without experience writing business plans may choose to outsource business plan preparation. However, ongoing and active involvement by entrepreneurs at each step of business plan preparation is absolutely essential. Table 11.6 details some of the sources and fees associated with engaging firms to prepare the business plan.

As discussed in this book, intellectual property and related legal discussions will be central issues for nanotech companies. As a result, companies must give them consideration in the business plan. For these discussions, entrepreneurs may seek legal advice and opinions to substantiate their own intuitions and to provide validation that the company's intellectual property and legal position can be defended. We provide a list of law firms with expertise in nanotechnology in Chapter 13.

CHAPTER 12

Early Stage Financing

This is the time to invest. I made most of my money from creating companies from the last nuclear winter in biotech that ended up ripening at the next IPO window.[1]
—Larry Bock (chairman of the board, Nanosys Inc.)

Nanoscale research and development is extremely costly, and nanotechnology start-ups usually require large capital investments to create successful companies. Nanotech companies need to purchase expensive tools and hire talented managers and researchers with specialized skills. Research is also usually very early-stage so that products may be many years away. Finally, regulatory scrutiny for some nanotech-based applications extends the time required for products to reach the market. As a result of these high expense hurdles, successful companies will need to develop comprehensive financing strategies that reflect all of their particular costs.

In this chapter, we describe different financing considerations for the different stages of company development. First, we discuss how to develop a financing strategy. We then provide an overview of how founders raise money from family and friends and "angel investors" in the "seed round." We also describe the process of obtaining federal grants available for nanotech companies. We then turn our focus to venture financing. We provide an overview of current venture capital investments in nanotech and profile several funds that are leading the way in investing in small-scale technologies. Finally, we explain the process of pursuing venture capital investment and summarize the key terms and conditions that accompany venture financing.

FINANCING STRATEGY

In developing a workable financing strategy, the company must first decide on short-term milestones involved in creating a profitable company. For example, companies may need to develop a prototype, integrate their technology with a partner's technology and obtain regulatory approval. Each step involves particular costs, and by separating the total investment into smaller, discrete amounts, the company can approach investors with the understanding that if it meets its goals at each stage, it can then obtain additional funds.

A comprehensive plan that takes into account all expenses necessary to meet goals is important for three reasons. First, it can prevent a "firesale" in which the company must sell equity quickly (usually at a discounted price) because it must raise capital to meet expenses. Second, planning prevents raising more money than the company needs and consequently prevents unnecessary dilution to existing stockholders. Third, identifying costs in the planning stage enables management to prioritize and eliminate expenses that are not essential to reaching its goals.

THE SEED ROUND

The seed round for nanotech companies is used to establish the business, obtain rights to necessary patents, and prove the technological feasibility of the product. The money is usually obtained either from founders, friends and family, angel investors, or occasionally venture capitalists.

In the early stages of the business, founders may need to finance operations with their own money. They may also choose to delay outside financing in order to preserve ownership, increase the company's valuation at the initial venture capital financing, and demonstrate to others that they are committed to the company. Some founders seek funds from their friends and family. While investment by friends and family can be a quick and effective means of obtaining capital, such dealings can also become complicated and strain personal relationships. Indeed, friends and family may not remain friends and family.

If the founders are unable or unwilling to raise their own capital, they should begin contacting angel investors. Angel investors are qualified individual investors who purchase debt or equity securities in a start-up company. They generally make smaller investments than venture capitalists, but may attach fewer conditions to their investments. Angel investors have demonstrated a growing interest in nanotech, and we expect this trend to continue. In 2002, a NanoBusiness Angels network was formed in New York to promote new companies and to "fuel the development of seed and early stage start-ups."[2] California's Central Coast Angel Network, which sponsors investor seminars, also recently started covering nanotechnology investment opportunity areas for its members.[3] A substantial angel investment can provide credibility to the company when seeking venture capital in the future.

It is also possible for founders to obtain seed funding from venture capitalists. In nanotech, venture capitalists will be reluctant to invest in very early-stage research. Nanotech companies receiving seed funding from venture capital companies involve more than just entrepreneurs with a good idea. Most have a long history developing their technology. In many cases this occurred by a professor at a university or as a result of many years of research at a national laboratory. Venture capital seed funding is thus the exception and not the rule.

GOVERNMENT FUNDING

Federal funds, which include grants, contracts, and purchase orders can be a valuable source of capital for early-stage companies. In addition to providing money to a company in its early days, federal funds can also increase the likelihood of obtaining venture financing later. According to one venture capitalist: "We use [federal funds] as an IQ test of the entrepreneur. Is she getting free capital before coming to ask us for capital? While we do not follow a sector simply because the government is putting money it, it is indeed a criterion for us."[4]

Indeed, government funds are widely considered "free" money. Although the actual terms of funds differ depending on the type of request and the agency involved, the intellectual property generated with public capital generally belongs to the company.[5] Conditions are usually limited and include submitting periodic reports to the funding agency. There can be two drawbacks to pursuing government funds. First, the process of writing a proposal, providing a panel with requested information, and waiting for a funding decision can be prolonged and tedious.[6] Second, because federal funds are such an attractive financing option, there may be many applicants competing for the same grants and contracts. The likelihood of success may therefore be low.[7]

Many folks in industry believe that the funding process is relatively objective. In some cases, however, funding decisions might rely on the credibility of the founding scientists and their relationships with the reviewers. If the founding scientists are reputable, they are more likely to succeed in obtaining federal funds. As one established professor put it, "For us, it was just a matter of writing."[8]

Federal agencies have several mechanisms to support start-up companies such as small business innovative research (SBIR) grants, small business technology transfer (STTR) grants, Advanced Technology Program (ATP) grants, and Defense Advanced Research Projects Agency (DARPA) grants. These grants have been crucial for many nanotech start-ups. As Table 12.1 illustrates, many nanotech start-ups have been able to survive without venture financing because of federal money.

TABLE 12.1 Examples of Nanotech Companies Receiving Federal Grants

Company	Technology	Agency/Type	Amount ($M)	VC
Npoint	Nanopositioning and control hardware	Various	$3.0	YES
Nanosys	Various, molecular electronics	NIH/SBIR; NSF/SBIR; various military agencies	$6.0	YES
Nano Group Inc. (Inframat and U.S. Nanocorp)	Coatings, water filtration, and fuel cells	DOD, NIH, NSF	$18.3	NO
Nanosphere	Nanoparticle probes for disease detection	NIH/SBIR	$1.5	YES
FeRx	Drug delivery	NIH/SBIR	$0.1	YES
NanoInk	Dip-pen lithography	NIH/SBIR	$1.3	YES
Nanomix	Carbon dioxide nanoelectric sensors	DOE/I&I	$0.1	YES
Espin Technologies	Polymeric nanofiber mats	NIST/ATP	$2.5	NO
Echo Technologies	Dendrimer-assisted DNA detection	Army	$0.1	NO
NanoBio	Emulsion against biological agents	DoD	$11.8	NO
Physical Sciences	Carbon nanofiber networks	Army/SBIR	$0.1	NO

SBIR/STTR Grants

First, companies can seek SBIR/STTR grants. These grants, which are offered by several agencies, are exclusively for small businesses.[9] The difference between SBIR and STTR programs is that STTR requires researchers at universities and other research institutions to play a significant intellectual role in each project. Different agencies have somewhat different procedures, but in general, there is a three-phase program for SBIR/STTR grants. Companies first apply for a Phase I award to test the scientific, technical, and commercial merit and feasibility of a proposed innovative research or activity. The Phase I award will be made for a maximum of $100,000 for SBIR and STTR. If Phase I proves successful, the company can apply for a two-year Phase II award of up to $750,000 (under both SBIR and STTR) to further develop the innovation. Phase III, which is rare, involves companies obtaining funding from the private sector and/or non-SBIR government sources to develop the concept into a commercial product.

A number of nanotech start-ups have received SBIR/STTR grants. In seeking to obtain an SBIR or STTR grant, a start-up must first identify a relevant grant. Agency solicitations can be found on the SBIR section of each agency's web site. Because nanotechnology is so interdisciplinary, companies should review the solicitations at numerous agencies including the National Science Foundation, (NSF), the National Institutes of Health (NIH), the Department of Defense (DOD), and the National Aeronautics and Space Administration (NASA). The applicant must then sift through the projects that the agency is soliciting to determine if it qualifies.

Defense Advanced Research Projects Agency

Nanotechnology companies may also be able to obtain funding from the Defense Advanced Research Projects Agency (DARPA).[10] DARPA manages and directs selected basic and applied research and development projects for the DOD. It focuses on funding projects where risk and payoff are both very high and where success may provide dramatic advances for traditional military roles and missions. Traditionally, DARPA has funded advanced research in military-related technologies like artificial intelligence, packet-switched computer networks, parallel processing, and semiconductor technology.

The primary responsibility of DARPA's Contracts Management Office (CMO) is to enter into and administer contracts, grants, and cooperative agreements.[11] DARPA solicits R&D work primarily through advertising in the FedBizOpps and the DOD SBIR Program Solicitation. DARPA also encourages industry to submit novel ideas for funding. DARPA seeks to fund "projects," which are collections of contracts and thrusts with a common theme. For example, DARPA is funding Phase I of a potential $7.2 million joint development project involving Nanosys, Sciperio Inc., the University of Texas at Dallas, and Penn State University. The contract is for the development of a new semiconductor technology for the fabrication of high-performance, large-area electronic systems on flexible substrates. The agency maintains that it funds one in three good project ideas.

Advanced Technology Program

Another source of federal funding for nanotech companies is the Advanced Technology Program (ATP). Administered by the National Institute of Standards and Technology under the Department of Commerce, the ATP is designed to stimulate "high-risk, high-potential" products, processes, and technologies.[12] ATP can fund up to 50 percent of research and development in technologies developed by industry. The government is entitled to a share of the licensing fees and royalties resulting from ATP projects.

Because ATP funds can match private funding, companies can obtain large sums of capital from NIST. ATP funds companies of all sizes, with approximately half of all ATP awards going to individual small businesses or to joint ventures led by a small business. ATP awards are selected through a rigorous peer review process. Technology-specific boards staffed with experts from particular fields evaluate the proposals. The agency accepts proposals only in response to specific, published solicitations.

Since 1991, ATP has made 39 awards, totaling nearly $142.5 million, to U.S.-based industry for high-risk R&D projects in the emerging area of nanotechnology.[13] Twenty-three awards have been made since 2000. Nanotechnology-related proposals are rapidly rising. A number of nanotechnology companies have received ATP funds, but perhaps the story of Zyvex Corporation best illustrates how massive and important these funds can be for nanotechnology companies.[14] Zyvex is a seven year-old start-up based in Richardson, Texas originally focused on developing the first molecular ("Drexlerian") assembler. The company partnered with Standard MEMS, Inc. and researchers from three universities in 2001 to secure a grant to develop low-cost, computer-controlled, microscale and nanoscale assemblers that operate in parallel to assemble three-dimensional microscale and nanoscale components. The project, which will take place over five years, has an estimated price tag of $24.5 milion. ATP will fund $12,170,000.

VENTURE CAPITAL

In order for a company to develop a marketable product, it will likely need to secure significant venture capital funding. After securing seed funding, venture capital funding usually proceeds in three stages called the "A Round,"

"B Round," and "C Round." Each round describes a different level of maturity for the company. In this section, we provide an overview of venture capital markets and describe investments in nanotechnology. We profile several venture firms that are focusing their investments in small-scale technologies and explain how companies can maximize the likelihood of securing venture capital. Finally, we summarize the key terms and conditions that accompany venture financing.

Venture Capital Investing

In exchange for investing in a company, venture capital firms obtain a share of the equity of the company and often play an active role in the company's affairs. In addition to assisting with business strategy, they may demand board representation, veto power over any major changes to the company's business operations or financial arrangements, the ability to recruit new managers, and the rights to liquidate their investment in preference to other investors. They try to focus on investments where the expected return is in the rage of 35 to 45 percent compounded annually. Since many of the investments will fail, they depend on a few star investments to bring up the overall return of the portfolio. Since rates of return start diminishing around the five-year point and most venture funds have a ten-year life span, venture capitalists favor investments with four- to seven-year life cycles.[15] As one investor explains: "[V]enture capitalists focus on the middle part of the classic industry S-curve. They avoid both the early stages, when technologies are uncertain and market needs are unknown, and the later stages, when competitive shakeouts and consolidations are inevitable and growth rates slow dramatically."[16] An exit vehicle, such as an initial public offering or the sale of the company to a larger company, provides a mechanism for investors to liquidate their shares of the company and lock in their outsize returns.

In the not-too-distant past, some venture capitalists (VCs) in emerging industries were prone to prematurely investing in companies. For example, VCs have funded different biotechnology companies attempting to commercialize pioneering breakthroughs despite the fact that it was likely they would not generate any revenue for at least a decade.[17]

Most venture funds are organized as limited liability partnerships. The investors are the limited partners, and the managers of the fund are the general partners. Approximately half of all of the money invested in venture funds comes from public and corporate pension funds.[18] Endowments and foundations, bank holding companies, and wealthy families and individuals each contribute about 10 percent of the capital to venture funds. The remaining investors include insurance companies, investment banks, nonfinancial corporations, and foreign investors.

Venture Investment in Nanotechnology

Many nanotech start-ups are not yet ready for venture capital investment. First, despite the enormous resources being dedicated to exploring and engineering at the nanoscale, the field is still in its infancy. Many start-up companies have just recently spun out of breakthroughs in universities and cannot reasonably expect to develop commercial products for at least seven to ten years. Even nanotech companies that can demonstrate functional products lack the ability to scale up for mass production.[19] Nanotech "cannot yet provide the customary tenfold returns on investment within five years, or fivefold returns within three years, which are usually expected by VCs."[20] Indeed, private equity investors do not have "deep enough pockets—or enough patience—to fund such a [nanotechnology] startup to profitability."[21]

Despite its nascency, many investors have identified nanotechnology as the next great technological wave. A trip down Sand Hill Road reveals that some venture capitalists are already proclaiming nanotechnology to be the "next big thing." At research universities across the country, venture capitalists are giving presentations to graduate classes in nanoscience and taking professors to lunch. As Stanford professor Mike McGehee notes on the first day of his nanoscience class, "Silicon Valley suddenly became interested in nanotechnology after the dotcom crash."[22] Steve Jurvetson, one venture capitalist fueling the excitement, declares, "We believe that nanotech is the next great technology wave, the nexus of scientific innovation that revolutionizes most industries and indirectly affects the fabric of society. Historians will look back on the upcoming epoch with no less portent than the Industrial Revolution."[23] According to an analyst, investors are "hypnotized" by nanotechnology.[24] Stanley Williams, head of Hewlett-Packard's quantum science research labs, notes: "I have many cold calls from venture capitalists displaced by the dot-com era and looking for the next big thing. A lot are focusing on nanotech as the next big payday."[25] Indeed, as nanoscience progresses, the "buzz is getting louder and louder."[26]

The preparation for future investment and the optimistic rhetoric has also been complemented by investment in some promising companies. In 2002, the NanoBusiness Alliance declared that 50 venture capital firms had made some type of nanotech investment.[27] A study conducted by the Venture Capital Journal showed that, in 2001, private equity investors funded 19 nanotech start-ups worth $190 million;[28] and in 2002, private equity investors funded 34 nanotech start-ups worth $427 million. According to the study, while there were three A rounds for $8.9 million in 2001, there were eight A rounds and three seed rounds totaling $73.3 million in 2002.

In order to better understand the nature of venture capital funding in nanotech, we conducted our own research. Our study differs from other studies in that we adopted a narrower definition of what constitutes a nanotechnology company. Specifically, we only evaluated companies that engage in R&D of technologies that depend crucially on some feature in the 1–100 nanometer range, excluding mainstream technologies that already exhibit this capability, such as semiconductor manufacturing, but including those that apply the current state-of-the-art to novel applications. As such, we excluded many companies in biotechnology and microtechnology that other studies have included. We estimate that private equity investors invested at least $807 million in 42 nanotech companies between 1999 and the end of 2003.[29] As shown in Table 12.2, companies completed 42 A rounds, 20 companies completed B rounds, and only 10 companies received C rounds. Total funding was evenly spread across A, B, and C rounds. This is because, although there are fewer companies reaching successive rounds of funding, the size of the financing grows in later rounds. As nanotech companies mature, they are likely to obtain larger sums of cash from investors.

Our study also attempted to identify what sectors of nanotechnology have received the most attention from investors. We placed companies into one of eight different categories. The categorization scheme is loosely based on the model that we developed in Chapter 2. Because companies in the "integrated systems" category are developing technologies with multiple applications, placing a company in a particular subcategory is often an arbitrary process. We attempted to identify a company's primary focus. As shown in Table 12.2, Life Sciences was the most active and advanced segment of nanotechnology, with 12 companies completing A rounds and eight completing C rounds. Optical components companies raised the most money in their first rounds, averaging $17.8 million in A rounds.

The pace of investment in nanotechnology will increase in coming years. Perhaps the most vivid illustration of the venture capital community's willingness to invest in this sector is the Series B financing secured by Nanosys, Inc. in 2003. Despite a lingering reluctance by venture capitalists to invest in early-stage companies, VCs were willing to pour $38 million in Nanosys, Inc. In early 2004 Nanosys filed for an initial public offering. If this offering provides a significant return to Nanosys' early venture capital investors, it is likely that more venture capital firms will be interested in investing in nanotech companies. In Table 12.3, we provide a detailed description of companies receiving funding in 2003.

As the field matures, many venture investors may make poor investments because they simply do not understand the fundamental limitations of the technologies in which they invest. Venture capital firms, which have expertise in fields such as biotechnology, software, and optical networking, have only begun to explore investing in nanotech. As Stan Williams, director of nanoscience at HP Labs, notes, "Those who excel at popularizing the field are not necessarily those who understand it."[30] Venture capitalists often "don't understand its physical limits or what's physically possible."[31] At one conference we attended, a partner at a fund focusing on nanotechnology had visible difficulty in explaining the basic technology of a company in which the fund had invested.

TABLE 12.2 Total Private Equity Investment in Nanotechnology between 1999 and Early 2004

No. Companies	A Round	B Round	C Round
12 Tools & Building Blocks			
No. Rounds	10	2	1
Average Round Size	$ 7.1	$10.7	$ 8.1
Total Funds	$71.1	$21.4	$ 8.1
8 Materials			
No. Rounds	7	2	0
Average Round Size	$ 5.2	$18.2	NA
Total Funds	$36.4	$36.3	—
19 Life Sciences			
No. Rounds	12	10	8
Average Round Size	$ 3.0	$ 9.8	$ 25.5
Total Funds	$35.6	$97.6	$203.7
20 Integrated Systems	13	6	0
Energy (3)			
No. Rounds	3	2	0
Average Round Size	$ 4.1	$23.8	NA
Total Funds	$12.2	$47.5	—
Nanostructures for MEMS (2)			
No. Rounds	2	1	0
Average Round Size	$ 4.2	$10.5	NA
Total Funds	$ 8.3	$10.5	—
Optical Components (4)			
No. Rounds	4	2	0
Average Round Size	$17.8	$ 5.4	NA
Total Funds	$71.2	$10.8	—
Electronics/Quantum Computation (7)			
No. Rounds	4	1	0
Average Round Size	$ 9.1	$ 4.0	NA
Total Funds	$36.5	$ 4.0	$ 30.0
Sensors (4)			
No. Rounds	0	0	0
Average Round Size	NA	NA	NA
Total Funds	$ 7.0	$24.3	$ 35.0
59 Total Private	42	20	9
Average Round Size	$ 6.6	$12.6	$ 30.8
Total Funds	$278.25	$252.4	$276.8

TABLE 12.3 Venture Financing in Nanotechnology in 2003

Company	Technology	Amount (M)	Round
Five Star Technologies	Nanomaterial synthesis	$ 4.5	C
D-Wave Systems	Quantum computing	$ 7.1	A
InMat	Nanocomposite coatings	$ 1.5	A
Molecular Imprints	Nano lithography	$30	B
Nanogram Devices	Nanomaterials for medical devices and batteries	$ 9.2	A
NanoInk	Dip-pen nanolithography	$ 6	A
NanoNexus	Components for testing integrated circuits	$14.5	C
Optiva	Nanomaterials for use in optical applications	$30	C
NanoOpto	Nano-fabrication technology for optical systems and networks	$ 7	A
Nanosolar	Self-assembled nanostructures for solar cells	$ 6.5	A
Nanosphere	Nanoparticle probes for nucleic acid and protein identification	$15	C
Nanostream	Microfluidics for drug discovery	$22	C
Nanosys	Inorganic semiconductor nanostructures for electronics, photovoltaics, sensors, and surfaces	$38	B
Nantero	Nanotubes for nonvolatile random access memory	$10.5	B
Novaled	Organic light-emitting diodes	$ 6.5	A
Nugen	Nucleic acid amplification and detection	$ 7.5	C
Santur	Tunable lasers	$10	C
Sionex	Chemical sensors chips and systems	$12.8	B
Solubest	Nanoparticles for drug delivery	$ 1.4	A
Nanomuscle	Products that displace small electromagnetic motors and solenoids	$16	C
Nanox	Nanostructured oxide materials for antipollution systems and alternate energy systems such as fuel cells, low temperature diesel oxidation catalysts	$ 4.1	B
Immunicon	Magnetic nanoparticles for cancer screening	$24.8	D
Quantomix	Imaging high-resolution wet samples in electron microscopes	$ 3.5	B

Profile of the Top Nanotech Venture Firms

Because many investors do not yet have a sophisticated understanding of nanoscience, it is useful to survey those venture firms that do have competence in nanotechnology. In this section, we provide summaries of the most active firms investing in small-scale technologies. The list was based on the number of deals completed by each firm as well as interviews and discussions with CEOs and investors in the field.

ARCH Ventures

ARCH Venture Partners invests in the development of seed and early-stage technology companies that focus on innovations in physical sciences, life sciences, and information technologies. ARCH operates nationally and works with leading academics to organize and build companies around fundamental, breakthrough research. ARCH has invested in the founding rounds of a number of companies involving nanotechnology, MEMS, and novel materials, including Nanosys, InnovaLight, Nanophase Technologies, Amberwave Systems, MicroOptical Devices, and Impinj.

Ardesta

Founded in October 2000, Ardesta is focused on investing in "small tech." The firm has created or invested in 15 companies, including Angstrovision, Discera, Handylab, Ion Optics, Konarka, MesoSystems, Micronics, Phoenix Bioscience, Sensicast, Sensicore, Therafuse, and Translume. In addition to financial capital, Ardesta supports its companies with business and technical services, including facilities, human resources, intellectual property, market research, government funding, and information technology. The firm's commitment to and focus on the field is best illustrated by its publication of *Small Times*. The magazine, which has a circulation of more than 25,000, provides daily news about the business of MEMs, microsystems, and nanotechnology.

Draper Fisher Jurvetson

Draper Fisher Jurvetson (DFJ) is one of the most dynamic and successful firms in the venture capital industry. DFJ has created a global network of affiliated venture funds with approximately $3 billion in capital commitments and offices in the major technology centers around the world. Headquartered in Silicon Valley, the firm can boast of a number of success stories, including Hotmail, Overture, Parametric Technology, and United Online.

The firm's focus on nanotechnology is driven by renowned venture capitalist Steve Jurvetson. Jurvetson, who was named one of the most influential people under 40 by *Fortune* magazine, is often referred to as "Mr. Nanotech." He is a cochair of the NanoBusiness Alliance and a prolific writer and speaker at nanotech conferences. Perhaps his influence in the field is best illustrated by his presence in the Oval Office for President Bush's signing of the nanotechnology legislation in December 2003.

The firm's involvement in the field began in 1985, when they invested in Molecular Solutions. Since then, DFJ has completed over 20 deals involving nanotechnology, MEMS, and novel materials, including Arryx, BinOptics, Coatue, D-Wave Systems, Flexics, HyperNex, Imago, Konarka, Luminus Devices, Microfabrica, Molecular Imprints, NanoCoolers, NanoOpto, Nantero, NeoPhotonics, Intematix, SiWave, Solicore, and Zettacore. DFJ can boast of two liquidity investments in nanotech: Molecular Solutions went public in 2000, and Coatue was acquired by AMD in 2003.

Harris & Harris Group, Inc.

Harris & Harris Group, Inc. is a publicly-traded venture capital company investing in "tiny technology." "Tiny technology" includes nanotechnology, microtechnology, microsystems, and microelectromechanical systems (MEMs). The company's focused approach to investment is motivated by the notion that both the miniaturization trend and exciting breakthroughs in nanotechnology will impact a variety of different products across multiple industries. Harris & Harris Group primarily focuses on early-stage investments, where being a public company with permanent capital affords it a certain level of patience and commitment to its portfolio companies.

Harris & Harris Group made its first investment in nanotechnology in 1994, when it invested in Nanophase Technologies Corporation. Nanophase is one of the only nanotech companies to have successfully completed an initial public offering. Harris & Harris Group has made significant tiny tech investments in Agile Materials & Technologies, Inc., Chlorogen, Inc., Continuum Photonics, Inc., NanoGram Devices Corp., NanoOpto Corporation, Nanopharma Corporation, Nanosys Inc., Nanotechnologies, Inc., Nantero, Inc., Neophotonics Corp., Optiva, Inc., and Questech Corp.

Led by a management team of experienced investors who are knowledgeable about the field, Harris & Harris may play an instrumental role in bringing products based on nanotechnology to market in the coming years.

Morgenthaler Ventures

Morgenthaler Ventures, which has been in existence for over 34 years, focuses on investing in emerging companies in the areas of enterprise IT, life sciences, semiconductors and components, and broadband communications. The firm has over $2 billion under management, including $850 million in its current fund (capitalized in the summer of 2001). Morgenthaler's interest in nanotechnology is primarily driven by general partner Greg Blonder, who led a number of research divisions at Bell Labs. Investments in microsystems and nanotechnology include LightConnect, Inc., NanoOpto, and Five Star Technologies, Inc.

Nanotech Partners Limited

Nanotech Partners Limited was launched in September 2001 to invest exclusively in nanotechnology. The fund has approximately $45 million, with commitments from Mitsubishi Corporation, Mitsubishi Chemical Corporation, and Honjo Chemical Corporation. The fund is managed by Nanotech Partners Limited in cooperation with Fullerene International Corporation, Materials and Electrochemical Research Corporation, Research Corporation Technologies, and Mitsubishi Corporation.

The firm will primarily focus on carbon materials such as fullerenes and carbon nanotubes. The firm's first investment was Frontier Carbon Corporation, a company developing mass-production facilities for fullerenes. The firm is preparing to invest in several nanotube projects. Nanotech Partners could also invest up to 20 percent of the fund in other areas of nanotechnology such as nanobio, nanomaterials (other than carbon), nanoelectronics, and other nanodevices.

NGEN Partners

Based in Santa Barbara, NGEN was formed to fund companies that are commercializing materials science technologies. Specifically, the firm focuses on polymers and organics (including molecular recognition, gene chips, and active organic coatings), energy and environmental technologies (including catalyst and sensor development and ceramics), displays and electronic technologies (including electrical, optical, magnetic, mechanical, luminescent materials, and dielectrics), and infrastructure and telecommunications (including photonics, simulation technologies, high throughput experimentation, informatics, processing, and manufacturing systems).

The fund's managers have a strong technology competence (including Nobel Prize–winner Alan Heeger) and a good understanding of this marketplace. They work closely with the fund's corporate investors, including BASF, DuPont, Canon, Boeing, Air Products and Bayer. This is useful both for reviewing technologies and understanding how the technology will be applied to markets. Further, because many of these investors, such as BASF, operate their own venture capital groups, NGEN can benefit from deal referrals by having them as investors in the fund. The LPs also enable reduced time to market through quicker valuation of the underlying technology and are likely to partner with portfolio companies.

NGEN's portfolio companies include Artificial Muscle, Konarka Technologies, Catalytic Solutions, Oxonica Ltd., Agile Materials, Nanosphere, InMat, Powerspan, Pionetics, Sensicore, PsiloQuest, Optiva and Fqubed.

Polaris Ventures

Polaris Ventures, which has offices in Boston and Seattle, invests in seed, first-round, and early-stage information technology and life science businesses. The firm has over $2 billion under management and current investments in more than 60 companies. The firm's investments in microsystems and nanotechnology include Acusphere, Advion Biosciences, Nanosys, and Optobionics.

Sevin Rosen Funds

Sevin Rosen Funds is a top-tier venture capital firm with a track record of funding successful companies since 1981. The partnership has consistently made early-stage investments in pioneering technologies and companies with the potential to create new markets. The most recent fund, Sevin Rosen Fund VIII, is capitalized at $600 million. Compaq, Lotus, Cypress, Citrix, CIENA, and Capstone Turbine are some of the firm's successful IPOs. Portfolio companies involved in nanotechnology, MEMs, and novel materials include Nanomix, InnovaLight, LightConnect, Luxtera, and Alien Technology, among others.

Seeking Venture Capital

To obtain venture capital, a nanotech start-up must have a viable strategy, protectable intellectual property (IP), and a strong business plan. We discussed these issues in the first two chapters of this section. We now describe the process of pursuing venture capital, from getting an initial meeting to negotiating a Letter of Intent.

The first major issue in nanotech fundraising is networking and leveraging a group of personal connections and contacts. During the Internet boom, entrepreneurs could expect to engage venture capitalists by simply submitting the business plan online to a number of different venture firms. In nanotech, however, obtaining venture capital will require deeper relationship building. Peter Grubstein, managing partner of NGEN Partners, notes, "We do actively review web submissions, but we are much more likely to follow up when we get a personal referral."[32] Who the referral comes from is also important. Referrals from financial investors are much less meaningful than referrals from scientists and industry experts because of the technology credibility scientific experts bring.[33]

Second, concluding a deal can be a difficult and time-consuming process. Although many investors are interested in nanotech, the venture community is taking a cautious approach to investing. If entrepreneurs can capture a venture capital firm's attention, it will engage in rigorous due diligence. A thorough review of the start-up's technology, intellectual property, and financial affairs may take several months. Completion times can be prolonged since most deals involve multiple firms contemplating investments. Although the lead investors will have primary responsibility for conducting due diligence, all investors will make some efforts to analyze the company and most will want to meet with management. This can be an intrusive process that diverts management time. After making exhaustive efforts to impress investors and

TABLE 12.4 View from the Inside: NGEN Partners

Founded:	2001
Purpose:	"To identify and fund, selectively, emerging businesses in [t]he new materials marketplace."
Fund Managers:	Peter Grubstein, Anthony Cheetham, and Steve Parry
Limited Partners:	14 Corporations
Scientific Board:	Alan Heeger, Jean Frechet, Edward Kramer, John Newsam, and Derek Stratham
Fund Size:	$ 70 Million
Nanotech Investments:	Approximately $15 Million in 11 Companies

Portfolio Overview:

	Stage	No. of Co-Investors	Total Round Size	NGEN Lead	Board/Other Representation
1. Catalytic Solutions	D	8	32,400,000		Yes
2. Agile	A	4	6,000,000	Yes	Yes
3. Konarka	B	5	13,500,000		Yes
4. Oxonica	A	4	6,240,000		Yes
5. Optiva	C	16	30,000,000		
6. Nanosphere	C	2	15,000,000		Yes
7. Inmat	B	3	1,500,000		Yes
8. Pionetics	B	4	3,850,000	Yes	
9. Powerspan	C	4	20,000,000		Yes
10. Psiloquest	B	6	7,000,000	Yes	Yes
11. Sensicore	B	4	10,000,000	Yes	Yes
Total			$145,490,000	4	9
Average	B-C	5	$ 13,226,364		

Data as of 1/31/03. Does not include all NGEN investments.
* For some of these investments, NGEN has invested in multiple rounds.
** Excludes other private investors not named in prospectus.

NGEN Investment Criteria:

1. Seeking to raise B and C Rounds
2. With realistic expectations
3. Not involved in toxic materials (i.e., cadmium)
4. With products that do not require regulatory approval or have already received regulatory approval
5. With qualified and experienced managers
6. With comprehensive IP portfolios

Markets where NGEN Believes Nanotech will have Impact:

1. Pollution Abatement Technologies (Catalytic Solutions)
2. Alternative Energy/Photovoltaics (Konarka)
3. Sensors (Sensicore)
4. Biomaterials (Nanosphere)
5. Displays/OLEDs (Optiva)
6. Microelectronics (Agile Materials)

enduring the due diligence process, companies could learn that venture firms are unwilling to invest. Often venture capitalists will closely analyze a company and then adopt a "wait and see" approach to monitor the company for up to a year before investing. If the company meets certain goals, they will decide to invest later.

During the process of raising venture capital, CEO's can wind up spending all of their time preparing presentations, meeting with investors, negotiating deal terms and accommodating due diligence requests. Raising capital can be so time-consuming that it detracts from business operations. For early-stage companies that need to quickly show progress, the distraction can be especially harmful. To deal with this issue, managers must be cautious to spend time only with serious investors and should be prepared for discussions by fully analyzing their own strengths and weaknesses in the business plan. Awareness and preparation can help turn a "wait and see" observer in to an active investor.

Third, when presenting the business, entrepreneurs should not expect investors to sign confidentiality agreements. There are several reasons for this. First, many venture capitalists receive too many submissions to be tied to confidentiality on every one. Similarly, because venture capitalists are actively investing in these areas, they will not agree to be tied down to non-compete provisions. Indeed, in many cases, a venture firm may have a portfolio company with related or identical technology to other companies it considers investing in. Consequently, entrepreneurs must either have patents, or must be careful only to disclose a minimal amount of information necessary to solicit the venture capital firm's continued interest.

Finally, the company must keep in mind that venture capital investment involves a long-term relationship. As equity holders in the company, venture capitalists demand board representation and involvement in major decisions. For example, at Sensicore, which develops environmental monitoring devices, Ardesta installed their vice president of finance to be Sensicore's CFO. This kind of hands-on involvement can present difficult issues since many technologists have differences of opinion with—and, in some cases, severe distrust of—venture capitalists. After breaking up with Celera Genomics, Founder Craig Venter explained the trade-off that technologists face in accepting venture financing: "[Y]ou don't enter into these Faustian bargains without holding up your end of it. They were putting up the money, so I had to play by their rules."[34] Although only naïve or very inexperienced entrepreneurs will enter deals expecting company management to remain unchanged, not enough entrepreneurs are fully ready and willing to be accountable to new personalities.

Terms and Conditions of Venture Capital Financing

In accepting venture capital, nanotech companies must agree to several terms and conditions. In this section, we provide a brief summary of the key contract issues in venture capital deals. Specifically, we discuss valuation, investment conditions, and operating conditions.

Valuation

First, venture capitalists and entrepreneurs must negotiate a "pre-money" valuation of the company. Indeed, the most important issue in the term sheet is the valuation of the company. The "pre-money" valuation is the value of the company before investors put their money in, and it is the basis for the "post money" valuation, or the value of the company after the investment. In each case, the valuation is determined by the fully diluted outstanding capital of the company multiplied by the price per share. For a more detailed discussion of valuation of nanotech companies, see Chapter 11.

Investment Conditions

In order to protect their investments, venture capitalists require risky companies such as nanotech companies to issue preferred stock, which provides holders with preferential rights over holders of common stock. Preferred stock can be an important way to ensure that selected investors have more control than others in the company. There are many different special rights that nanotech entrepreneurs might grant investors through preferred stock:

Liquidation Preference: The first protection venture capitalists seek in preferred stock is its liquidation preference over common stock. In the event the company is liquidated or consolidated with another company, preferred stock holders receive certain predefined payments before any assets are distributed to holders of common stock. When a company is not in dire need of money and can obtain the money from different sources, the liquidation preference is equal to amount invested plus accrued dividends.[35]

However, when a company needs an immediate infusion of capital to avoid collapse, investors may require very beneficial terms. For example, the agreement may require a payment of up to three times the investment amount before any other investors receive anything.[36] In most of the nanotech deals we have seen, investors insist on "participating" preferred stock, which entitles the preferred stock to participate with the common stock in the distribution of any assets left after payment of the liquidation preference.

Dividend Preference: Preferred stock is generally given a dividend preference over common stock. Because nanotech companies are unlikely to declare dividends, dividend preferences are not an important consideration.

Redemption Provisions: Most technology deals include certain redemption provisions or terms under which investors can redeem their stock for cash. A "mandatory" redemption provision allows the investors to force the company to repurchase the investors' preferred stock at its purchase price plus a redemption premium.

Conversion Rights: In almost every deal, preferred stock is convertible into common stock upon demand by the holder. With a long and potentially turbulent road to profitability for most nanotech start-ups, companies may be forced to raise money at a lower price in a future round of financing. To encourage initial investors to return to the table for future financing, nanotech companies may want to try to include "pay to play" provisions in their deals. This term results in the automatic conversion of preferred stock to common stock if the holder declines to purchase his or her pro rata share of a lower-priced offering.[37]

Antidilution Protection: Nanotech deals usually require antidilution protection. An antidilution provision results in an increase in the conversion ratio if the company sells its stock in the future at a price lower than that paid by the preferred stockholder. There are two primary methods of calculating the new conversion price: "weighted average" and "full ratchet." The weighted average formula uses the sale price and number of shares sold to adjust the conversion price.[38] In contrast, the ratchet formula reduces the conversion price to the most recent lower price at which stock was sold, regardless of how many shares were sold at that price.[39]

Registration Rights: Nanotech deals should include a registration rights agreement, which gives investors the right to require the company to register their shares with the SEC when the company goes public.[40] Certain registration rights also allow investors the right to sell their shares or to force the company to make a public offering.

Right of First Refusal and Co-Sale: Many venture deals include a "right of first refusal," which provides investors with the right to purchase the founder's stock before it is sold to any third parties. Investors may also include a "co-sale" agreement in the deal, which provides them with the right to participate in any proposd sale of the founder's stock to third parties.

Operating Conditions

In exchange for their capital contribution, venture capitalists will also impose operating conditions on management. First, management must provide investors with access to company information. As mentioned, VCs are also likely to demand certain voting rights attached to their preferred stock and board representation. They might also include specific provisions that prevent the company from taking certain actions without investor approval.

FUTURE ROUNDS OF FINANCING

Most nanotech companies will require multiple rounds of financing before they are self-sustainable. As soon as a company completes its first round of private equity financing, the race toward the next round begins. The terms and conditions associated with future financing turn on the success of the company in following its business plan and developing its technology. Investors will evaluate whether the company meets certain goals. If a company progresses according to plan, subsequent rounds will be easier to obtain. First-round investors will become active partners in soliciting additional investment for a next and higher-valued round. They do this both to help the company and to increase the value of their first-round investments. The company must negotiate the deal terms of the "new money" with the lead investor.

If a company has not progressed according to plan, it will be costly to raise new capital. The company may receive a lower valuation, allowing the original investors to invoke anti-dilution protections. The company will be forced to accept "tougher" deal terms. Further, venture capitalists may demand changes in the management and structure of the company. We have also seen the worst case scenario when companies are forced to shut down because they miss key milestones and are unable to raise additional capital.

Chapter 13

Intellectual Property

As products come to market, there are certain to be a number of disputes. The chaotic and uncertain environment presents a number of strategic challenges for nanotech companies. A company must have a thorough understanding of the patent landscape, develop concrete IP strategies, and constantly monitor patent publications, issuances, licenses, and litigation.[1]

—Vivek Koppikar (PTO examiner),
Stephen Maebius (patent lawyer),
and J. Steven Rutt (patent lawyer)

Intellectual property considerations are critical to a successful nanotechnology business. A compulsion to patent has swept nanotechnology researchers and corporations. Thousands of patents are being filed on different tools, materials, systems, and devices in nanotechnology. In Chapter 5, we discussed general patent issues facing the entire nanotechnology industry. In Chapter 10, we discussed intellectual property issues that must be waded through in starting a nanotech company. Specifically, we explained how start-ups forming out of universities or government labs obtain licenses and how companies spinning out of other companies address potential trade secret litigation. In this chapter, we explore intellectual property issues faced by growing nanotechnology companies.

We first explain how maturing nanotech companies can generate intellectual property through means other than their own internal R&D efforts. We then address the most important form of IP protection for nanotechnology companies: patents. We begin by explaining the importance of patents to commercialization of nanotechnology. We then discuss issues and strategies related to obtaining, enforcing, licensing, and litigating nanotechnology patents. We then analyze when trade secret law might be used by nanotech companies as an alternative to patents as well as how trademark law impacts nanotechnology companies. Finally, we provide a brief summary of copyright law and how it might be relevant to nanotech companies.

GENERATING INTELLECTUAL PROPERTY

As firms mature, they can generate intellectual property from their own research and development efforts. They can also continue to leverage R&D efforts at universities and government labs. In Chapter 10, we identified universities and government labs that are most active in nanotechnology and discussed issues involved in concluding license agreements. In this section, we identify ways for nanotech firms to outsource R&D by pursuing collaboration with university and government labs.

Universities

In addition to direct licensing agreements, there are a number of different mechanisms by which nanotech firms can collaborate with research universities.[2] First, nanotech companies can form research contracts with universities. These arrangements involve a company identifying a specific research problem and funding university research intended to address the problem. The company is entitled to license the results of the research.[3]

Second, firms can pay some universities for the opportunity to follow the ongoing work of researchers. A firm might "sponsor" a particular lab to get exposure to its research and development. Depending on the sponsorship agreement, the firm could have rights to license inventions that take place in the lab.[4]

Nanotech companies can also use university research facilities to conduct their own research. For example, the National Nanofabrication Infrastructure Network is a collection of research facilities located on a dozen campuses.[5] The Network enables users to "fabricate advanced nanostructures within weeks of initial contact."[6] Rates and conditions of use vary between facilities. In general, firms own intellectual rights to inventions.[7]

Government Labs

Companies can work with the federal laboratories by entering into Cooperative Research and Development Agreements (CRADAs). A CRADA is a written agreement to work jointly on a project with the laboratories toward predefined goals. The government can grant or waive rights to intellectual property covering inventions discovered through joint research.[8] CRADA's generally involve "march-in rights," in which the government maintains the right to license the technology to a third party in the event that its partner abandons its commercialization efforts.[9]

Another way to work with the federal laboratories is to enter into a Work For Others (WFO) agreement, in which the private company pays the laboratory to conduct the desired work.[10] A number of requirements must be met to conclude a WFO agreement. The work must be consistent with the laboratory's mission, the government must maintain a nonexclusive, royalty free license to any invention, and the work must "use a unique capability of the Laboratory and not place the Laboratory in direct competition with the private sector."[11] In a WFO arrangement, the company can also negotiate exclusive ownership of intellectual property.

A User Facility Agreement (UFA) permits private industry to conduct research using the laboratory's facilities and equipment.[12] Typically, a partner pays upfront fees and establishes a work schedule with the laboratory that does not interfere with the laboratory's other users. The companies that use the facilities typically own any resulting intellectual property.

Finally, companies wishing to work with the federal laboratories may choose to enter into a Personnel Exchange Agreement (PEX).[13] A PEX is an agreement in which employees are exchanged between the laboratory and the company. Either the company sends its employees to work at the laboratory or the laboratory sends its employees to work at the company's facilities. The company generally pays for either the laboratory time or for 50 percent of the employee's salary.

PATENTS

In Chapter 5, we showed why protecting intellectual property with patents is critical to commercialization of nanotechnology. In order for a start-up to bring a product to market, it needs a dominant intellectual property portfolio. If it does not have rights to intellectual property covering its technology, it could find itself embroiled in a ferocious patent dispute. A determination of patent infringement can result in a start-up being enjoined from commercializing its product.[14] Even if a start-up can prevail in court, litigation can be extremely costly and distracting. Some start-ups may be fortunate enough to avoid litigation if other patent holders are willing to license their technology or enter into cross-licensing agreements. Nevertheless, start-ups can still be financially drowned by a range of different patent holders demanding excessive royalties in licensing negotiations.[15] Finally, intellectual property disputes can scare away potential investors or ruin an acquisition or initial public offering.[16] Perhaps the importance of patents on the valuation of nanotech companies is best illustrated by the relationship between stock prices and patent issuances. When the Patent and Trademark Office granted a new nanotech patent to Nanogen in December 2003, the company's stock immediately jumped by 52 percent.[17] Similarly, the issuance of a patent covering "spintronic" magnetic technology to NVE Corp. in June 2004 caused its stock price to rise. And as this book was going to press, Nanosys sought to go public by emphasizing the value of its IP portfolio.

Filing Nanotech Patents

In the quest to build strong IP portfolios, many nanotech companies are filing as many provisional patent applications as possible. A provisional application is merely a description of the invention—it is relatively inexpensive to file, does not require claims, is not examined, and cannot mature into a patent. Nevertheless, priority can still be claimed from the date of the provisional application by filing a nonprovisional application within one year from the provisional application's filing date. In the United States, patent protection is given to the "first to invent." (Most other countries have adopted a "first to file" system). Generally, priority is awarded to the inventor that first reduces the invention to practice.[18] Constructive reduction to practice takes place when a provisional is filed, even if the invention has not been fully tested. Thus, in order to establish priority over competitors working on similar technologies, it might be desirable to file provisional applications on new inventions before they are fully reduced to practice, their patentability assessed, and their market applications explored. Filing a provisional on an invention before it has been fully tested, however, might also increase the likehood of non-enablement of the claims of the patent.[19]

A nanotech company must be more selective in filing nonprovisional patent applications and pursuing issuance of patents. Prosecuting a nanotechnology patent can be an expensive and time-consuming process. Depending on the complexity, claims, and length of the application, filing fees can go as high as $1000.[20] Attorneys fees for prosecuting nanotech patents are usually between $25,000 and $35,000.[21] Therefore, in order to file a nonprovisional application and pursue issuance of a patent, a company must: (1) believe that the patent would be a valuable asset; and (2) be confident that it would be able to obtain and enforce the patent.

First, a nanotech company must consider the value of the patent. In nanotechnology, firms are racing to develop large arsenals of intellectual property. An arsenal of patents can provide substantial leverage vis-à-vis competitors. If a company's IP weaponry is more powerful than its competitors', it can become offensive and seek to squash its competitors or extract large royalties.[22] A sizable and credible patent portfolio can also deter competitors from making charges of infringement or force competitors to enter into cross-licensing arrangements. In this competitive IP race, it may be advantageous to file as many patents as possible. The most valuable patents claim inventions that will be widely needed. Nevertheless, in this field, it appears that firms are filing patents claiming nearly everything—from compositions to small, incremental improvements in manufacturing.

Second, the company must be confident that filing would result in the issuance of a patent that could be enforced. In Chapter 5, we provided a brief overview of the requirements for obtaining a patent from the Patent and Trademark Office (PTO). For the sake of convenience and clarity, we repeat this discussion here. In order to obtain a patent, an inventor must file an application with the PTO within one year of the first offer for sale or public disclosure of the invention. The application must meet the disclosure requirements in Section 112, which states that the specification must "contain a written description of the invention, and of the manner and process of making and using it, in such full, clear, concise, and exact terms as to enable any person skilled in the art to which it pertains, or with which it is most nearly connected, to make and use the same[.]" The patent will issue if the PTO determines that several conditions are satisfied. First, the invention must be patentable subject matter under Section 101. Second, the invention must display some utility. Third, the invention must be novel. Under Section 102(a), an invention is novel unless the invention was known or used by others in this country or patented or described in a printed publication in this or a foreign country before the date of invention by the applicant. Fourth, the invention must not be obvious. An invention is obvious if the prior art would have suggested to one of ordinary skill in the art that this process should be carried out and would have a reasonable likelihood of success.

Determining the patentability of an invention generally involves conducting a thorough prior art search. One nanotech patent agent recommends the following databases to search prior art for nanotechnology inventions: Thomson Derwent (World Patents Index, Patents Citation Index); Thomson Delphion; issuing authorities' websites (USPTO, European Patent Office, Japanese Patent Office); IFI CLAIMS (U.S. Patents/Citations, Current Patent Legal Status); assignee websites; INPADOC (family and legal status); Dialog (Dialindex); JAPIO (patent abstracts of Japan); science and engineering databases (INSPEC, EiComendex, SCISEARCH, CAS); and markets and business databases (Factica or PROMT).[23]

Ultimately, it may be difficult to determine whether the PTO would issue a patent on a particular invention. As argued in Chapter 5, the PTO has failed to adequately prepare for the filing of large numbers of nanotechnology patent applications. Patent applications are being directed to different centers for review, and most examiners do not have a solid understanding of the science and technology involved in nanotech inventions. As a result, some patent applications are being unfairly rejected while others are issuing with overly broad claims. Although the PTO Customer Partnership Meeting in September 2003 was launched to help the agency prepare for nanotechnology, much work remains to be done. Indeed, as demonstrated by the questions asked and issues raised by lawyers and businessmen at the meeting, there is a great deal of uncertainty involved in filing nanotechnology patents. Many patent lawyers fear that the agency is unlikely to take any substantive steps toward improving review of patents in this field in the near future.

Even if a patent does issue, it may not be enforced by the courts. Although patents are presumed valid,[24] they are subject to review and invalidation by courts. The burden of establishing invalidity of a patent or any claim rests on the party asserting such invalidity.[25] As will be discussed later in the chapter, it is extremely difficult to anticipate the outcome of patent litigation. This uncertainty is likely to be particularly acute in nanotechnology litigation, where most federal judges and their clerks have little understanding of nanoscience.

If a patent on a particular invention would be valuable and it is likely that the patent can be obtained and enforced, then the patent should be filed and prosecuted. Even if a patent is unlikely to issue or be enforced, however, a nanotech company may still wish to file the application. Simply filing the application allows a company to waive "patent pending" status,[26] which can demonstrate legal and technical sophistication to would-be investors, collaborators and competitors.

Generally, nanotech companies engage outside counsel to assist in the prosecution process. It may be desirable to retain legal counsel in Washington D.C. due to the proximity to the Patent Office. Many patent lawyers in Washington have personal relationships with the examiners and can meet with examiners in person to resolve patentability issues. There is also a good case to be made, however, for having patent counsel located nearby. Face-to-face meetings can be extremely valuable in understanding and fleshing out complex issues. The following firms are a few that have substantial experience in working with nanotech companies in filing patents:

- Burns, Doane, Swecker, & Mathis, L.L.P.
- Finnegan, Henderson, Farbow, Garrett & Dunner, L.L.P.
- Fish & Richardson P.C.
- Foley & Lardner
- Howrey Simon Arnold & White, LLP
- Lathrop & Gage L.C.
- Winstead, Sechrest, & Minick P.C.

Drafting Nanotech Patent Applications

Once the decision has been made to file a patent, nanotech companies must work with their counsel to draft the application. There are two primary components of a non-provisional patent application: the written description and the claims.

Written Description

The written description includes a description of the "Background of the Invention," a "Summary of the Invention," and a "Detailed Description of the Invention." The Background section should use simple and concise language. Although law holds that an applicant should use language directed to a person of skill in the art to which the invention pertains, nanotech applicants should draft the Background section with an eye to the judge, jury, and investment banker, as well as the patent examiner.[27] A good example of a well-written Background section can be found in Patent 6,593,731:

> Background of the Invention
>
> 1. Field of the Invention
>   This invention generally relates to displacement transducers, and, more specifically, to high sensitivity displacement transducers for MEMS and NEMS.
>
> 2. Background
>   A displacement transducer is a device which senses displacement of an object, and, responsive thereto, provides an electrical signal representative of the displacement of the object. Known displacement transducers sense macrolevel displacements. However, these transducers are not readily scalable to micro- or nanoscale dimensions, although many applications exist for micro- or nanolevel displacement transducers. System-level applications for such transducers are generally referred to as microelectromechanical systems (MEMS) or nanoelectromechanical transducers (NEMS). Consequently, there is an unmet need for micro- or nanoscale displacement transducers.[28]

The Summary of the Invention broadly describes the technology and distinguishes it from the prior art. The Detailed Description of the invention must be sufficiently full, clear, concise, and exact as to enable any person skilled in the art to which the invention pertains to make and use the invention. The Detailed Description also contains the best mode of carrying out the invention known to the inventor at the time of filing. Generally, drawings are provided in the Detailed Description and are required "where necessary for the understanding of the subject matter sought to be patented."[29]

Claims

Patent claims, which are comprised of elements (or limitations), determine the scope of patents. As stated by the Federal Circuit, it is "the claims that measure the invention" and "[c]laims are infringed, not specifications."[30] There are four primary types of patent claims in nanotechnology: (1) "composition" claims; (2) "device, apparatus, or system" claims; (3) "process or method" claims; and (4) "product-by-process claims."

Composition Claims: Claims covering chemical compounds or combinations are characterized as "compositions." Examples of composition claims are claims to nanomaterials such as carbon nanotubes, silicon nanowires, nanocomposites and metal nanoparticles. These claims are generally the broadest and, to that extent, the most likely to be infringed. A competitor that uses a different process to manufacture the nanomaterials would still be infringing the claim.

Device, System, or Apparatus Claims: Examples of apparatus claims are claims to electrical, mechanical, and optical devices incorporating nanomaterials, or tools used to prepare, characterize, and position nanomaterials.

Process or Method Claims: Process or method claims describe ways to synthesize nanomaterials or construct devices or systems. Process or method claims can also involve new uses for nanomaterials and devices.

Product-by-Process Claims: Product-by-process claims allow applicants to claim complex products whose structure or other characteristics are insufficiently known to permit adequate description of the product itself. The law is conflicting as to whether "claimed products" are limited to the process described and claimed in the patent.[31]

Most nanotech patent applications have more than one claim. Claims can be independent or dependent. An independent claim is completely self-contained while a dependent claim refers back to an earlier claim. Generally, the broadest claims are presented first, with more limiting claims following.

Applicants must make several strategic decisions in drafting nanotech claims. First, applicants must carefully decide how broadly to draft the claims. A broad claim might enable a patent holder to assert that a range of different technologies infringe the patent. The PTO has granted a number of broad and overlapping claims on nanomaterials, devices, systems. Examples discussed in Chapter 5 include the following:

- Hyperion Catalysis International's claim to a "cylindrical discrete carbon fibril"
- IBM's claim to single-walled carbon nanotubes
- Advanced Technology Materials Inc.'s claim to a "microelectronic or microelectromechanical device, comprising: a substrate, wherein the substrate includes an oxide layer and an etch stop layer for the oxide layer, and a fiber formed of a carbon-containing material"
- Stanford University's claim to "an apparatus comprising: (a) a substrate with a top surface; (b) a catalyst island disposed on the top surface of the substrate; (c) a carbon nanotube extending from the catalyst island"
- Rice University's claim to a "composition of matter comprising at least about 99% by weight of single-wall carbon molecules"
- Starfire Electronic's claims to silicon nanocrystals and germanium nanocrystals
- UC Berkeley's claim to "particles of III-V semiconductor...between 1 nanometer and 6 nanometers across"
- MIT's claim to a "coated nanocrystal capable of light emission..."

- UC Berkeley's claim to "luminescent semiconductor nanocrystal compound capable of linking to an affinity molecule and capable of emitting electromagnetic radiation in a narrow wavelength band when excited..."

These broad claims provide the patent holders with sweeping powers. Based on the PTO's willingness to grant broad claims in this field, many applicants may wish to "shoot for the moon" and claim as broadly as possible in the hope that examiners will not object.

There are, however, risks associated with a strategy of drafting overly broad claims. If the PTO rejects a claim and the applicant is forced to make an amendment, the scope of the patent could become narrower than if the applicant had filed a narrower claim in the first place. Understanding how filing an overly broad claim might serve to restrict the scope of the claim requires a detailed explanation of how courts determine patent infringement.

Infringement of a claim takes place when there is literal infringement or infringement under the doctrine of equivalents.[32] Literal infringement occurs when every element in a claim is found in the accused device. Under the doctrine of equivalents, infringement takes place when every element in a claim or its equivalent is present in the accused device. An element is equivalent if it performs substantially the same function in the same way to obtain the same result. Prosecution history estoppel limits the reach of the doctrine of equivalents by preventing the patent holder from charging infringement when the accused device contains a feature that was given up by the patent holder through amendment or during prosecution. In 2002, the Supreme Court laid out specific rules governing infringement of claims that have been amended.[33] When a patent holder asserts that a product infringes an amended claim under the doctrine of equivalents, there is a rebuttable presumption that there is no infringement. To overcome this presumption, the patent holder must show that, at the time of the amendment, one skilled in the art could not reasonably have been expected to draft a claim literally encompassing the alleged equivalent. Thus, submitting an overly broad claim could have the perverse effect of limiting the reach of the claim more so than if the applicant had simply submitted a defensible claim.

Although the PTO has been willing to grant some broad claims in nanotechnology, as is discussed in the following pages, the agency is likely to more carefully scrutinize the scope of the claims in the future. As such, applicants should carefully draft the scope of their claims.

In addition to making a strategic decision about the breadth of the claims that will be filed, applicants must also consider what language will be used to describe the subject matter in the claims. The nature of language makes it impossible for an inventor to describe an invention with complete precision. Different applicants, working with their patent attorneys, could draft very different claims to describe the same invention. As one court stated, "often the invention is novel and the words do not exist to describe it. The dictionary does not always keep abreast of the inventor. It cannot. Things are not made for the sake of words, but words for things."[34]

This is particularly true in nanotechnology, where different words are used to describe the same nanomaterials. For example, quantum dots, semiconductor nanocrystals, nanodots, colloidal crystals, and nanoparticles might all be used to describe a cluster of atoms with certain properties. Similarly, carbon fibers, carbon fibrils, carbon nanostructures, carbide nanomaterials, carbon whiskers, and molecular wires could all refer to what are commonly known as carbon nanotubes. Different types of dendrimers might also be described as dendritic molecules, polymers, or starburst conjugates.

In most circumstances, applicants will want to use claim language that will clearly flag their ownership rights. In other words, applicants will want others to discover the patent in conducting searches. In some cases, however, an applicant may intend to conceal the patent from others conducing prior art searches. Firms generally survey the intellectual property landscape before investing in a particular technology. If a critical patent on the technology is not detected, the firm may invest in developing a product, only to later discover that it needs a license to market the product. In this case, the patent holder is in a strategic bargaining position; it can demand excessive royalties from the product developer. Attempting to "hide" a patent can also ensnare the patent holder in an intense intellectual property battle that might otherwise be avoided. In many instances, a firm wishing to develop a product is willing to license the necessary patents. If a critical patent is not detected, the firm may file for a new patent without citing the hidden patent. If the examiner does not discover the patent, it could approve a patent that should otherwise be amended or rejected.

In addition to choosing claim language that clearly flags the subject matter of the invention, applicants should carefully define terms used in the claims. In construing claims, courts generally evaluate the plain and ordinary meaning of the language.[35] Courts consult dictionaries, encyclopedias, and treatises as well as the specification and prosecution history to ascertain the ordinary meaning of claim terms.[36] A patentee can, however, be his or her "own lexicographer."[37] In other words, an applicant can define the terms used in the claims in any manner he or she wishes.

Many terms used in nanoscience might have a range of different definitions depending on the dictionary, treatise, or encyclopedia used. Some terms have no definition. One nanotech patent lawyer describes how different definitions of the word "crystal" might lead to different claim interpretation.[38] A dictionary defines "crystal" as a "3-dimensional structure made up of atoms, molecules or ions arranged in basic units that are repeated throughout the structure."[39] In contrast, a more technical reference provides a more narrow definition of the term:

> A solid in which the atoms or molecules are arranged periodically.... In scientific nomenclature, the term crystal is usually short for single crystal, a single periodic arrangement of atoms.... In electronics the term is usually restricted to mean a single crystal which is piezoelectric.[40]

If a court chose the second definition of "crystal," a claim might be limited to a "single crystal" with piezoelectric properties. By contrast, if a court chose the first definition of crystal, neither the "single crystal" nor the "piezoelectric" limitations would be read into a claim term. Thus, if applicants do not clearly define the words used in the claims, they are at risk of having their claims more narrowly interpreted later. A good example of patentees serving as their own lexicographers is U.S. Patent No. 6,500,622, where the words "semiconductor nanocrystal" and "quantum dot" are described as follows:

> The terms 'semiconductor nanocrystal,' 'SCNC,' 'quantum dot' and 'SCNC nanocrystal' are used interchangeably herein and refer to an inorganic crystallite of about 1 nm or more and about 1000 nm or less in diameter.... SCNCs are characterized by their uniform nanometer size. An SCNC is capable of emitting electromagnetic radiation upon excitation (i.e., the SCNC is luminescent) and includes a 'core' of one or more first semiconductor materials, and may be surrounded by a 'shell' of a second semiconductor material.[41]

Responding to PTO Rejections

Once the application is drafted and filed with the PTO, it is given an application number, processed, and assigned to a patent examiner who evaluates whether the claimed invention is patentable. The PTO examiner conducts a search for prior art relevant to the claimed subject matter. As explained in Chapter 5, a nanotech patent application might be assigned to an examiner in several different centers for review, and the PTO is only beginning to develop databases necessary for effective prior art searching in nanotechnology. The examiner can allow claims, reject claims, object to claims, or object to the written description.

In the process of prosecuting a particular patent, applicants often file additional patents based on the original, "parent" application. For example, if the applicant wishes to change the written description to include a new variation of the invention, he or she may file a "continuation-in-part" (CIP) application, which has some subject matter in common with the parent but also has new subject matter. A continuation-in-part can claim the parent's filing date as to any subject matter in common, but only its own filing date as to new subject matter. Most nanotech companies control several different patents all relating to the same basic invention.

There are three primary issues that applicants might face in obtaining issuance of nanotechnology patents. PTO examiners might object to claims based on the doctrines of inherent anticipation, obviousness, and enablement. Nanotech applicants may also have to prove that they were the first to invent in interference proceedings in order to obtain patents. Even if applicants can successfully patent their inventions, they could confront challenges to their patents based on these doctrines in litigation.

Sections 102 and 103: Inherent Anticipation and Obviousness

The primary issues facing patent prosecutors in this field involve rejections of claims based on prior art. Developing a nanoscale form of a material or device that already exists at the micro scale presents Section 102 and Section 103 issues. As explained earlier, Section 102 bars patenting inventions that have been anticipated by prior art, and Section 103 requires inventions to not be obvious in order to be patentable. Under the doctrine of inherent anticipation, a prior art reference may "inherently" anticipate a claimed invention, even if the reference does not expressly disclose the later invention.[42] In one presentation given at a Customer Partnership Meeting, the PTO identified several cases holding that scaling down an existing technology does not give rise to a patentable claim.[43] Since many nanotechnology patents claim to make an existing material, device, or system smaller, many applications are facing inherency problems.

In countering an examiner's objections based on inherency, a nanotech company can make several arguments. First, the cases supporting rejection can be distinguished in that they involve devices that did not perform differently from prior art devices. For example, in *Gardner* (one of the cases cited by the PTO), the patent claimed dimensional limitations on a device for drying ink such as making part of the device as "small as is mechanically practicable." In upholding the trial court's decision to invalidate the claims on obviousness grounds, the Federal Circuit held that, other than the dimensional limitations, the patent holder failed to point out "a feature of [the device] which performed any differently from prior art [devices]."[44] Most inventions in nanoscience perform differently from similar technologies with micro dimensions. Indeed, the revolutionary potential of the field stems from the ability to leverage new and unexpected properties that do not exist at the micro scale. Additionally, applicants can argue that, because the prior art did not enable production of the nanoscale version of an existing substance, the invention is patentable. For example, in *In re Hoeksema*, the court held that a claimed chemical compound could be nonobvious, even though its structure is suggested, when no process existed at the time that would have enabled its production.[45] In other words, if the process for making a nanoscale material, device, or system is not disclosed in the prior art, the invention should be patentable. Finally, applicants can point to a number of nanotech patents that the agency has issued claiming smaller versions of existing technologies.

Section 112: Enablement

Applicants might also face PTO rejections of patent applications or challenges to their issued patents based on the enablement doctrine. The enablement doctrine requires the inventor to provide sufficient information to enable a person skilled in the relevant art to make and use the claimed invention without "undue experimentation."[46]

There are two opposing lines of cases dealing with application of the enablement doctrine when the specification describes one of several different processes for obtaining a broad claim. The first line of cases uses the enablement doctrine to narrow the scope of the claims. *O'Reilly v. Morse*[47] represents the starting point for this analysis. In that case, the Supreme Court narrowed the patent on the pioneering invention of the telegraph. Morse's patent claimed "the exclusive right to every improvement where the motive power [was] the electric or galvanic current, and the result [was] the marking or printing [of] intelligible characters, signs, or letters at a distance."[48] The specification only contained one method of printing signs and letters at a distance. The Court held that, because Morse was claiming the right to use a process that he had not described, the claim was too broad.[49] Granting such a broad claim would prevent future inventors from developing more advanced telegraphs.[50]

Courts have used the enablement doctrine to limit the scope of patent claims in biotechnology. They have generally reasoned that the unpredictable nature of the science warrants narrowed claims to genes, proteins, and host cells. For example, in *In re Goodman*[51], the patent claimed a "method for producing a mammalian peptide in plant cells," and the specification contained a single example of producing gamma-interferon in the dicotyledonous species, tobacco.[52] The court held that the single example could not "enable a biotechnician of ordinary skill to produce any type of mammalian protein in any type of plant cell."[53] Similarly, in *Enzo Biochem, Inc. v. Calgene*[54], the claims were directed to antisense constructs, methods of regulating gene expression in a cell using antisense constructs, and cells containing antisense constructs. The court found that claims to genetic antisense technology in the "entire universe of cells" were not enabled, because the disclosure only discussed *E. coli* cells.[55]

As shown in Chapter 5, the PTO has generously issued claims of considerable breadth in the first wave of nanotechnology patents. As time passes, however, the PTO can be expected to more carefully scrutinize claims in nanotech patents, and parties are certain to raise the enablement issue in litigating nanotechnology patents. The initial biotechnology patents contained extremely broad claims. As the field matured, and many of the broad biotech patents were found invalid in court, examiners began to require applicants to narrow the scope of the claims.[56] Similarly, as the PTO and courts become more familiar with nanoscience and the prior art, they can utilize the broad and nebulous doctrine to narrow the scope of overly broad claims in nanotechnology. Producing nanostructures is a dynamic and unpredictable area of research. A single example of producing a protein in a certain plant species could not sustain a claim to proteins produced using alternative plant species and examples of genetic antisense technology in *E. Coli* cells could not enable claims to genetic antisense technology in the "entire universe of cells"; in the same way, single methods of synthesizing nanostructures do not enable claims to nanostructures with slightly different morphologies produced through other means. Examiners and courts can decide that the specifications in patents such as those identified in this chapter do not enable the claims to nanomaterials and devices, regardless of how they were produced. Arguably, the claims should only extend to nanomaterials and devices derived using the methods outlined in the patents.

Nanotech applicants can fight PTO rejections and litigation challenges based on the enablement doctrine by arguing that the specification does provide a "representative" group of examples in relation to the scope of the claim based on the relative predictability of the area in question.[57] Further, they can cite cases where courts have been reluctant to use the enablement doctrine to narrow patents in dynamic industries. This line of cases blooms from *In re Hogan*.[58] In that case, the court upheld the validity of a patent claiming a genus of homopolymers of 4-methyl-1-pentene, including both low- and high-molecular weight homopolymers. The patent only disclosed how to make low-molecular weight homopolymers, and the process for producing high-molecular weight homopolymers was disclosed after the filing of the patent.[59] The court held that subsequent prior art cannot be evidence of lack of enablement and noted that a policy against broad protection for pioneer inventions would be "shortsighted and unsound from the standpoint of promoting progress in the useful arts..."[60] A recent application of the *Hogan* principles can be found in *Amgen v. Hoechst Marion Rousel*.[61] The patent claimed certain gene sequences while only discussing exogenous gene sequences in the disclosure. The court held that the claim covered exogenous and endogenous gene sequences. "[W]here the method is immaterial to the claim, the enablement inquiry simply does not require the specification to describe technological developments concerning the method by which a patented composition is made that may arise after the patent application is filed."[62]

Interference Proceedings

Pending nanotech patents could also be subject to "interference" proceedings. As argued in Chapter 5, many nanotech patents filed by different inventors claim substantially the same patentable inventions. The purpose of an interference proceeding is for the Board of Patent Appeals and Interferences to determine which party invented first. In general, the inventor who proves to be the first to conceive the invention and the first to reduce it to practice will be held to be the prior inventor.[63] Issued patents can also be subject to interference proceedings if the patent has not been issued for more than one year prior to the filing of the conflicting application, and provided that the conflicting application is not barred from being patentable for some other reason.[64] The decision can be appealed to the Court of Appeals for the Federal Circuit, or the losing party can file suit in U.S. district court.[65]

Issuance of the Patent

A patent application is published after the expiration of 18 months from the date of filing, unless applicants make a nonpublication request. Publication is not always beneficial as it could provide competitors with information on inventions and product strategy that may be better off kept secret until the patent issues. Additionally, if the application does not ultimately issue as a patent, publication will nonetheless disclose material that may otherwise have been subject to trade secret protection. Publication can only be avoided, however, if the invention disclosed is not and will not be the subject of an application filed in a foreign country that requires publication.[66]

If the examiner determines that the invention is patentable, a U.S. patent is issued. Currently, it takes approximately three to five years from the date of filing until the patent issues.[67] Once a patent is issued, the holder must pay issue fees and maintenance fees throughout the life of the patent. The inventor or his or her assignee has the right to exclude others from practicing the invention for twenty years from the filing date of the application.[68] As noted earlier, issued patents are presumed valid. However, they are subject to review and invalidation by federal courts. The burden of establishing invalidity of a patent or any claim rests on the party asserting invalidity.

Filing Foreign Patent Applications

Obtaining a U.S. patent only protects the invention in the United States. In order to protect its technology in the global marketplace, a nanotechnology company must file patents in other countries. After filing with the USPTO, an inventor has one year to file an "international" patent application under the Patent Cooperation Treaty (PCT).[69] Under PCT rules, inventors specify in which countries they intend to file patents. The international application is then subjected to an "international search," and an "international search report" is sent to the applicant. The report lists the citations of published documents that might affect the patentability of the invention claimed in the international application. If the applicant elects to proceed, the PCT provides inventors up to 30 months from the date of filing the international application to file the patent in other countries.[70]

Filing international patents can be complicated and present unexpected hurdles. For example, publishing an article about an invention before filing a patent can extinguish patent rights in other countries. (In the United States, the applicant is given one year to file after publication.) Additionally, some countries where patent protection might be highly desirable for nanotech companies are not parties to the PCT. For example, in order to establish protection in Taiwan, companies must file directly with the Taiwan Patent Office in a timely manner.

Infringement, Enforcement, Licensing, and Litigation

In many different areas of nanotechnology, the intellectual property landscapes are fragmented. A large number of patents held by different entities cover similar inventions and improvements to the same invention. At the time of this writing, the intellectual property battlefield in nanotechnology is relatively quiet. As materials, devices, and systems based on nanotechnology begin to have a commercial impact, there are certain to be a number of patent disputes. For example, as demonstrated in Chapter 5, a company attempting to market a product based on carbon nanotubes will confront a number of upstream patents, including IBM's patent on the single-walled carbon nanotube. The chaotic and uncertain environment presents a number of strategic challenges for nanotech companies. In order to stay alive, a company must have a thorough understanding of the intellectual property landscape across which it travels. Companies should develop concrete IP strategies and constantly monitor patent publications, issuances, licenses, and litigation.

Infringing Patents Held by Others

If a start-up recognizes that it may be infringing a patent held by another company in its early phases of development, it must carefully analyze its strategic options. As discussed above, infringement of a claim takes place when there is literal infringement or infringement under the doctrine of equivalents. When infringement occurs, the patent holder may seek enforcement of the patent in a federal district court and obtain preliminary or permanent injunctions, damages for lost profits, attorneys' fees, and treble damages.[71]

First, the company should seek an opinion of counsel that supports a conclusion that the patent(s) at issue are either not infringed (a noninfringement opinion) and/or are invalid (an invalidity opinion). Such opinions are generally sought from independent sources (that is, outside counsel, not in-house counsel). If litigation does take place later, such an opinion can prevent claims of willful infringement and enhanced damages.[72]

Second, the company can attempt to engineer its product to avoid infringing the patent. While this may be possible in certain circumstances, it is extremely difficult. There are a number of upstream patents around which firms cannot design; broad patents have issued on compositions of nanomaterials as well as on the uses of nanostructures in different applications. Additionally, there are lists of patents covering improvements on processes for synthesizing, characterizing, and manipulating nanostructures. Even if a company can design around a patent, it may still find itself embroiled in litigation at some point. Patent holders often file suits even when they are likely to lose the case on the merits.[73] This is particularly likely to be the case in nanotechnology, where patents are extremely complex and cluttered with obscure language. Because courts will have difficulty detecting frivolous suits without engaging in a detailed analysis of the technology, litigants will be forced to spend large amounts of money proving their case. The sheer costs of defending against the suit could force the start-up to agree to a license.

Third, instead of attempting to design around a patent, a start-up can decide to litigate the validity of a patent. This strategy requires the start-up to be prepared to challenge the validity of the patent when the patent holder seeks to enforce its rights. Generally, patent holders will wait until a company is earning revenue before alleging infringement. As we argue above, there are sound legal tools available to narrow the scope of several broad patents in the field. Indeed, a number of start-ups that will infringe IBM's patent on single-walled carbon nanotubes when they bring products to market believe that the patent is invalid. As noted below, however, the costs of patent litigation could render this strategy inferior to simply obtaining a license from the patent holder.

A cheaper alternative to full-blown litigation is reexamination. If a company can find prior art that the original patent holder did not review and the prior art is material to the claims of the issued patent, the company can petition the PTO to review the validity of the patent. If the patent was filed after November 9, 1999, the infringing party can elect to participate in the reexamination proceeding or remain anonymous.[74] If the patent was filed before November 9, 1999, the party challenging validity cannot participate in the reexamination proceeding. If the company chooses to participate in the reexamination proceeding, it cannot litigate issues that were raised or could have been raised during reexamination. Further, the company cannot appeal the reexamination beyond the Board of Patent Appeals and Interferences.[75]

Fourth, a start-up might decide that it should seek a license from the patent holder rather than attempt to design around the patent or litigate the matter. We discussed difficulties associated with concluding license agreements in detail in Chapter 5. A company must decide whether it will seek a license from the patent holder before developing a commercial product. If the patent holder is a potential competitor, the company should probably not seek a license prior to launching a product. In order to conclude a license agreement, the company would have to disclose that it is likely infringing as well as sensitive information related to the technology. If the patent holder is a university, government lab, or another company in a different industry, however, this option may be desirable. Since there are large risks associated with product development and the value of the final product is highly uncertain, the company may be able to obtain the license on more favorable terms at an early stage. At the same time, the uncertainties may dissuade a patent holder from entering into a license agreement at this stage. Additionally, a start-up may not have the capital necessary to conclude a license agreement.

Finally, a start-up can generally deal with claims of infringement by pursuing a strategy of deterrence. If a company can develop its own offensive weaponry against a potential competitor, it can deter the competitor from launching an infringement suit. Indeed, as explained earlier in the chapter, numerous companies have adopted strategies of building large patent arsenals.

Enforcing Patents

At this stage, it is difficult to detect infringement of nanotechnology patents. Most companies are still experimenting with different nanomaterials and techniques for synthesizing nanomaterials, and there are few products on the market. A survey of patent infringement law suits as this book went to press revealed that there are no pending nanotech patent wars.

If a company can prove that another company is infringing its intellectual property during research and development, the patent holder has several options. First, if the infringer is a potential competitor, the patent holder can wield its intellectual property as a sword to prohibit the company from conducting research and development using its technology. Second, the patent holder can force the company to obtain a license. If the patent holder is confident that the company will fail, it might make sense to attempt to squeeze a license, with a large upfront fee, out of the infringer. If, however, the infringer is likely to succeed, the patent holder should wait to enforce its patent rights. After a product has been developed, the patent holder will have a good understanding of the value of the patent to the infringer and the market value of the infringing product. Additionally, the patent holder will have more leverage to exact greater royalties from a company that has already developed a product.

Licensing Nanotechnology Patents

Concluding license agreements will be integral to a nanotech company's quest to commercialize its technology. We engaged in a detailed discussion of negotiating licensing agreements, types of licenses, and the important terms of license agreements in Chapters 5 and 10.

Nanotechnology Patent Litigation

If nanotech companies cannot successfully conclude licensing agreements, they may ultimately be forced to resort to litigation. One company might launch its legal attack by seeking a preliminary injunction against another company. A preliminary injunction will be issued based on several factors, including whether the moving party is likely to prevail and whether the moving party would be subject to irreparable harm if the injunction was not granted.[76] A court can issue a preliminary injunction relatively swiftly and thereby order the infringing party to not practice the invention. As a result, the infringer will likely seek to conclude a license agreement in order to continue operating. If a court does not issue a preliminary injunction, however, a patent holder must decide whether it will pursue full-blown litigation.

Companies should carefully consider the implications of patent litigation. First, nanotechnology patent litigation will be extremely costly. In addition to exorbitant attorneys' fees, there are excessive discovery costs. Further, litigating nanotech patents will involve high fees for expert witnesses. Experienced patent litigators estimate that it could cost up to several million dollars to litigate a single nanotechnology patent. Patent litigation will also be time-consuming and distracting for management. Generally, cases take at least two years to get to trial and one to two more years on appeal.[77] Additionally, nanotechnology patent litigation will present tremendous uncertainties. The scope and meaning of patent claims are determined by federal judges.[78] Although federal judges and their clerks are bright and hard working, they may encounter difficulties in defining terms such as nanocrystals, diblock copolymers, and quantum confinement. Claim infringement, which is determined by juries, is even more unpredictable than claim construction. As two patent lawyers warn, "attempting to plan or predict the course of patent infringement litigation is often regarded as futile."[79] Depending on the circumstances of the case, either party might be found to be infringing or have their patents declared invalid. Patent litigation can ruin a company's chances of obtaining additional financing. Investors are unwilling to invest in firms that are engaged in court battles over intellectual property. Finally, patent litigation can risk revealing trade secret and confidential business information.

PROTECTING TRADE SECRETS AND CONFIDENTIAL INFORMATION

We discussed trade secret issues related to employees leaving their former employer to start a nanotech company. In this section, we focus on how the nanotech start-up should protect its trade secrets and other confidential information. While patent law requires a company to disclose its inventions to the public, trade secret law provides protection for concepts that are kept secret within the company. Nanotech companies primarily rely on patents instead of trade secrets to protect their important inventions. In addition to being critical to attract investors, patents provide a more sturdy form of protection. Nevertheless, in certain circumstances, trade secrets may be preferable to patents. Nanotech companies can also use other doctrines such as the duty of loyalty and unfair competition to protect ideas and concepts that are not trade secrets. Finally, nanotech companies have several contractual tools to protect their confidential information.

Trade Secret

Unlike patent, trademark, and copyright law, trade secret law is primarily governed by state law. Trade secrets are of potentially infinite duration since they last as long as secrecy can be maintained.[80] The Uniform Trade Secrets Act (UTSA), which has been adopted in some form by most states, defines a trade secret as:

> Information, including a formula, pattern, compilation, program, device, method, technique, or process, that: (i) derives independent economic value, actual or potential, from not being generally known to, and not being readily ascertainable by proper means by, other persons who can obtain economic value from its disclosure or use, and (ii) is the subject of efforts that are reasonable under the circumstances to maintain its secrecy.[81]

In making reasonable efforts to assure secrecy, a company is allowed to provide employees and other important individuals with limited access to the information. Confidentiality agreements with employees, collaborators, and sources of capital are essential to demonstrating that reasonable efforts have been made to preserve trade secrets.[82] Companies should also identify and catalog trade secrets, clearly flag confidential information, and restrict access to information by maintaining an access log.

Nanotech companies can use trade secret law to protect concepts and ideas that cannot be patented or to protect inventions that do not justify the expense of filing patents. For example, at this stage of development, some of the most valuable information generated at start-ups involve failed experiments. This "negative information" can be protected by trade secret. Information such as customer and investor lists are also ripe for trade secret protection. Finally, many nanotech companies prefer to maintain the details of their manufacturing methods as trade secrets.

Once a company proves that a trade secret exists, it must then prove that misappropriation occurred. Under the UTSA, a trade secret is misappropriated if it is (1) acquired by "improper means," or (2) disclosed or used by an individual who either used improper means to acquire it or knew that he or she was under a duty to protect its secrecy.[83] "Improper means" includes theft, bribery, misrepresentation, and espionage through electronic or other means. The most typical case of misappropriation takes place when an employee lawfully acquires the information, but then violates a duty not to disclose the information. Thus, if information or technology subject to trade secret protection is reverse-engineered by proper means or developed independently, there is no misappropriation.

Even when there is no evidence of intentional misappropriation, when an employee leaves a company, the former employer can argue that the employee should be restrained from working for the new company under the doctrine of inevitable disclosure. This doctrine creates an evidentiary presumption that an employee cannot help but rely on knowledge of the parent company's trade secrets in his or her new job. As argued in Chapter 10, courts typically invoke the doctrine only when a departing employee acts in bad faith.

Companies should also be aware that, even if they cannot seek redress under state trade secret law, there may be a remedy under the Economic Espionage Act (EEA).[84] EEA broadens both the kind of information covered and the type of conduct prohibited. Under the Act, a company can call the FBI when it thinks its trade secrets have been misappropriated. When a competitor steals a trade secret through a new employee, the employee and company's management team could be subject to jail sentences and large fines.

Other Common Law Tools

Even if information does not qualify as a "trade secret" and even if the employee did not sign a nondisclosure agreement, a company can still prevent an employee from disclosing information under other common law doctrines. An agent's duty of loyalty includes confidentiality of all information related to the principal's business "unless the information is matter of general knowledge."[85] The tort of "unfair competition" can also be asserted against individuals who obtain confidential information by improper means for purpose of advancing a rival business interest.[86]

Contractual Tools

Several contractual tools can also be used to protect confidential information. As explained in the preceding section, contractual arrangements can be necessary to secure trade secret protection. If trade secret law cannot be utilized, contract law might serve as a basis for legal action. Four types of agreements might be used by nanotech companies to secure their proprietary information.

First, all nanotech companies should have invention assignment agreements, which require all technical personnel to promptly disclose all inventions conceived or learned by the employee during employment and stipulate that all such inventions will belong solely to the company. Invention assignment agreements are generally upheld by courts. Many employers include "trailer" or "holdover" clauses in employment contracts, which state that inventions made after the employee leaves the company are still owned by the employer. Courts construe these provisions narrowly. As explained in Chapter 10, some courts have completely voided the provision when the period of time is unreasonable while other courts have held trailer clauses only apply to inventions made using the parent company's trade secrets.

Second, companies should have nondisclosure agreements with their employees. Nondisclosure agreements involve the employee's commitment not to disclose any trade secrets or other confidential information that he or she acquires during employment by the company to any third parties. Nondisclosure agreements are usually upheld as long as they are reasonable, but some courts have modified clauses that are clearly overly broad.[87]

Third, companies might establish non-compete agreements with their employees. A noncompetition agreement prevents an employee from working for a competitor for a specified period of time following his or her departure. The company can condition its offer of employment upon execution of a noncompetition agreement by the prospective employee. As explained in detail in Chapter 10, different states have different rules regarding the enforceability of noncompete clauses.

Finally, companies might negotiate nonsolicitation agreements. Nonsolicitation agreements prevent a former employee from recruiting key employees of the former employer. Although nonsolicitation agreements are generally upheld, they are strictly construed.[88]

TRADEMARK

Trademark issues will be important to nanotechnology companies. A trademark is a word, name, symbol, device, or any combination thereof, which is used to distinguish the goods of one person from goods manufactured or sold by others. The term "service mark" is used to refer to marks that are used in connection with services. For purposes of this chapter, the term "trademark" will be used to refer to both trademarks and service marks. Nanotech companies may wish to trademark the name of the company, the company's logo, and the names of specific products and/or services. For example, Nantero sought to register the name "NANTERO" for a long list of products that might be marketed under the company's name as well as "NRAM" for computer memory chips. Similarly, Quantum Dot Corporation claims trademarks to the "Q LOGO," "QDOT," and "QCELL." On the services side, Quantiam Technologies has filed for service mark protection of the mark "QUANTIAM" for custom manufacture of advanced materials services, and Nanoink, Inc., is attempting to register "NANOINK" for use in connection with technical consultation, research, and support in the field of nanolithographic printing.

Establishing Trademark Ownership

The creation of rights in marks occurs automatically under the common law when a mark is used in trade. Registration of a mark is not necessary to acquire ownership rights and enforce those rights. Despite the fact that a company can acquire a trademark without registering the mark, companies should register their marks for several reasons. First, registration flags the mark to the public and reduces the chance that the PTO will register a similar mark. Second, registration provides a right to assistance from the U.S. Customs Service in preventing the importation of infringing goods. Third, registration creates a presumption of mark ownership and validity that increases the likelihood of prevailing in trademark disputes. For example, after five years of registration, certain challenges to the mark cannot be raised.[89] Finally, registration enables the mark owner to obtain rights in a greater geographical area than would be possible under the common law.

In order to register the mark, the PTO must determine that two primary conditions are met. First, the mark must be distinctive. Marks are placed into one of four categories:

1. Arbitrary and fanciful
2. Suggestive
3. Descriptive
4. Generic marks

Arbitrary and fanciful marks and suggestive marks are considered distinctive. Arbitrary and fanciful marks do not describe the products they identify in any way. Examples of an arbitrary or fanciful mark might be "ZYVEX" for "laboratory robots for use in the field of nanotechnology," "ZETTACORE" for "chemical processes for use in the manufacture of integrated circuits and electronic, molecular, and semiconductor memories," and "INFRAMAT" for "nanostructured materials." Suggestive marks allude to or indirectly refer to the product. The consumer must use imagination and a multistage reasoning process to reach the conclusion that the product is associated with the mark.[90] Examples of suggestive marks might be "DENDRITECH" for "chemicals," "NANOINK" for "electrical and scientific apparatus ... for use in the manufacturing, design, and operation of a wide variety of microscale and nanoscale electronic products," and "NANOPHASE" for "ceramic particles."

Descriptive marks are not inherently distinctive, but can become so if they acquire secondary meaning. A mark is descriptive when little or no "imagination, thought, or perception is required to reach a conclusion on the nature of the goods or services."[91] Secondary meaning arises when the relevant consuming public has been exposed to the mark enough to automatically associate it with the product or service. Five years' exclusive presence of a product in a specific market is prima facie evidence of secondary meaning under Section 43(a) of the Lanham Act.[92] An example of a descriptive mark might be "CARBON NANOTECHNOLOGIES" for carbon nanotubes. CNI registered the mark "CARBON NANOTECHNOLOGIES, INCORPORATED." as a design mark for the custom manufacture of carbon nanotubes for others. Perhaps due to the fact that "carbon nanotechnologies" is descriptive of the product, CNI had to disclaim the right to use "CARBON NANOTECH-

NOLOGIES, INC." apart from the mark as shown. Having to disclaim, in essence, all of the components of the mark makes the protection that CNI received on this mark extremely weak.

Generic marks can never become distinctive. A generic mark is the common descriptive name of the good or service it is used to identify. For example, "NANOTECHNOLOGIES" might be considered a generic mark. Indeed, Altair Nanotechnologies, Inc. had to disclaim the right to use "NANOTECHNOLOGIES, INC." in registering their mark. Similarly, Quantum Dot Corporation was forced to abandon its application pending for "QUANTUM DOT," because the term was a generic term used to describe semiconductor nanocrystals.

Second, to register a trademark, an applicant must convince the PTO that its mark is not likely to be confused with a previously registered mark.[93] Examiners generally evaluate the following eight factors to determine likelihood of confusion:

1. Strength of the registered mark
2. Similarity of the marks
3. Relatedness of the goods
4. Evidence of actual confusion
5. Marketing channels used
6. Likely degree of purchaser care
7. Intent in selecting the mark
8. Likelihood of expansion of the product lines

As start-ups in the field proliferate and the number of trademarks incorporating the word "nano" rises, there is an increased chance that an examiner will determine that a proposed mark is likely to be confused with a registered mark.

Use of the Mark

Trademarks must be used correctly and consistently for an owner to maintain its rights. Widespread misuse of a trademark may lead to the permanent loss of all trademark rights in the mark, regardless of whether such misuse is made by the trademark owner, the trade, the press or the consuming public.[94]

It should be remembered that marks are adjectives, not nouns. Therefore, a mark should always be used with an initial capital letter and in its adjectival form (that is, modifying a noun or other descriptive language). For example, a proper use of the mark "NRAM" in connection with computer memory chips is "the NRAM computer memory chips are considered to be...," not "NRAMs are considered to be...." One trick is to use some descriptive or generic language and/or "brand" following the use of a mark (for example, "Kleenex-brand tissue," not just "Kleenex"). Not following such guidelines can result in loss of rights in a mark. For example, terms like "aspirin," "escalator" and "elevator" were at one time registered marks. However, use of the terms in a generic sense over time resulted in loss of exclusive rights in the marks.[95]

If a trademark is not registered with the USPTO, or if a company is uncertain of the registration status, the superscript™ should be used to identify the mark. If the mark is registered with the PTO, the registration symbol ® should be used to identify a mark. This distinction should be carefully followed since use of the registration symbol ® on an unregistered mark is actionable under the law. The trademark symbol should be placed next to the most prominent use of the trademark in each publication. The symbols should also be placed next to the first use of the mark in any text or body copy. The trademark symbol does not need to be repeated next to the trademark throughout the publication.

Trademark Litigation

Trademark infringement suits turn on whether two marks are confusingly similar. Courts use the same multifactor test used by PTO examiners in making this determination. Litigants also quarrel over the validity of each mark. Thus far, with few products on the market, trademark lawyers have not been called to battle by nanotech companies. As the field develops, however, there are certain to be a number of trademark disputes. For example, while U.S. Nanocorp has a registered trademark for "NANOX" for battery, fuel cell, and supercapacitor applications, there are a handful of companies using "NANOX" for different products. At one time, there were even rumors that Quantum Dot Corporation might sue anyone who capitalized the "Q" and the "D" in Quantum Dot unless the words were used to refer to their company.

COPYRIGHT

Copyrights are granted to original works of authorship in any fixed, tangible medium of expression.[96] Copyright protection extends only to the work's expression and not the ideas embodied in the work. In addition to protecting literary and musical works, copyright protection also extends to software.[97] Copyright provides the owner with certain exclusive rights, including the right to reproduce the work, distribute the copies, create derivative works, publicly perform the work, and display the work.[98] Generally, in order to win a claim of copyright infringement, a plaintiff must show (1) that he or she holds a valid copyright in the copied work and (2) that the defendant copied the work's protected expression.[99] Under the fair use doctrine, courts can avoid finding infringement when doing so "would stifle the very creativity which that law is designed to foster."[100] For example, courts have held reverse-engineering of software to constitute fair use[101] under certain circumstances.

For the most part, copyright protection will have limited importance for nanotech companies. However, copyright protection is often available for documents of original authorship, such as technical manuals, training materials and the like. Additionally, as explained in Chapter 1, computational nanotechnology is critical to development of technologies. Nearly all research and development programs involve sophisticated software. In many cases, the company will want to protect its software. Although software programs are eligible for patent protection, the costs of filing patents combined with the disadvantages of publishing may dissuade companies from patenting their software. Therefore, many nanotech companies will seek to copyright software. A company is not required to register copyrights in order to be protected under copyright law.[102] A company cannot sue for infringement, however, until the copyright has been registered.[103] To register a copyright, the copyright owner submits an application and fee and deposits the required copies of the work with the United States Copyright Office.

CHAPTER 14

Corporate Partnering and Globalization

> *[N]anoparticles are far from straightforward to work with and require tremendous chemistry knowledge in order to tap into the much desired "nano-properties." Most companies cannot be fully integrated on their own, so we have focused on finding partners in each of our target market applications.*[1]
>
> —Randy Bell (CEO, Nanotechnologies Inc.)

Corporate partnerships and strategic alliances are terms used to describe the "relationship between two companies for developing and exploiting technology, products and markets..."[2] These relationships will be essential for the commercialization of nanotechnology. Each type of collaboration carries certain responsibilities and risks, and the success of a partnership is dependent on the nature of deal terms and the effectiveness of project implementation. While a successful arrangement can pave a road to commercial success, a poor deal can cripple a fledgling start-up.

This chapter explores corporate partnering in nanotechnology. We first explain why companies should pursue partnering arrangements and identify the risks associated with partnering. This is followed by a discussion of when start-ups should begin to seek corporate partners and how they should go about finding a good partner. We then identify the issues raised in partnering agreements and discuss how different partnerships should be structured and implemented. Since many corporate partnerships involve foreign corporations, we take advantage of the opportunity to discuss globalization issues facing nanotech companies. In the last section, we highlight four issues that nanotech companies must consider in approaching the global marketplace.

THE CASE FOR PARTNERING IN NANOTECHNOLOGY

Partnering arrangements are an increasingly common strategy by which companies get products to market. They will be an important consideration for nanotech companies. As one commenter put it, if the 1980s were the "Decade of the Merger/Acquisition," the 1990s was the "Decade of the Strategic Alliance."[3] Indeed, corporate partnerships are "a permanent feature of [t]he business world."[4] The growth of partnerships has been especially strong in emerging technology areas where new companies require help in turning technologies into products. Their growing importance has been seen most clearly in the biotechnology industry, where start-ups rely on strategic relationships with large pharmaceutical companies to obtain regulatory approval and market new drugs.[5] Large companies also benefit from these relationships since they access new technologies, sell products that are complementary to their existing products and target new groups of customers. Partnerships are such an essential part of their business that large companies such as Eli Lilly have even created special groups whose sole focus is to "establish best practices for working with partners."[6]

While some academics continue to debate whether these arrangements are value-enhancing or simply a "faddish panacea,"[7] most executives, consultants, and investment bankers view partnerships as valuable tools that can be used to help accomplish business goals. Kathryn Harrigan detailed the different motives that two companies might have when forming a corporate partnership:

A. Internal suses
  1. Cost- and risk-sharing (uncertainty reduction)
  2. Obtain resources when there is no market
  3. Obtain financing to supplement firm's debt capacity
  4. Share outputs of large, minimally efficient scale plants
    a. Avoid wasteful duplication of facilities
    b. Utilize byproducts, processes
    c. Shared brands, distribution channels, wide product lines, and so forth
  5. Intelligence: obtain window on new technologies and customers
    a. Superior information exchange
    b. Technological personnel interactions
  6. Innovative managerial practices
    a. Superior management systems
    b. improved communications
  7. Retain entrepreneurial employees
B. Competitive uses (strengthen current strategic positions)
  1. Influence industry structure's evolution
    a. Pioneer development of new industries
    b. Reduce competitive volatility
    c. Rationalize mature industries
  2. Preempt competitors ("first-mover" advantages)
    a. Gain rapid access to better customers

b. Capacity expansion or vertical integration
   c. Acquisition on advantageous terms, resources
   d. Coalition with best partners
3. Defensive response to blurring industry boundaries and globalization
   a. Ease political tensions (overcome trade barriers)
   b. Gain access to global networks
4. Creation of more effective competitors
   a. Hybrids possessing owners' strengths
   b. Fewer, more efficient firms
   c. Buffer dissimilar partners

C. Strategic uses (augment strategic position)
  1. Creation and exploitation of synergies
  2. Technology (or other skills) transfer
  3. Diversification
     a. Toehold entry into new markets, products, or skills
     b. Rationalization (or divestiture) of investment
     c. Leverage related owners' skills for new uses[8]

As of this writing, dozens of strategic alliances have been formed between start-ups and large companies. Table 14.1 illustrates that the nanotechnology community is already embracing corporate partnerships. As the field begins to blossom, this trend is likely to continue. Indeed, partnering may play an even greater role in nanotechnology than other industries.

There are several reasons why partnerships are critical to start-up companies. First, partnering provides a valuable, alternative means of financing the development and commercialization of nanotechnology. Developing tools, materials, devices, and systems based on nanoscience is cost-intensive and takes many years. As shown in Chapter 12, the costs of obtaining venture financing are greater than ever before. Despite much hype, venture capitalists are still resistant to invest in the field. Even when they are willing to invest, the terms of investments are onerous—company ownership is seriously diluted and venture capitalists often demand substantial control. Large corporations can make substantial equity investments in fledgling start-ups without stripping the founders of ownership and control. As one professor puts it, the returns sought by large companies "are primarily strategic, not financial."[9]

Second, partnerships are critical to the commercialization of nanotechnology. Most nanotech start-ups do not expect to develop end products. Rather, they are focused on developing materials, devices, and systems that will be incorporated into existing products and manufacturing processes. Thus, in order to succeed, start-ups must leverage the existing technology platforms of large corporations. Randy Bell, CEO of Nanotechnologies, Inc. explains, "[N]anoparticles are far from straightforward to work with and require

TABLE 14.1 Examples of Corporate Partnerships in Nanotechnology

Company	Company	Technology	Terms
Nantero	ASML	Integrate Nantero's nanotube-based NRAM device into ASML's semiconductor equipment	Joint development
Nanotechnologies, Inc.	Air Products	Pulsed plasma technology that produces nanoscale oxides as well as nanoscale metals	Joint development (exclusive development for selected segments) Equity investment
Nanotechnologies, Inc.	Essilor	Use of nanoparticle manufacturing technology for eyewear applications	Joint development
Nanocor	PolyOne	Nanocomposites based on polyolefin, polyvinyl chloride, and other polymers	Joint development and marketing agreement
NanoGram Devices	EaglePicher Technologies	Nanomaterial technology for batteries in implantable medical devices	Joint development and manufacturing agreement
NanoOpto	Enplas	Nanofabrication for rapid design and high-volume manufacture of nano-optic devices for optical systems and networks	Manufacturing and licensing agreement
Carbon Nanotechnologies	Performance Plastics Products Inc. (3P)	Single-wall carbon nanotubes incorporated into 3P's custom manufactured components	Joint development agreement
Nanocor	Mitsubishi Gas Chemical	High-barrier nanocomposite plastics for consumer and industrial packaging applications	Manufacturing, sale, and joint development agreement
Nanomix	Dupont Electronic Technologies	Field-emitting thick film materials containing carbon nanotubes for use in flat panel displays	Exclusive license to make and sell
Nanosys	Matsushita Electric	Nanocomposite solar cells into building materials	Joint development
Nanogate Technologies	Holmenkol (Loba)	Innovative sports coatings, including ski wax, that exploit Nanogate's chemical nanotechnology	Joint venture (formation of new entity: Holmenkol Sport-Technologies GmbH & Co. KG)
Carbon Nanotechnologies	Sumitomo	Carbon nanotubes	Joint development (Asia)

tremendous chemistry knowledge in order to tap into the much desired 'nano-properties.' Most companies cannot be fully integrated on their own, so we have focused on finding partners in each of our target market applications."[10] Partnering also provides start-ups with access to large-scale manufacturing operations as well as domestic and international distribution channels and customer bases. As one nanotech venture capitalist explains: "There is no way start-up companies can be selling a fuel additive in Asia and Europe with a three-person sales force. So they are forced to go to Mobile, Exxon or one of the large corporations to give them an area. Or you might, for example, break up part of Asia; India, China, Japan and the rest of Asia and there might be four different agreements with local companies in each one."[11] In the nanobio arena, partnering can also be critical to navigate the regulatory landscape. A large company can lead a start-up focused on a new drug delivery device through the process of obtaining FDA approval.[12]

Third, partnering can provide credibility to nanotech start-ups. A number of start-ups we interviewed enter negotiations for corporate partnering to enhance their credibility. Validation from large corporations such as Intel or BASF make investments from venture capital firms more likely.

Just as start-ups need large companies to commercialize nanotechnology, large companies can leverage nanotechnology by partnering with start-ups. Many large companies in established industries do not have the flexibility to develop their own research and development programs. Further, at this point in time, acquisitions are not common. Despite substantial interest on the part of large companies, they are choosing to wait and see rather than actively acquire nanotechnology companies because it is still a new and unproven area. Partnership arrangements provide large corporations with an opportunity to get involved in the field without having to undertake the risks associated with acquisitions.

THE RISKS OF PARTNERING

While strategic alliances with larger companies offer many potential advantages, there are also risks associated with such arrangements. Companies should carefully consider several potential pitfalls to partnership. First, collaborations historically carry a high risk of failure.[13] For example, the anticipated technology outcome might not be realized, integration of the start-ups technology might not go as planned, or the larger company might abandon the project for its own strategic reasons that have nothing to do with the project itself. If the deal collapses for any reason, the start-up could find itself with personnel, equipment, and facilities for which it no longer has capital to support. The perception of the start-up in the business and investment communities will also suffer. Outsiders will have many concerns about the loss of a larger partner and the company may encounter difficulty in raising additional capital and finding other corporate partners.

Second, even if the partnership succeeds from the standpoint of product commercialization, a poorly drafted agreement can strangle a start-up. If the scope of the alliance is too broad or the terms of the deal too restrictive, then the start-up's growth could be stifled. For example, in some cases developing and integrating a product for the larger partner may divert resources from other areas, and the effort required might significantly limit the ability of the company to pursue other projects. In effect, the partnership could be an "inadvertent acquisition of the smaller company by the larger company."[14] The start-up might become the research and development arm of the larger company.

Finally, even when technological development proceeds smoothly and the agreement is drafted properly, there is always a risk that the parties' goals and incentives will not be aligned. For example, a large company may have an incentive to suppress the value of the project in order to obtain a lower valuation in a future acquisition of the company.

While corporate partnerships are usually less consequential for large companies than for start-ups, they also carry certain risks. If the agreement is frustrated by technological delay, those within the large corporation who backed the partnership may want to back out of the deal to protect their personal credibility. If the deal is too successful, the large company might find itself in several difficult situations. The large company might become dependent on the revenue they are realizing from the start-up's technology. Depending on the terms of the original agreement, the start-up might be able to exert a great deal of influence over the large company by being in a much stronger negotiating position for an extension of the deal.

SEEKING A CORPORATE PARTNER[15]

Notwithstanding the risks associated with strategic alliances, corporate partnerships will be a requirement for most nanotech companies. From the company's inception, managers should begin to evaluate potential partners. One CEO remarks that a nanotech start-up should begin to consider corporate partners "as early as possible."[16] However, a company should not actively seek partners unless the conditions are ripe for a deal.

First, the start-up must be able to demonstrate its technological capability before being able to secure a corporate partnership.[17] The level of technological development needed to lure a strategic partner depends on the particular technology, the potential partners, and the type of arrangement contemplated. However, almost all corporate partners will at a minimum need some proof of concept and at least minimal proof of technology potential. Interestingly, most venture capital–backed nanotech startups that we have evaluated have successfully secured partnerships within two years of being formed.

Second, a start-up is probably not ready to pursue a partnership if it is starving for cash. The process of forming an alliance may take longer and be more complicated than securing venture capital. Many large companies interested in strategic partnerships are interested because they want access to research and development without incurring all of the costs. Therefore, they will be seeking a discount to developing the technology themselves and may not be interested in being a full funding partner for the start-up.

Pursuing a Deal

If a start-up decides that the time is ripe to establish a corporate partnership, managers should first identify what the company is seeking from a partner and what it can provide to the partner. For example, Table 14.2 describes some of these essential qualities.

Second, the start-up must establish a strategy for engaging large companies and develop a plan to implement the strategy. Management should work with legal and financial counsel to develop alternative deal structures and to analyze the potential outcome of different deal types. By properly framing the situation, managers "avoid wasting resources by focusing on too many players or the wrong ones."[18] It is also very important that the start-up put together a comprehensive list of potential partners. The company should compile information packets that may be derivations of the business plan, designed to send to potential partners to generate interest in the company. Finally, a timeline must be established for implementing the plan. Even though the final deal outcome will likely be very different from the initial plan considerations, preparation is an essential component of deal preparation.[19]

TABLE 14.2 Selected Large Company and Start-Up Contributions to a Partnership

Ideal Partner for Start-Up Nanotech Company	Bargaining Chips of Start-Up Nanotech Company
Market leader in a particular application area	Nanotech with high growth and market potential
Broad chemistry capability for targeted applications	Nanotech that complements large company's business
Interest in developing nanotech applications	Intellectual capital and specific technology competence
Culture and history of technical innovation	Patents to block large company from developing technology
Willingness to invest significant resources to project	Threat of partnership with competitor

Third, it may be valuable for the start-up to be the aggressor in initiating deal discussions. A company can contact several potential partners at the same time in order to maximize bargaining leverage and to gauge which opportunities make the most sense. In the case where the start-up desperately needs a partner, being aggressive and keeping multiple lines of discussion open can ensure alternatives in the event of negotiation failure. Quite obviously, the most attractive targets will have objectives that complement the needs of the start-up. A target that has some familiarity with the start-up should be given higher priority than one that does not, since the familiar companies are in a better situation to judge whether the partnership is likely to be of interest. Management should investigate whether the large company is likely to undergo any major internal changes or other events that might disrupt its stability. It is not uncommon that large companies in nanotech begin discussions with start-ups as a means to learn more about the industry. Ultimately, these companies are more curious than serious. As such, some discussions can waste valuable time. Finally, the start-up should focus on targets that have employee and work atmospheres that are compatible with the start-up. One of the greatest risks to any partnerships is the inability for two different kinds of companies to see eye to eye. Different company cultures can destroy the ability to effectively partner.[20]

Fourth, a start-up must identify key decision makers to effectively manage deal execution process. While the start-up's management can quickly go forward with a proposed deal, staffs at large corporations may not have the authority to approve a deal. They will likely need to propose a deal structure to their own managers and senior executives within their company. Therefore, the start-up must work with the corporate staff in negotiating the deal, but must also make sure that the corporate staff is properly engaging the final decision makers throughout the deal process. During the negotiation process, it is important for managers to develop a strong relationship with the figure in the large company that is determined to see a deal consummated. Many start-ups have been the unfortunate victims of being misled by corporate staff who were unable to persuade their own companies to enter into a deal even though they had finalized their own negotiations.

During the initial contact, management must present general information regarding the start-up and how a partnership might be mutually beneficial. If, after the introduction, the parties elect to pursue further discussions, a carefully drafted confidentiality agreement will be necessary. There may also be a need for an agreement that the large company will not attempt to lure away key employees from the start-up.

Finally, the start-up should take the lead and present a proposal to the large company. The initial proposal should be in outline form, leaving details to be clarified in negotiations. The goal of taking the initiative in this way is to reach a letter of intent specifying the key terms of the deal. After reaching a letter of intent, negotiations begin to flesh out the final documents. Throughout the negotiation process, the start-up should have experienced counsel. The next section describes how a well-crafted agreement is crucial to successful implementation.

DOING THE DEAL

There are several different types of partnership agreements. As stated in the introduction of a leading book on corporate partnering: "[Corporate partnerships] come in a nearly infinite number of variations and can be adapted to almost any set of mutually agreed-upon objectives. This often makes structuring and negotiating alliances complex and time-consuming. It demands flexibility and creativity."[21] Certain arrangements may take the form of equity investment by the large corporation along with distribution rights. Others might involve the two companies' jointly developing a technology. Each deal will be different based on the unique objectives of each party. As shown in Table 14.1, nanotechnology start-ups seek equity or development assistance, and large companies are interested in manufacturing, distribution, and licensing rights to the start-up's technology. In this section, we discuss the different terms of different partnership arrangements. Specifically, we review the key issues that must be addressed in partnerships involving equity investments, technology development agreements, manufacturing agreements, distribution agreements, and licensing agreements. This section is only intended to provide a brief overview of the terms that must be addressed in order to conclude a partnership agreement. For checklists and sample agreements with which to build a customized partnering arrangement, more detailed texts and guides should be consulted.[22]

Equity Investment

Some nanotech partnerships involve an equity investment by the large company in the start-up. For example, the strategic alliance between Air Products and Nanotechnologies, Inc. involved a substantial equity investment by Air Products. Generally, the large company receives convertible preferred stock in exchange for its investment. The partners must agree on the rights and protections of the preferred stock. Liquidation preferences are common while dividend, redemption preferences, and voting rights are less common. The price of the preferred stock must also be negotiated based on the rights and privileges of the stock and the value of the contract. In most cases, large companies invest less than 20 percent of the equity in order to avoid recording losses from the start-up on financial statements.[23] The start-up must also decide whether the large company will be given a seat on the board. In negotiating this provision, nanotech start-ups should carefully analyze the goals and interests of the partner. A board seat would make the partner privy to sensitive information about the company and could generate serious conflicts of interest. Large companies might also seek to gain power over the start-up's business decisions. For example, a large company may insist on a right of first refusal to participate in future equity offerings or transfers of technology by the start-up. Even more drastic, a large company may push for an option to purchase the start-up. These provisions can reduce flexibility and lock a start-up into an undesirable situation. Finally, the agreement should include an exit mechanism in case the collaboration fails. For example, the start-up could be given an option to repurchase the large company's shares if certain milestones are not met.

Technology Development

Most partnership deals we evaluated involved some form of technology development. The collaboration between Nantero and ASML Holding NV is illustrative of a typical technology development agreement in the field. In 2003, Nantero and ASML began working to integrate Nantero's nanotube-based NRAM device into ASML's semiconductor equipment. The project took place in ASML's research facilities with a joint team from Nantero and ASML's Special Applications Group.

Effective technology development agreements must clearly define each partner's commitment to the project in terms of technology, personnel, facilities, equipment, funding, and services. The large company often provides funding for the project in stages, marked by development milestones. The partners must establish acceptance tests to determine if a milestone has been reached. Confidentiality provisions are essential in the agreement; each party must guarantee that the other party's proprietary information will only be used in the joint development. The agreement must also carefully define two different ownership rights. First, the agreement should define ownership of the base technologies brought by each side to the project. Second, the agreement should define ownership of technologies resulting from the joint efforts. Most agreements specify that the large company will have rights to a particular application, vertical market, or geographic territory.

In some cases, the alliance may involve the formation of new entity to conduct research and development. For example, a new entity, Holmenkol Sport-Technologies GmbH & Co., was created to carry out a collaboration between Nanogate Technologies and Holmenkol to develop new coatings. The primary motivation for a separate entity is ensure managerial and operational independence. There may also be other benefits such as tax, accounting, cultural, or liability limitation.

Manufacturing

Many nanotech partnerships involve joint efforts to develop materials, devices, and systems that will be integrated into technology platforms of established companies in different industries. These types of partnerships generally include provisions relating to manufacturing rights. Often, the large company has exclusive manufacturing rights to the resulting technology. In some cases, however, start-ups have already developed a technology to the point where it can be directly incorporated into a manufacturing process. Thus, some partnerships do not involve joint development efforts and are formed for the primary purpose of manufacturing. For example, Nanogram Devices Corporation signed an agreement with EaglePicher, Inc. for the manufacture of nanomaterial-based medical power sources. Similarly, NanoOpto entered into a manufacturing agreement with Enplas Corporation. The agreement involves NanoOpto's integrating a portion of their manufacturing technology into Enplas' facilities for production of optical devices.

In exchange for exclusive manufacturing rights, large companies often pay the start-up a lump sum payment, amortized over an agreed-upon number of units to be manufactured by the large company. The agreement should include provisions allowing the start-up to terminate the deal if the large company cannot meet certain quality standards. A start-up should reserve the right to require product revisions in certain circumstances, and a large company should be barred from changing the product without approval. The agreement should also clearly establish that proprietary information regarding the product and manufacturing process belongs to the start-up.

Distribution

Partnership agreements often include provisions relating to product distribution. A large company will usually acquire certain exclusive rights to a particular market segment or geographic territory. For example, in the strategic alliance between Nanosphere and Takara Bio, Takara has exclusive distribution rights for certain Nanosphere products in Asia and nonexclusive rights in European countries.

A large company generally pays for distribution rights with a large, up-front payment. Exclusivity may also be conditioned upon a large company attaining agreed-upon sales quotas. A large company must use its best efforts to market the product, and the start-up usually provides some marketing assistance. A start-up should consider insisting on the ability to alter its prices in the future. The agreement must flesh out other issues such as trademarks, packaging and notices, maintenance, support and training, information and reports, and policing infringement. Finally, the agreement must specify a period of time before it expires. The terms of many distribution agreements are between two and five years. Either party should be allowed to terminate the deal in the event the other fails to perform its duties.

Licensing

Some partnering arrangements involve license agreements. The start-up licenses its intellectual property to the corporate partner. Therefore, licensing agreements necessarily include manufacturing and distribution rights.

The most important feature of the license agreement is the scope. Generally, a large company acquires an exclusive license to a particular application in which the company has special expertise and market power, and the start-up maintains rights to all other applications. A good example of a licensing agreement is the agreement between Nanomix and DuPont Electronic Technologies. DuPont obtained an exclusive license to make and sell field-emitting, thick-film materials containing carbon nanotubes for use in flat-panel displays.

Most licensing agreements involve the large company's paying royalties on each unit of product manufactured and sold. In some cases, partners may agree to a lump sum payment for the license. If the agreement includes royalties from a company in a foreign country, there are complicated issues associated with withholding taxes and exchange controls that must be decided. Finally, indemnification provisions generally designate the burden of litigating patent disputes.

IMPLEMENTING A SUCCESSFUL PARTNERSHIP

At the time of this writing, it is difficult to discuss factors that are indicative of successful partnerships in nanotechnology. In many cases, the success of nanotech partnerships will turn on technological factors. Based on our conversations with managers and research of partnerships in other industries, we can make several predictions about factors that are likely to accompany successful partnerships.[24]

First, strong personal relationships between managers and employees at each company are critical to an effective partnership. Feelings of dislike or distrust between key players on each side can derail the entire alliance. Establishing early involvement from people of different levels within the company involved in the collaboration can be very helpful. During the planning, due diligence, and negotiations process, active involvement can facilitate positive personal relationships.

Clarity and precision in deal terms are also likely to enhance the likelihood of a partnership succeeding. Each party must accurately identify the other party's objectives and goals for the partnership. The terms of the deal must also be clearly spelled out, and each party must understand its rights and obligations.

Another key ingredient of successful partnerships is installing incentives for each side to perform throughout the duration of the partnership. According to one nanotech CEO, "Good partnerships are based on an understanding of win-win arrangements."[25] The deal should be structured such that each party's obligations are consistent with its business strategies. Some deals build in mechanisms such as penalties for failure to perform.

Finally, effective communication channels and procedures for dispute resolution are also important. Disagreements between newly formed partners are almost certain to arise at some point. Regular formal communications such as quarterly status meetings can prevent differences from becoming conflicts. When controversies do arise, pre-litigation dispute-resolution procedures can thwart escalation. Specifically, parties can reach fair and efficient outcomes by establishing mechanisms that require the parties to discuss the conflict until resolution is reached at the highest management levels. [26]

GLOBALIZATION

Economies around the world are increasingly linked through trade, integrated financial markets and the movement of people and information across borders. Globalization offers nanotech companies access to more and larger markets for exports, capital, labor, and ideas. However, in order to reap the benefits of globalization, nanotech companies must monitor legal, business, and policy developments in other regions of the world. It is beyond the scope of this chapter to show how companies should approach the new global marketplace.[27] In this section, we provide a brief overview of four of the most important globalization issues for nanotech start-ups to consider:

1. foreign partnerships and investors
2. foreign ideas and talent
3. manufacturing abroad
4. intellectual property protection in the global marketplace

Foreign Corporations and Investors

Foreign corporations can serve as strategic partners for nanotech start-ups. Many Asian and European firms perceive investing in American nanotech start-ups as a means to familiarize themselves with the technology, American market, and distribution channels. CEOs at startups that have formed such relationships warn that geographic distance, language barriers, and cultural differences can complicate the negotiation and implementation process. Further, partnerships with foreign entities can involve complex legal issues.[28]

Foreign investors might also serve as valuable sources of capital. Although venture capital is not as entrenched in foreign countries as it is in the United States, it is increasing. In Asia, Innovation Engine, Apax Globis Partners & Co., and Juniper Capital Ventures Pte. Ltd. have all made investments in nanotech start-ups. Venture firms interested in nanotechnology in Europe include Technostart, Ariadne Capital, and 3i.

Foreign Ideas and Talent

Other countries can also be a valuable source of ideas and skilled labor for nanotech start-ups. As discussed in Chapter 8, governments around the world

TABLE 14.3 Asian Nanotechnology Research Institutions to Watch

Country	Research Institutions
Japan	AIST, University of Tokyo, Kyoto University, Osaka University
China	Chinese Academy of Sciences, Beijing University, TsingHua University, Hong Kong University of Science and Technology
South Korea	Seoul National University
Taiwan	National Taiwan University

have followed the US lead and are now pouring billions of dollars into nanotechnology research and development. Nanotech companies should keep a careful eye on developments at key research institutions in Asia and Europe. Further, as mentioned in Chapter 8, the US is likely to face a shortage of skilled labor in nanoscience. As nanotech companies grow, they could encounter difficulties in hiring scientists and engineers with necessary skills. Companies may find talented workers abroad. However, in light of the current national security environment, hiring foreign scientists involves navigating through complex regulatory procedures.[29] (See Tables 14.3 and 14.4 for examples of nanotech research institutions in Asia and Europe to watch.)

TABLE 14.4 European Nanotechnology Research Institutions to Watch

Country	Research Institutions
Belgium	Katholieke Universiteit Leuven
Finland	Helsinki University of Technology, University of Jyvaskyla
France	CNRS, MINATEC, University Paris-Sud
Germany	University of Karslruhe, TU Dresden
Israel	Technion, The Weizmann Institute of Science
Netherlands	TU Delft (DIMES), BioMaDe, University of Twente (MESA+)
Sweden	University of Lund, University of Linkoping, Chalmers University of Technology
Switzerland	Swiss Federal Institute of Technology, Paul Scherrer Institute

Manufacturing Abroad

If a nanotech company prefers to take products to market without the help of a large corporate partner, it must consider establishing its manufacturing facilities in a foreign country. Large American technology companies are increasingly shifting manufacturing operations to less-developed countries. China, for example, offers cheap labor, workers with technical education, inexpensive land, and tax breaks. The advantages of locating manufacturing operations abroad are illustrated by Intel's decision to build a factory in western China to convert silicon wafers into semiconductors for sale worldwide.[30] Despite the apparent economic advantages, manufacturing nanotechnology materials, devices, and systems abroad involves wading through complex export control laws.

Protecting Intellectual Property in the Global Marketplace

Finally, nanotech companies should be mindful of intellectual property issues in the global marketplace. The obsessive compulsion with intellectual property rights in the United States has not been contagious. Many other countries do not place the same emphasis on enforcing intellectual property (IP) rights as the US.[31] Because most companies rely on patents instead of trade secrets to protect their inventions, foreign firms may be able to replicate patented technology and market it in countries that do not have solid IP regimes. Under the General Agreement on Tariff and Trade's 1995 TRIPS Agreement, WTO member countries must make patents available and enforce patent rights.[32] Although TRIPS bodes well for the future of intellectual property protection, nanotech companies should be prepared to accept piracy in the international marketplace for the foreseeable future.

CHAPTER 15

Consolidation and Standardization

It is this inability to identify standard measurement and characterization methodologies for nanomaterials and nanodevices that may signal that nanotechnology development is still more weighted toward science rather than commercialization... Without anticipatory standards, integration and large-scale manufacturing could remain elusive.[1]

—Edward Rashba (IEEE), et al.

A survey of the hundreds of companies engaged in nanoscale research reveals that there are large numbers of companies competing to develop similar technologies. For a variety of reasons, fragmentation might inhibit commercialization of nanotechnology. In this chapter, we explore strategies that enable coordination in research and development of nanotechnology. First, we argue that companies should consider pursuing mergers with other companies developing similar technologies. In certain circumstances, consolidation might eliminate destructive patent disputes, eradicate inefficient, duplicative research, and increase the likelihood of developing commercial products within a shorter time horizon. Second, even without consolidation, companies engaged in nanoscale research would benefit from standardization. Standards are currently needed for consistent characterization of nanomaterials and will be required for the future interoperability of devices and systems.

CONSOLIDATION

The Case for Consolidation

When particular sectors have become "hot," large numbers of start-ups with similar business models spring up and are funded.[2] For example, during the biotech boom of the early 1990s, several dozen firms adopting similar approaches to tackling the same types of diseases received financing.[3] Similarly, when the Internet and telecommunications sectors sizzled between 1998 and 2000, venture capitalists funded a large number of similar firms. At one point, there were nine different Internet pet food suppliers.[4]

As nanotechnology becomes "the next big thing," many companies are likely to jump on the bandwagon and rush to bring products to market. Already, several groups of companies are doing research in similar areas. For example, at the time of this writing, there are a variety of companies working with carbon nanotubes. Generally, these companies are focused on developing a process for large-scale production of nanotubes or constructing materials and devices such as composites, sensors, and field emitters. Table 15.1 provides examples of start-ups competing to develop carbon nanotube technology.

Generally, competition between firms fosters technological development. Under the threat of competition, different companies pursue different strategies, different researchers come across independent discoveries, and employees work hard.[5] In some circumstances, however, a large number of similar start-ups might hurt development of the sector. For example, most of the similar biotech start-ups financed during the roar of the early 1990s "yielded very disappointing returns for their venture financiers and modest gains for society as a whole."[6] According to one scholar, competition between these start-ups stifled development of the sector due to "highly duplicative research agendas, intense bidding wars for scientific and technical talent culminating with frequent defections from firm to firm, costly litigation concerning intellectual property rights and misappropriation of ideas across firms, and sudden termination of funding for many of these concerns."[7]

Similar to what happened in the biotech sector, in some instances, it is possible that having too many small nanotechnology companies could be antithetical to the development of the industry. First, it is arguable that similar research conducted by competing nanotechnology companies is inefficient.[8] For example, several different firms researching and developing processes for synthesizing carbon nanotubes could be wasteful since a single firm might succeed in developing an effective process by itself. In this case, overlapping development money might have been better used in a number of ways such as developing complementary technologies.

Second, having too many small companies competing for the same customers may deter large investments necessary to bring nanotechnology products to market. The research and development costs for nanoscience endeavors are very high, as are the resources required to commercialize products. When many firms pursue similar strategies, there is a risk that none will obtain sufficient funding to develop and commercialize products. Instead of capital being devoted to a single company, it is spread out over several companies. In various markets, no single firm can obtain the economies of scale necessary for efficient R&D and mass production.[9] Further, when

TABLE 15.1 Companies Developing Carbon Nanotube Technology

Company	Technology
Applied Nanotechnologies	Nanotube-based field emission x-rays
	Nanotube-based microwave amplifiers
	Nanotube-based gas discharge tubes
Applied Nanotech	Carbon nanotube cathodes for television, new lighting devices, x-ray, and microwave generators
	Carbon nanotubes for sensors
Bucky USA	Carbon nanotube production
CarboLex	Carbon nanotube production
Carbon Nanotechnologies	Carbon nanotube production
Carbon Solutions	Carbon nanotube production
Frontier Carbon	Carbon nanotube production
Hyperion Catalysis	Carbon nanotube production
	Nanotube composites
Iljin Nanotech	Carbon nanotube production
Materials and Electrochemical Research	Carbon nanotube production
	Nanotube composites
	Energy conversion systems
Microtechnano	Carbon nanotube production
Molecular Nanosystems	Carbon nanotube production
	Nanotubes for AFM tips
	Nanotubes for field emission devices
	Nanotubes for sensors
Nanocarblab	Carbon nanotube production
Nanocyl	Carbon nanotube production
Nano Lab	Carbon nanotube production
	Field emission
	Photonic devices
Nanoledge	Carbon nanotube production
	Nanotube composites
Nanomix	Nanotube sensors
	Field emission
	Hydrogen storage
Pyrograf Products	Carbon nanotube production
Rosseter Holdings	Carbon nanotube production
	Field emission
	Hydrogen storage
Southwest NanoTechnologies	Carbon nanotube production
Sun Nanotech	Carbon nanotube production
	Hydrogen storage
	Nanotube composites

competing start-ups seek future financing, the large number of other companies working on the same product may deter investors from making additional investments.[10]

The need to exceed a minimum size threshold is also important when it comes to finding customers. The confusion caused by having many small entrants means customers will delay purchases. This is especially true for certain nanotechnologies, where switching costs for customers may be exceedingly high. For example, building a new semiconductor fabrication facility for nanoimprint lithography would cost billions of dollars for a prospective customer. It is a difficult sale to begin with, and having ten different companies all with variations on this theme would make it unlikely that the customer would choose one. Customers would be more likely to adopt a "wait and see" approach, delaying the adoption of new nanotechnologies.

Finally, the most harmful impact of a fragmented landscape of nanotechnology companies could be the resulting intellectual property squabbles. As we argued in Chapter 5, the compulsion to patent anything and everything has swept nanotechnology scientists and entrepreneurs. Hundreds of broad, overlapping patents on the tools and processes for synthesizing, manipulating, and characterizing nanostructures are held by different universities, large corporations, and start-ups. In order for a start-up to bring a product to market, it might need a handful of different licenses from different entities. Start-ups competing to bring similar products to market might each hold patents needed by the other. If one company is likely to win the race to launch a product, the other company might waive its blocking patents to hinder or delay the introduction of the product. Patent litigation can be devastating for fledgling start-up companies. A determination of patent infringement can result in a start-up being enjoined from commercializing its product. Even if a start-up can prevail in court, litigation can be extremely costly and distracting. Some start-ups may be fortunate enough to avoid litigation if other patent holders are willing to license their technology or enter into cross-licensing agreements. Nevertheless, start-ups can still be financially drowned by a range of different patent holders who demand excessive royalties in licensing negotiations. Finally, intellectual property disputes can scare away potential investors or ruin an acquisition or initial public offering.[11]

In addition to start-ups waging intellectual property (IP) wars on each other, a large number of start-ups competing against each other may provide additional advantage to entrenched corporations. For example, if a start-up is attempting to develop a nanotube device, it may need to obtain a license from IBM.[12] Even if a start-up holds intellectual property that IBM needs, the computer systems giant is unlikely to enter into a cross-licensing agreement. Since protracted patent litigation can ruin a start-up without having any impact on a large corporation, the latter's threat of litigation can shut down the start-up or dictate very favorable terms for itself in an acquisition. Indeed, established corporations seeking to prevent start-up companies from disrupting their market dominance in nanotechnology markets should perceive the fragmented landscape as a favorable condition. They can slowly starve their start-up competitors into extinction by waging a protracted battle on the IP front.

Rather than attempting to overcome the technical, market, and legal roadblocks as separate entities, in some cases, investors and managers might consider combining their intellectual and financial strengths.

Issues Involved in Mergers Between Nanotech Start-ups

Corporate consolidation can take place through mergers and acquisitions. A merger "collapses two corporations into one, which is either the surviving corporation, or a new, 'resulting' corporation."[13] By contrast, an acquisition is the unification of two firms by one of several "non-merger" techniques, such as the purchase of the assets or shares of one firm by another.[14] In both mergers and acquisitions, the acquiring firm can use cash, its own stock, or any other agreed-upon form of consideration as payment.[15]

Consolidation of nanotechnology start-ups should take place through stock-for-stock mergers, with the formation of new companies. Mergers provide the most favorable tax treatment and the simplest means of obtaining necessary shareholder approval.[16] Mergers between private companies are becoming more common.[17] There is little literature, however, on private-private mergers. The mergers of nanotechnology start-ups could be extremely difficult to coordinate and consummate for a variety of reasons. As one experienced Silicon Valley deals lawyer declares, "These are hard deals to get done."[18] There are three primary issues that must be fulfilled in order for a merger between nanotech start-ups to succeed. First, the merging parties must agree to core terms of the deal. Second, the key players at each company must agree to the merger. Third, the merger must not give rise to antitrust scrutiny.

Agreement Between Merging Companies

In order for a merger between two nanotech companies to take place, the investors with substantial ownership and control of each company must agree to the deal. At the outset, it must be noted that each merger presents a unique set of circumstances and issues. In some instances, neither company will have accepted venture capital financing, and the relevant parties will be the founders and managers of each company. In other instances, two companies, each having six different investors and two rounds of financing, may seek to merge. The level of complexity greatly increases when there are a larger number of investors. In any case, there are several issues on which both sides must agree.

First, the merging companies must agree on relative valuations. Because a higher relative valuation for one company entails a lesser ownership stake for the shareholders in the other company,[19] negotiations to determine agreeable valuations can be contentious. When publicly traded companies merge, the market provides a measure of the value of both companies.[20] When a publicly traded company acquires a private company, the value of the acquirer is known, and the negotiation focuses on the valuation of the target company. When two private companies merge, however, there is no market to measure the value of either company. Parties must negotiate the value of both companies.[21] In Chapter 11, we discussed why valuing nanotechnology start-ups is a difficult and subjective process. In addition to the reasons provided in that discussion, valuations in the context of mergers are further complicated by the fact that parties may be reluctant to fully disclose all relevant information necessary for accurate valuations in merger negotiations. If two competitors are contemplating a merger, each will be reluctant to disclose trade secrets relating to product development or business strategy.[22]

When one or both companies have venture backing, the valuation process is further complicated by antidilution protections. Antidilution protections are triggered when a company sells shares of stock at a price per share less than that paid by the preceding venture investor.[23] Although there are several different types of antidilution protections, they generally involve an adjustment of the conversion price of the preferred stock into common stock. If the exchange value at which the merger takes place is less than the price of preferred stock purchased by earlier investors, the antidilution protections will go into effect. The resulting change in the number of shares of common stock into which each share of preferred stock is convertible can complicate the deal "so as to make it impossible."[24] Thus, valuations must be large enough to avoid triggering antidilution protections.

Even if two companies can agree on valuations, a host of other ownership issues must be resolved before the conclusion of such deals. Valuation of each company does not necessarily provide a method for determining ownership stakes in the newly merged entity. The parties must agree on how different types of stock will be woven together into the capital structure of the new company. For example, if a company has received venture financing, it probably issued convertible preferred stock. As explained in Chapter 12, convertible preferred stock is voting stock that provides its holders with certain preferences and privileges that are superior to the rights afforded holders of common stock. The specific terms that govern venture investments differ from company to company and from venture fund to venture fund.

To understand how different rights accorded to different preferred stock can complicate the process of determining ownership, consider the following scenario. Venture Capital Fund A invested in Company A, and Venture Capital Fund B invested in Company B. Each company has the same valuation and relative amounts of common and preferred stock. At first glance, it appears that an acceptable merger would provide the venture capitalists (VCs) with equal amounts of preferred stock in the new company. But a deeper analysis of the terms of the preferred stock might reveal that one of the venture funds is entitled to a greater ownership stake in the new company. If the rights provided by the preferred stock held by Venture Capital Fund A were far superior to the rights provided by the preferred held by Venture Capital Fund B, Fund A might argue that its stock is more valuable than the stock held by Fund B. As such, the value of Fund B's stock should be discounted in determining the relative ownership of the new company.[25] Because there is no formula for comparing relative values of preferred stock, VCs must arbitrarily establish their relative ownership ratios of the new company.

The task of determining ownership of the merged entity becomes even more arduous when the merging companies have different valuations, different rounds of financing by different venture capital funds, different amounts of issued stock, and different ownership structures. Further, the consolidations may be accompanied by a new round of financing, something that raises a host of other complicated issues. Because investors are struggling to maximize their ownership, determining who owns what percentage of the new company can be extremely difficult.[26]

Even if two companies can agree on the relative ownership of the new company, the marriage of the companies is still a long way off. Companies must also agree to other critical issues such as the terms of the new stock, the employee rights, the new management, and the composition of the board of directors. Intractable disagreement on any of these issues can frustrate the merger.

Despite these apparent complexities, however, skilled venture capitalists and deal lawyers can conclude a "nano deal." When companies have received venture financing, the valuations used in the previous rounds of financing can be an approximate starting point for determining valuations. Further, valuations for merger purposes need not reflect the actual economic values of the companies—they are only necessary to establish an exchange value for two companies. Skilled negotiators should "be able to squint and agree on some overall relative valuations."[27] Willing parties should also be able to negotiate a satisfactory capital structure and agree on other relevant terms.

Agreement within Companies

Perhaps a more difficult challenge facing companies in facilitating consolidation in nanotechnology is enlisting the support of all the key employees at each company. These include founders, managers, and other employees whose technical expertise is critical to R&D and product development. Obtaining the support of all key employees is critical for two reasons. First, the merger agreement must be approved by the vote of a majority of all issued and outstanding stock entitled to vote on the merger.[28] Thus, an unwilling founder might be able to exercise his or her voting power to veto the merger. Second, even if a merger could be initiated, the merged entity could fail without the support of a single key employee. Many nanotechnology start-ups were founded by pioneering scientists in the field. The continued participation of a founder with the technical expertise and human capital needed to transform the technology into marketable products could be critical to success of the enterprise after the merger.

However, key employees might resist consolidation for a number of reasons. First, a number of studies demonstrate that scientists are overly optimistic about the likelihood of success of their personal projects.[29] They are likely to reject the notion that merging with another outfit would increase the likelihood of successfully bringing a product to market. Second, even if scientists recognize the benefits of collaboration, they are likely to oppose the changes that might accompany the merger. Changes in company name, location, and management are not likely to be well received by strong-willed founders and key employees. Third, key employees are likely to cringe at changes in the ownership structure. Not only might they perceive the deal as severely diluting their ownership, but they are also likely to be highly suspicious of all the complexities associated with the resulting changes. In these cases, discussions of mergers are likely to be greeted with contempt and suspicion.

Proponents of the merger might employ several tools to convince key employees of the need for consolidation. As part of the consolidation agreement, cash or stock bonus plans can provide incentives for key employees to support the deal.[30] When venture capitalists are pushing for consolidation, they can rely on their most powerful financial weapon: further financing. When start-ups begin to need a new infusion of capital, VCs can offer financing on the condition of consolidation.

Regulatory Concerns

Even if all relevant parties agree to consolidation, regulatory barriers might frustrate mergers between nanotechnology start-ups. Section 7 of the Clayton Act prohibits mergers that substantially lessen competition or tend to create a monopoly.[31] A merger that threatens competition in the form of higher prices or reduced output should be found unlawful.[32] Enforcement authorities will challenge mergers under Section 7 if they reduce competition in "innovation markets." An "innovation market" is defined as the research and development directed to particular new or improved goods or processes, and the close substitutes for that research and development.[33] Close substitutes are those R&D efforts, technologies, and goods that would "significantly constrain the exercise of market power with respect to the relevant research and development, for example by limiting the ability and incentive of a hypothetical monopolist to retard the pace of that research and development."[34] Thus, if a nanotechnology merger restricts the number of competitors in a particular innovation market without promoting innovation, the merger might be prohibited.

The determination of whether a merger should be regulated depends on how the innovation market is defined. Nearly every case involving innovation markets has been in biotechnology or the pharmaceutical industry.[35] An example of how the innovation market is defined is the Federal Trade Commission (FTC)'s review of the merger between CibaoGeigy and Sandoz to form Novartis. The FTC argued that the merged firm's combined patent protection on upstream gene therapy procedures could be used to prevent potential competitors from producing downstream products.[36] Although it ultimately approved the merger, the consent order required the licensing of certain key patents.[37]

There are several reasons why mergers between nanotechnology start-ups should not trigger antitrust scrutiny. The analysis of a merger of two or three companies that develop nanotube-based devices is illustrative. First, as argued earlier, consolidation might substantially increase the efficiency of nanotechnology research and development and the likelihood of developing downstream products. The efficiency benefits of the merger would arguably outweigh the anticompetitive effects.[38] Second, it is likely that the merger would still allow competition between different companies using different techniques to develop nanotube devices. Third, even if consolidation restricts competitors in companies working with carbon nanotubes, it is arguable that nanotubes are competing with other nanomaterials to become a reliable "building block" for devices and materials. In other words, the innovation market is the entire R&D on the "building blocks" and not simply the R&D on carbon nanotubes. Thus, companies focused on nanotube synthesis are competing with companies working with nanowires, quantum dots, semiconducting polymers, and other nanomaterials. In the final analysis, antitrust authorities should not object to most nanotechnology mergers.

STANDARDIZATION

Even if companies do not pursue horizontal consolidation, they should explore industrial standards. Standards in different areas of nanotechnology could be crucial to commercializing materials, devices, and systems. We provide an overview of standards, explain why they are necessary in nanotechnology, summarize efforts taken to develop nanotechnology standards, and identify legal issues that could arise in developing standards.

Defining Standards

Standards are defined as "documented agreement containing technical specifications or other precise criteria, to be used consistently as rules, guidelines, or definitions of characteristics, to ensure that materials, products, processes and services are fit for the purpose."[39] There are different types of standards. Some standards allow different players to have consistent measurement and characterization techniques. For example, there are standards on the measurement of various electromagnetic fields relevant to telecommunications. Other standards allow parties to develop technology and products that are interoperable, or compatible. For example, standards govern the size, shape, and pin assignment configuration of computer chips. While performance requirements and aspects of product designs are standardized, the hardware and software architectures of the products themselves are not standardized. Though they adhere to applicable standards, different manufacturers compete to differentiate their products—on the basis of, for example, price, quality, attributes, or added features—and to attract customers to their particular offerings.

Several benefits are derived from standards. By normalizing a technology and enabling economies of scale in production, they enhance the efficiency of labor, energy and materials in manufacturing. Standards also can support competition, simplify business processes, aid quality, and improve health, safety and the environment. Other benefits attributed to standards include:[40]

- Building confidence among end users that products are safe, reliable and perform as intended. They set consistent expectations and ensure that those expectations are met.
- Creating a common language so producers and users communicate on such issues as quality and safety.
- Promoting product compatibility and interoperability.
- Lowering trade barriers and opening global markets.
- Fostering the diffusion and adoption of new technology.

Generally, the private sector is responsible for developing and promulgating voluntary standards for adoption by industry at the national, regional, or international level. In contrast with most other nations, the voluntary standards system in the United States is highly decentralized and aligned with industrial sectors. Currently, there are more than 450 private-sector standards developing organizations ("SDOs") in the United States and about 150 consortia that also develop technical specifications. Approximately 20 SDOs develop about 90 percent of U.S. standards, many of which are used internationally.[41] A recent study identified 43 SDOs and consortia in the "telecommunications and computer-networking industries" alone.[42]

The competition involved in setting standards should not be surprising, given the high stakes that may be involved. Participants in limited-consensus standards efforts (i.e. those taking place in SDOs or consortia that may limit membership to a select group of companies or organizations) may be successful in crafting the standard in a way that unfairly confers advantages to particular companies, industries, or economies. Likewise, testing requirements and other specifications can be devised to deter competition from exporters by impeding access to markets. To address these potential barriers to trade, the Technical Barriers to Trade Agreement ("TBT") requires nations to give preference to international, full consensus standards as a basis for their technical regulations. In addition, the agreement encourages national and regional standards developers to defer to international standards in their activities.[43] In the United States, consensus based SDOs that are certified by the American National Standards Institute ("ANSI") may submit their standards to international standards organizations, such as the International Organization for Standardization ("ISO") and the International Electrotechnical Committee ("IEC"), for adoption. The IEEE is the first organization to sign a "Dual-Logo" agreement with the IEC, facilitating the adoption of existing IEEE standards by the IEC.

In support of U.S.-based standards organizations and their industrial customers, the National Institute of Standards and Technology ("NIST"), a U.S. Department of Commerce agency, provides specialized assistance, developing measurement methods, testing tools, calibrations, and other underpinning technical elements of standards.[44]

Anticipatory Standards Efforts in Other Fields

In the past, standards were primarily retrospective. They were chosen from among candidate solutions that had already been developed to resolve well-recognized problems and needs. A review of recent, successful standards efforts in other fields reveals that it is advantageous to promulgate "anticipatory" standards that precede the products and procedures whose performance they proscribe. Such forward-looking standards can facilitate the evolution of technologies, serving to "jump-start" an emerging field. When a standard is developed ahead of the release of new technology platforms, producers, users and the general public gain confidence in the new technology and rapid adoption is promoted. For example, the publication of IEEE 802.11b, or "Wi-Fi" as it is popularly known, demonstrated the disruptive potential of high-speed, wireless Internet to the telecommunications industry.[45] Because the IEEE 802 working group was able to reach consensus on the goals of the next IT generation and avoid the delay and market fragmentation competing technologies can create, the wireless industry has blossomed.

In contrast, some experts have argued that the microelectromechanical systems industry ("MEMS") industry has struggled because of a lack of anticipatory standards. The director of research and development at one company argues that the lack of standards is "considered the number-one barrier to greater acceptance of MEMS technology."[46]

The Need for Anticipatory Standards in Nanotechnology

In the same way that implementation of an anticipatory standards strategy accelerated the adoption of and broadened the market for a new generation of wireless devices, so can such a strategy significantly advance the adoption of nanotechnology-based products. One of the difficulties facing the successful commercialization of nanotechnology is the ability to consistently characterize the properties and benefits of the basic building blocks that underpin much of the theoretical work; different researchers use different tools for the characterization of nanomaterials and methods of reporting performance and other data. For example, in 2002 and 2003, a number of researchers argued that a quantum phenomenon known as ballistic magnetoresistance was responsible for substantial increases in electrical resistance in ultra-small magnetic sensors. Several studies sparked speculation that storage capacity might be increased by a factor of a thousand or more. After conducting its own studies, NIST maintained that the changes in electrical resistance were unrelated to the quantum effect. NIST researchers concluded that the observed electrical properties were caused by the shortening of wires, which severely distorted the cluster of atoms that formed the nanometer-scale contact between them. According to NIST research chemist William Egelhoff Jr., "At this point, it is inconclusive as to whether a real BMR effect will be found and, if so, whether it will prove large enough to be of much interest to the magnetic-sensor community."[47] Indeed, agreement on the more advanced properties of nanomaterials is impossible until fundamental consensus on the underlying structures is achieved.

It is this inability to identify standard measurement and characterization methodologies for nanomaterials and nanodevices that may signal that nanotechnology development is still more weighted towards science rather than commercialization. This may result in companies experiencing difficulties selling their products to customers and raising capital from investors. Further, when system architectures based on nanotechnology come to market in the more distant future, standards will be needed to ensure inter-operability. Without anticipatory standards, integration and large-scale manufacturing could remain elusive. By taking steps to develop measurement and characterization standards now, scientists and engineers involved in nanoscale research will be able to create a framework that will establish the foundation for future integrated nanotechnology products.

Nanotechnology Standards Efforts

Efforts to develop standards that address various nanotechnology-related needs are under way at a growing number of organizations, including the Institute of Electrical and Electronics Engineers ("IEEE"), American Society For Mechanical Engineers ("ASME"), and European Committee for Standardization ("CEN"). In China, the National Accreditation Board of Laboratories has reportedly established a committee for accrediting nanotechnology-related R&D laboratories and inspection bodies.[48] Additionally, many of the world's national metrology institutes, such as NIST, are focusing on measurement and instrumentation needs. In this section, we focus on efforts underway at IEEE and NIST.

The IEEE

In fall 2003, the IEEE Standards Association, an organization responsible for standards in technologies ranging from telecommunications to computer software, launched the Nanotechnology Standards Initiative.[49] The standards are intended "to facilitate the movement of nanotechnology innovations from a research to a market environment and to establish fundamental nanotechnology platforms that support accelerated growth of the sector."[50] Generally, the initiative will seek to identify:

- Nanoelectronic technologies likely to generate products and services having high commercial and/or societal value
- Areas where new standards can aid rapid commercialization, technology transfer and diffusion into the market
- People and institutions to lead and support IEEE nanotechnology standards projects

The first project promulgated by the initiative is IEEE P1650,™s titled "Draft Test Methods for Measurement of Electrical Properties of Carbon Nanotubes."[51] The project aims to develop a standard to define electrical testing procedures and suggest characterization tools for carbon nanotubes. Several IEEE Study Groups are being created to identify additional areas for nanotechnology standards development: electrical characterization of nanoscale materials for fuel cells and interoperability for nanotube devices to silicon devices. This activity is part of a broader nanotechnology effort at the IEEE driven by the IEEE Nanotechnology Council (NTC), an interdisciplinary group whose members are drawn from 19 IEEE Societies.

Research Efforts at NIST

NIST will play a crucial role in developing measurements and data for nanotech standards efforts. One of the "Grand Challenges" outlined in the U.S. National Nanotechnology Initiative emphasizes measurement and instrumentation needs. In 2002, NIST funding for nanotechnology-related projects in the NIST laboratories totaled almost $40 million.[52] Nanoscale research at NIST cuts across five important areas: (1) fundamental science and basic measurement capabilities; (2) characterization of nanostructured materials; (3) nanoscale electronics, optoelectronics, and magnetics; (4) nanochemistry and nanobiotechnology; and (5) quantum computing and communications. NIST scientists and engineers have already made a number of pioneering breakthroughs in tools and processes and will be useful in developing standards in nanotechnology. The pace of progress should increase with the opening of NIST's Advanced Measurement Laboratory ("AML") in June 2004. Achieving an unprecedented combination of temperature, vibration, and other environmental controls, the AML's almost "noise-free" conditions will enable NIST scientists and their collaborators to focus directly on key nanotechnology challenges, such as teasing out cause and effect, definitively linking structure and function, or simultaneously achieving high levels of specificity and resolution in chemical analyses. Of the more than 100 projects to be housed in the new facility, many will concentrate on furthering efforts to explore, measure, rearrange, formulate, or fabricate at the nanoscale. The results will translate into clearer guidance on the road to commercialization of nanotechnology.

Issues in Setting Nanotechnology Standards

IEEE, NIST, and the nanotechnology community will encounter several challenges in developing nanotechnology standards. The experience of other standards efforts, such as IEEE 802, can shed light on important issues for those involved in nanotechnology standards.

First, those involved must ensure that the standards-setting process is democratic. The most successful standards efforts garner input from a broad range of interests and represent both minority views and the views of end users. The standards are based on a consensus process that distills agreement from a broad range of constituencies; often balancing the interests of industry, government, consumers, trade unions, environmental groups and others. IEEE working groups are open to anyone to participate—participants do not have to be IEEE members. Nanotechnology meetings have attracted a diverse group of representatives from academia, government labs, start-ups, and large corporations.

In some cases, however, nanotechnology players may have an incentive to exclude potential competitors from the process. Participation in standards efforts can provide companies with competitive advantages through early access to information and know-how. Participants can also have an impact on the content of the standards and are likely to favor standards based on their technology.

IEEE 802 was successful in preventing standards from becoming competitive weapons by adopting one of two approaches. If the technological approaches are similar enough, the Committee seeks to foster compromise. If, on the other hand, technological approaches are so different that compromise is unlikely, it allows competing standards to be created so the market can choose between them.

Second, in the more distant future, standards on devices and systems at the nanoscale are likely to read on the large number of patents being filed. IEEE standards may include the use of patents, including patent applications, provided the IEEE receives assurance from the patent holder or applicant with respect to patents essential for compliance with both mandatory and optional portions of the standard.[53] When this happens, there could be fierce disagreements over what standards should be chosen. Additionally, there could be disputes over licensing terms for patents that claim technology embodied in the standards. The IEEE requires that licenses will be made available without compensation or under reasonable rates, with reasonable terms and conditions that are demonstrably free of any unfair discrimination. However, determining what royalties are appropriate for patent holders whose technology is used is an extremely difficult task.

In addition to disputes over patents and licensing rates, when standards read on patents, there is also the risk of opportunistic behavior. The story of Rambus, a company developing memory chips, is illustrative. Rambus participated as a member in JEDEC, an industry standard setting organization, without disclosing that it was actively developing patent rights with claims encompassing technologies adopted in the standards. The company filed amendments, divisionals, and continuation applications to ensure the claims covered technology in the standards. When the standards became widely adopted by the DRAM industry. Rambus enforced such patents worldwide against companies manufacturing memory products in compliance with the standards.

The nanotechnology standards community will also need to carefully choose which standards projects it devotes times and resources to. On the one hand, it is important to pay attention to future needs. Part of the reason IEEE 802 is so effective is that working groups rapidly reset goals for the next generation of technology, which allowed the group to address issues that would impede growth, such as security in LANs. These goals pave the path for new working groups, which form the foundation for a successful standards effort. On the other hand, strict scrutiny should be applied to the selection of new working groups. Any new standard must have broad market potential, be technologically and economically feasible, not duplicate other standards, and provide a unique solution.

Another potential issue that could impact the development and acceptance of nanotechnology standards is international harmonization. The EU nations might view ISO as a unique source of international standards, whereas U.S.-based organizations, such as IEEE and NIST, may not necessarily agree. These latter organizations believe that international standards are preferable to a bevy of conflicting or duplicative nationally or regionally promulgated standards. This situation could give rise to nations jockeying to gain some standards-related advantage for their nanotechnology industries in the making.

Finally, an important activity for the working group focused on nanotechnology electronics standards involves coordination with the Compact Model Council ("CMC"), which prepared the current silicon design software. The CMC includes semiconductor and electronic design automation vendors and promotes standards for compact model formulations that aid the diffusion of next-generation technologies. The CMC has had a great deal of experience in fostering the market diffusion of silicon-based technology and will be a useful guide in helping the working group create nanotechnology-based electronics standard.

CHAPTER 16

Exit Opportunities

> *The IPO market is like the weather: very hard to predict and constantly changing. When times are good, the sun shines and seems that it will never end. Yet when the clouds roll in, we wonder if the sun will ever come back. Giant waves of good IPO weather are driven by industry upheaval, yet some academic studies show those who sell on the first day do better than everyone else. What in the heck can we do about this situation that seems almost too wild even for Wall Street?*[1]
>
> —James E. Schrager (Chicago Graduate School of Business)

As a company matures, its founders and investors may want to sell some or all of their interest in the business through an initial public offering (IPO), or sale to another company. In addition to providing capital, going public or getting acquired can also serve as an "exit opportunity," enabling equity holders to profit by selling their interest in the business. There is no simple formula to determine exactly when investors and founders should pursue exit opportunities. Entrepreneurs must instead consider many different benefits and challenges involved in exit. Usually, exit opportunities are only available to companies that have met a minimum threshold of growth and financial stability.[2] At the time of this writing, however, nanotechnology is so new that it is unclear what exit opportunities will be available to companies. Early signals suggest that investors will respond favorably to nanotech companies and grant generous valuations. This chapter will describe what entrepreneurs can do to increase the likelihood of a successful exit.

We first provide an overview of the most common transactions used by maturing companies to raise money. We discuss the advantages and disadvantages of debt, IPOs, and acquisitions. Next, we describe what nanotech companies should expect to endure on the path to exit. Finally, we focus on how nanotech companies can best position themselves for acquisitions and initial public offerings. Specifically, we make predictions about what types of companies will be eligible for such transactions and describe the deal process for each transaction.

MECHANISMS THAT RAISE MONEY

There are a variety of different mechanisms that maturing companies can use to raise capital. It is beyond the scope of this chapter to explore all of the different financing options available.[3] The most common transactions available to growth firms, such as nanotech start-ups, are debt or bond offerings, initial public offerings, and mergers and acquisitions.[4] (See Table 16.1.) Unlike debt, an acquisition or an initial public offering can serve as an opportunity for entrepreneurs and investors to realize some return on their investment. Each of these alternatives is described below.

TABLE 16.1 Summary of Different Financing Mechanisms Available

Transaction	Description
Initial Public Offering (IPO)	The sale of company stock in public equity markets. An IPO provides an opportunity for the company to raise money in return for giving up some ownership in the firm. The primary advantage of an IPO is the ability to provide liquidity to investors. An IPO also carries a number of requirements such as enhanced regulatory scrutiny, giving up equity in the company, and underwriting fees.
Bond Offering	Debt issued for a period of more than one year. Public and private organizations sell bonds. Like any loan, a bond is an agreement to repay the loan at a specified time with specified interest.
Convertible Debt	General debt obligation of a corporation that can be exchanged for a set number of common shares of the issuing corporation at a prestated conversion price.
Mergers and Acquisitions (M&A)	A combination of two companies. The transaction can take many forms including one company purchasing all of the assets and liabilities of the other, or a combination of the assets of the two companies.

Debt

Debt is a common mechanism by which a company can raise additional capital. Debt and bond offerings enable companies to raise funds without giving up control of their ownership of the business. They may also provide certain tax advantages above equity offerings.[5] For high-growth companies, debt may entail a significantly lower cost of capital than equity. For example, during the Internet boom, debt was far cheaper than equity.[6] However, the exact cost of debt for any individual firm depends on the probability of company default.[7] Cash-strapped companies, such as early-stage nanotech companies, are more likely to default and will have a higher cost of debt than more financially sound companies. Certain forms of debt are designed to meet short-term capital constraints and help companies survive until they can sell, go public, or otherwise raise more funds. For example, a "convertible bond" (also known as "bridge financing") is a coupon-paying bond that can be converted into stock at predetermined terms at the discretion of the holder.[8] Generally, convertible debt is used as a last resort and carries the risk that investors will perceive the company as weak. Investors in convertible rounds usually have already invested in the company and are enabling the company to maintain solvency while seeking longer-term sources of capital.[9]

Initial Public Offering

An initial public offering (IPO) is the sale of company stock in public equity markets. An IPO provides an opportunity for the company to raise money in return for giving up some ownership in the firm. An IPO might offer a number of advantages above an all-out sale of the company.[10] First, an IPO provides greater flexibility than a sale. Management can determine how much equity they would like to sell. IPOs for early-stage companies usually involve a sale of 15 to 20 percent of the equity of the company.[11]

Second, there may be higher technology valuations in public equity markets than in acquisitions.[12] At the time of this writing, euphoria for nanotechnology is beginning to grip investors. As illustrated by the stock rallies of companies with the terms "nano" in their names in late 2003 and early 2004 (Table 10.5), investor excitement could translate into high valuations for start-ups seeking to go public. Large acquirers may be unwilling to pay the same valuation premiums that equity market investors will pay.

Third, an IPO allows management to maintain control of the company.[13] If the company is sold, positions can change and certain employees may be eliminated. In an acquisition, employees from the acquiring company will often assume the most important positions in the company. Finally, there are customer considerations. Customers feel more confident in a public company since it it must disclose information about its operations through Securities and Exchange Commission filing requirements.[14] Customers could also become suspicious of the company if it is purchased by one of its competitors. An IPO may avoid such issues of ownership.

Mergers and Acquisitions

Mergers and acquisitions (M&A) involve the combination of two companies. In Chapter 15, we discussed legal and business issues associated with mergers of two nanotech start-ups. In this chapter, we focus on start-up companies that are purchased by large corporations.

There are several reasons why a start-up might prefer an outright sale to an IPO. First, a sale often provides insiders with more liquidity than an IPO. The Securities and Exchange Act of 1933 prohibits company insiders from selling stock for at least 180 days after a company's IPO.[15] In an acquisition, shareholders can obtain substantial liquidity if they receive cash or freely tradable stock. Additionally, complying with the numerous regulatory and financial disclosure hurdles can make an IPO an arduous and lengthy process.[16] Finally, for small companies, an IPO might result in the sale of only a few shares due to scant market trading.

Second, there are privacy considerations for managers and investors. Companies filing IPOs are subject to numerous financial and company disclosure requirements. In an acquisition by another private company, there are virtually no reporting requirements. In an acquisition by a public company, certain disclosures may be necessary, but they are far less significant than those required to make a public offering.

Third, an acquisition may provide greater immunity from market instability. While IPOs can be threatened by general changes in the market, acquirers with an interest in the technology or a belief in the value of the business may not care that the broader equity markets are slowing down. In fact, many studies have indicated that there is an increase in acquisition activity when equity markets experience downturns.[17]

A fourth consideration is company credibility. For nanotech companies, it is likely that acquisition by a larger company will add credibility to the technology and a belief in the long-term sustainability of the business.

An acquisition may also be more efficient for product development, marketing, and distribution. As we argued in Chapter 15, an acquisition can bring together two former competitors and eliminate the many costs involved in competition. These include costs of duplicative research and development as well as reductions in product price brought on by competition. Further, nanotech companies seeking IPO must establish manufacturing and distribution capabilities, financing mechanisms, and effective administration and management. The process of developing the necessary infrastructure can take a long time and cause substantial dilution. Selling to a larger organization may eliminate the need to go out and build these capabilities.

Finally, the process of going public is costly.[18] There are legal expenses associated with registering the necessary forms with the Securities and Exchange

Commission (SEC). Investment banking fees consume a significant portion of the capital raised by the IPO. Finally, under the Sarbanes-Oxley Act of 2002 and the SEC rules that have been promulgated in its wake, it is becoming more costly to take a company public.[19] This is because the process of preparing to go public is increasingly arduous and lengthy. Public companies must maintain a majority of independent directors, and executives must certify the company's financial statements. Recruiting the necessary number of independent directors and managers skilled enough to certify the financial statements can take a great deal of time. Further, once a company does go public, it must comply with myriad new regulations such as the composition and qualification requirements for audit committees, limitations on certain services by companies' auditors, prohibitions on personal loans to executives, implementation of codes of conduct, and a variety of disclosure requirements.

THE PATH TO EXIT

The path to acquisition or IPO can be long and difficult. Most companies have to go through several rounds of financing over the course of a decade before an exit opportunity arises. The company may have to change its strategy and R&D focus several times. It is also likely to experience changes in the composition of the board and management. The firm that is ultimately sold might look nothing like the original company. Throughout the journey, nanotech companies should make efforts to keep potential investors updated. For example, Nanosys has developed large databases of people from industry, investment communities, and the media. The company's management sends out emails announcing research breakthroughs and intellectual property licensing agreements. Their communication efforts have made Nanosys a popular topic in casual discussions among public equity investors. While many of these investors have little understanding of the technology being developed at Nanosys, they are more likely to purchase securities of Nanosys when it makes an offering than other similar nanotech start-ups that are less visible.

GETTING ACQUIRED

At the time of this writing, little M&A activity has taken place in nanotechnology. As shown in Table 16.2, the limited activity that has occurred has largely been confined to the market for nanotechnology "tools." As explained in Chapters 1 and 2, these companies develop tools for synthesizing, positioning, characterizing, and manipulating nanomaterials. For example, Veeco Instruments, Inc. has acquired over a dozen start-up companies in the last six years. These tools companies are purchased for anywhere from $16 million to several hundred million dollars.[20]

TABLE 16.2 Examples of Nanotech Acquisitions

Acquiror	Acquiree	Technology	Date	Price ($)
VEECO	Applied Epi	Molecular Beam Epitaxy	9/17/01	136M; 30M cash balance in stock
VEECO	Thermo Microscopes	Atomic Force / Scanning Probe Microscopes	7/16/01	Undisclosed
VEECO	CVC	Physical Vapor Deposition (PVD), Atomic Layer Deposition (ALD), E Beam, and Metal-Organic Chemical Vapor (MOCVD) Deposition Systems	5/5/00	370.8M in stock
VEECO	Advanced Imaging Systems	Micromachining tools for storage devices	11/19/03	60M in cash plus up to 9M over the next three years if certain performance goals are met
VEECO	IBM SXM	Atomic Force Microscopy	4/5/00	Purchase of certain assets (undisclosed amount)
VEECO	Monarch	Quasi-Static Testing Systems	1/28/00	15.8M in stock swap
VEECO	Optimag	Automated Optical Defect Inspection Systems	10/15/99	Undisclosed
VEECO	Ion Tech Inc.	Ion Beam Deposition Process Equipment	10/14/99	37M in stock

Acquiror	Acquiree	Technology	Date	Price ($)
VEECO	Digital Instruments	Atomic Force / Scanning Probe Microscopes	5/19/98	150M in stock
VEECO	MRC Process Equipment	Physical Vapor Deposition Systems	4/10/97	Purchase of certain assets (undisclosed amount)
VEECO	Wyko Optical Metrology	Laser Interferometers, Optical Profilers, and Inspection Systems	7/25/97	9M in stock (3M shares at mid-March 1997 prices)
VEECO	NanoDevices	AFM Probes	6/5/03	Undisclosed
Cabot	Superior MicroPowders	Proprietary nano- and micro-powder production systems	6/2/03	16M
Konarka Technologies	Quantum Solar Energy	Plastic-based solar cells	2/10/03	Undisclosed
Nanopowders Industries	Aveka	Metal nanoparticle products for electronics (inks and coatings)	11/02	No info
AMD	Coatue	Plastic-based memory	8/6/03	Undisclosed

Becoming an Acquisition Target

Unlike companies seeking to go public, which must be prepared to issue public financial statements and for whom profitability and revenue are generally very important qualities, companies interested in being acquired may have a different set of factors that determine their success. This is because acquisitions may be motivated by strategic rather than financial considerations. Large companies may seek to acquire technology that benefits, or directly competes with, their existing lines of business. These acquirers are less interested in direct financial return, but are focused on the strategic value of the underlying technology.[21] Consequently, the hurdles that companies face will be related more to technology due diligence than with financial due diligence involved in IPOs. The acquisition of Aveka by Nanopowders Industries represents an example of a strategic acquisition in which the primary consideration related to the value of the target's technology to the acquirer's product portfolio. Nanopowders, an Israeli technology company seeking to enter the American defense market, viewed Aveka's technology as useful in developing next-generation rocket fuel.[22]

Preparing for Acquisition

When management feels that the company is ready for an acquisition, it may consult an investment bank.[23] The bank will supply valuation estimates and provide schedules outlining the process of getting acquired. The bank will also help the company identify and contact potential acquirers. In Chapter 14, we discussed considerations that made large corporations a good strategic "fit" for partnering with a start-up. In many cases, strategic partners are the most likely candidates to acquire a company since they have a history of involvement with the company and a demonstrated prior interest in its development.

The Deal

Once a relationship with a potential acquirer is established, the parties must negotiate the terms and conditions of the deal. The process of concluding an agreement can take several months and generally involves negotiating a letter of intent, conducting due diligence, negotiating and signing the definitive agreements, and closing the transaction.[24] In this section, we provide a brief overview of how a deal is concluded and implemented.

First, the parties should agree to a letter of intent (LOI). The LOI is a short document that summarizes the form of the transaction, the valuation, the negotiating period, and other proposed provisions in the final deal.[25] As explained in Chapter 15, corporate consolidation can take place through mergers and acquisitions. A merger collapses two corporations into one, which is either the surviving corporation, or a new, resulting corporation. By contrast, an acquisition is the unification of two firms by one of several non-merger techniques, such as an asset purchase agreement or a stock purchase agreement. In both mergers and acquisitions, the acquirer can use cash, its own stock, or any other agreed-upon form of consideration as payment. There are different rules regarding shareholders' rights and assumption of liabilities for mergers,[26] asset purchase agreements,[27] and stock purchase agreements.[28] It should also be noted that there are a variety of tax considerations that often have a large influence on what transactional form is chosen.[29]

One of the most important issues in negotiating the letter of intent is determining valuation. We discussed valuation issues in nanotechnology in Chapter 11. Small companies should be ready for a barrage of questions designed to identify flaws in the business plan, technology, and strategy of the company. In addition to the fundamentals of the business, entrepreneurs should be aware of recent nanotech acquisitions to have a gage for determining if what is being offered is in line with market norms.

Conclusion of the letter of intent will be followed by an exclusive negotiating period to conduct due diligence and conclude a definitive agreement. Due diligence efforts will be focused on the following areas: technology, marketing, financial, and legal/intellectual property. The parties must then negotiate the details of the definitive agreements, which include the precise terms of the transaction, the break-up free,[30] representations and warranties,[31] and how labor issues will be addressed.[32] The company's board of directors must approve the definitive agreements before signing and closing.

Regulatory Issues

Acquisitions can invoke regulatory issues involving securities laws and antitrust laws. Generally, companies will not be forced to register their shares with the SEC when being acquired.[33] In Chapter 15, we summarized Section 7 of the Clayton Act, which prohibits mergers that substantially lessen competition or tend to create a monopoly. Acquisitions of nanotechnology companies should not trigger antitrust scrutiny. Most acquisitions will involve a large company acquiring a start-up that is competing with other start-ups. As such, the acquisition is not likely to give rise to monopolies over upstream technology.

INITIAL PUBLIC OFFERINGS

At the time of this writing, it is uncertain whether public markets will serve as desirable exit strategies for nanotech companies in the near future. As shown in Table 16.3, few nanotechnology start-ups have been able to successfully transform into a public company. Just as this book was going to press, Nanosys filed its IPO registration statement with the SEC. The unfolding of

TABLE 16.3 Examples of Nanotech IPOs

Company	Market	Technology	Price	Date
Nanophase	Nasdaq: NANX	Nanocrystalline materials / nanopowders	$8 per share; 4M shares; $32M raised; approx. 83% of total post-IPO market cap	11/26/97
Nanogen	Nasdaq: NGEN	Molecular analysis and assaying (genetics)	$11 per share; 3.6M shares; $40M raised; 20% of total post-IPO market cap	4/14/98
Acusphere	Nasdaq: ACUS	Drug delivery	$14 per share; 3.75M shares; $52.5M raised; 26.25% of total post-IPO market cap	10/8/03

this transaction in the coming months will have a major impact on the nanotech community. A successful IPO with a strong valuation could pave the path for other nanotech offerings in the near future while a poor valuation could turn cause Wall Street to shy away from nanotechnology.

Even if the market does open up in the near future, however, we don't expect IPO conditions similar to those in the late 1990s to reappear soon. This is in light of the backlash against Wall Street for behavior during the Internet boom. Investigations of major investment banks culminated in a $1.4 billion settlement and an agreement that firms would fund independent research for five years and separate analysts from underwriters.[34] Further, the Sarbanes-Oxley Act increases regulation of analysts as well as companies seeking to go public.[35] As such, Wall Street is likely to be much more careful in approaching nanotech IPOs. Unlike many Internet companies, nanotech companies going public may be required under Sarbanes-Oxley to demonstrate strong management, consistent growth, and future revenue projections. Most nanotechnology start-ups will not be able to boast of consistent growth and clear evidence of future revenue and growth within the next two to three years. Still, the possibility of a speculative bubble remains.

Nanosys' attempt to go public has unleashed a fierce debate in financial communities about what requirements should be met by nanotech companies seeking to go public. In a speech given at Stanford University, prominent venture capitalist Vinod Khosla, of Kleiner Perkins Caufield & Byers, stated:

Personally, I think it's the wrong model for a company, and I think it's a shame that they're going public because I don't think they are in a position to be predictable enough. And whether they are doing it knowingly or unknowingly, there is a reasonable likelihood that they will defraud the public market, and I think that's a shame.[36]

Nanosys does not have a specific product that that is generating substantial revenue and is unlikely to have one in the immediate future. Although Nanosys does have collaborations with Intel, DuPoint, Matsushita, and SAIC, the timing and ultimate success of product lines in each of these areas is uncertain. Nanosys maintains that the value of the IP portfolio furnishes a key consideration for the offering. Whether this model will succeed both in the near term and long term remains to be seen.

In the more distant future of five to seven years, nanotech companies should be more equipped to raise money through public offerings. Although we are reluctant to make predictions about such a dynamic and unpredictable sector, we will not be surprised if nanotechnology companies do begin to flood the public markets before the end of the decade.

Becoming an IPO Candidate

There are a number of factors that must be considered to determine whether an IPO is a viable alternative for a nanotech company. The characteristics of the industry, current cost of debt to private companies, the state of the market, the size, revenue, and the growth nature of the company are all significant factors in determining a company's suitability for IPO.[37] At the time of this writing, it is difficult to speculate about what will be required for a nanotech company to be eligible for a public offering. In general, a nanotech company pursuing an IPO should focus on three broad themes.

First, a company must strive to demonstrate consistent revenue and earnings growth. Earnings visibility is reflective for a mature company and provides investors with reassurance that management understands and is able to adjust to changes in its business. Very few early-stage and high-growth companies are able to deliver the visibility required to go public.

Second, a company seeking to go public should focus on large and diverse target markets. Nanophase (NANX: Nasdaq) is an excellent example of an early-stage nanotech company whose IPO was due in part to the success of the company in convincing investors that its technology addressed a large market need. The company produces engineered nanomaterial products for a wide variety of diverse markets: personal care, sunscreens, abrasion-resistant applications, environmental catalysts, antimicrobial products, and a variety of polishing applications, including semiconductors, hard disk drives, and optics. Similarly, Nanosys maintains that its semiconductor nanostructures have applications in the energy, defense, electronics, healthcare, and information technology industries. Just as investors flocked to stocks of Internet companies that promised to transform numerous existing industries, investors might be likely to buy stocks of nanotech companies with technology that can impact many markets.

Walking through an IPO: The Traditional Route

If a company opts for the traditional route to going public, an investment bank will guide it through the process and serve as an intermediary between the company and the public. While investment banks do not yet have significant practices dedicated to nanotech, they will likely be active players in the near future. Merrill Lynch has stated that it believes "nanotechnology could be the next growth innovation." Other banks are devoting resources to understanding nanotechnology and developing strategies for approaching it. As one financier states, "The markets are a strange beast, and if there is demand for nanotechnology companies in the public markets, one can be assured there will be bankers that are there to take firms public."[38]

The investment bank engages in a due diligence process and negotiates with the company the amount of capital to be raised, the type of security to be issued, the price of the security, and the cost of the issuance to the company.[39] There are two types of agreements between the bank and the company: a Firm Commitment and a Best Efforts agreement. A Firm Commitment is the bank's promise to purchase the entire issue from the corporation and sell the stock to the public. Because the bank must pay the corporation for the stock no matter what, the bank assumes the risks of the offering. In contrast, with a Best Efforts agreement, the bank promises to use its best efforts to sell the securities, but does not guarantee a sum of capital. The underwriter is compensated out of the proceeds of the offering. The corporation sells the stock to the underwriter, and the underwriter then sells the stock to brokers and the general pubic at a higher price. The difference between the price at which the stock is purchased by the underwriter and the price at which it is sold is the latter's compensation.

In many cases, several investment banks want to underwrite the same deal. When several banks wish to make Firm Commitments, they form a syndicate that participates in selling the issue. A syndicate manager is designated to enter into the agreement with the company, determine the offering price and effective date, and work with the SEC to ensure compliance with securities laws. When several banks want a Best Efforts agreement, they form a Selling Group. Each underwriter is allotted a share of the issue and must do its best to sell the share.

The process of going public is regulated by the Securities Act of 1933. The Act prohibits the offer to sell or buy any security before a registration statement has been filed. An "offer" can include anything that may contribute to public interest in the securities of an issuer such as any unusual publicity about the issuer's business or projections for the industry.[40] An issuer is permitted to publish a brief notice of a proposed offering prior to filing the registration statement.[41] The registration statement includes background information on the corporation and its officers and directors, comprehensive financial statements, how the proceeds of the offering will be used, and any legal proceedings involving the company. The filing of the registration statement is followed by a "cooling off" period.[42] During this period, the SEC reviews the company and the registration statement to ensure that full disclosure has been made. The only materials that can be distributed by the issuer relating to the offering are a preliminary prospectus[43] and the "tombstone ad."[44] The preliminary prospectus, also known as the red herring, generally contains the information found in the registration statement as well as more detailed market information, but does not contain the offering price or the effective date. The tombstone ad contains similar specified information. Management also stages "roadshows," which are meetings between senior management and investors to solicit interest in an offering. During this period, interested buyers cannot yet order stock but can express their interest in the stock. Once the SEC approves the issue, the stock can be sold to the general public. Prior to this "effective date," the company must meet with the investment bank to flesh out any remaining issues. On the effective date, the Final Prospectus is issued. The Final Prospectus, which includes the information contained in the red herring as well as the price and effective date, must accompany the sale of any securities.[45]

The Future: Issuance of Securities through Electronic Means

By the time the wave of nanotech companies begins to break on to the capital markets, the public offering of securities may primarily take place online. Electronic distribution of securities can enhance the traditional public offering by enabling underwriters to reach large numbers of potential investors. The Internet could even allow nanotech companies to circumvent traditional underwriters and make direct public offerings of securities.

Since the mid-1990s, the SEC has issued several rules and interpretive releases on application of federal securities laws to offering activities by electronic means.[46] The commission has approved of the issuance of securities through the Internet and crafted specific requirements for such issuances.[47] First, unlike the issuance of traditional securities, the issuer must obtain an investor's informed consent to receipt of the information through the Internet. Consent can be obtained through written communications, electronic communications, or by telephone. Second, the issuer must provide adequate and timely notice of the electronic information to the investor. Issuers can satisfy this requirement by directly emailing potential investors. Third, investors must have access to the information that is "comparable" to postal mail. Information must remain accessible and updated, and investors should be able to download the information. The SEC has approved the use of Portable

Document Format (PDF) to deliver documents. Finally, issuers must ensure that delivery of the information actually took place. Delivery confirmation can be achieved by "return receipt" systems, in which the e-mail recipient hits a reply button upon receipt of the electronic document. Other than the consent, notice, access, and delivery rules outlined by the commission, the process of conducting an Internet offering is similar to a traditional offering.

Traditional underwriters have begun to use the Internet as a means to sell stock.[48] In the future, the Internet could enable nanotech companies to make direct public offerings (DPOs) without an intermediary. In other words, nanotech companies might sell securities directly to investors. Although DPOs have been available to companies wishing to raise capital for years, they are only now beginning to represent an effective tool to raise money. The first nanotech company to conduct a DPO online was California Molecular Electronics Corporation, now renamed Accelyrs. The company raised approximately $721,000 between February 2000 and February 2001.[49]

Much progress must take place before nanotech companies that seek large sums of capital can seriously consider bypassing traditional underwriters in favor of direct public offerings. Sophisticated, online financial intermediaries must become capable of facilitating the sale of large quantities of securities. At the time of this writing, these intermediaries are developing their business models, technologies, and databases of potential investors for new stock offerings. In the coming years, nanotech companies may be able to work with an online intermediary to send direct e-mail alerts to millions of potential investors interested in nanotechnology and sell large quantities of securities online.[50]

CHAPTER 17

Conclusions

In 1952, the post–World War II economic boom stirred the American imagination. A new American president was elected and the nation's bright prospects for a higher standard of living were embodied in its burgeoning plastics industry. Americans celebrated Tupperware that could keep food fresh longer and Formica that was a lighter and more durable alternative to wood for many household construction projects. Today, plastics is a $393 billion industry that has forever changed the foods we eat, the cars we drive, the homes we live in, and the air we breathe.

Only seven years after the advent of Tupperware, Richard P. Feynman gave his now famous 1959 speech at Caltech entitled "There's Plenty of Room at the Bottom"—a speech that presaged the advent of nanotechnology. The parallels and lessons between plastics and nanotechnology are striking. Just as plastic was a lighter and more durable material than wood, nanotechnology will create materials with new properties that offer advantages over those in use today. Computers that weigh pounds today could weigh ounces tomorrow. Solar panels that are bulky and costly now could be small and efficient in the future.

Nanotechnology has the potential to be an even more disruptive economic force in this century than plastics were in the last one. The shift to nanotechnology could revolutionize a number of different industries. But the difference between what a technology can do and what will make money is a more relevant distinction for readers of Part III of this book. History is replete with examples of promising technologies that were underfunded, ultimately infeasible, or simply so misunderstood that they never realized their potential. When historians look back on the development of nanotechnology, they could see both smashing successes and colossal failures.

As this book goes to press, there are many looming business considerations about the business of nanotechnology. Will science produce technologies that meet the needs of the market? What business models are most likely to succeed? Will companies with promising technology and skilled entrepreneurs be destroyed by costly intellectual property disputes? Will the capital markets witness another exciting wave of initial public offerings or will acquisition be the most common exit strategy for nanotech start-ups?

Part III of this book is a first attempt to explore these questions. Certainly the answers to these questions will become increasingly clear in the coming years. Lawyers, managers, and investors in 2010 will know more than we can today. In 2004, we can offer five broad themes about the business of nanotechnology.

1. Initial Business and Legal Decisions Will Have the Biggest Impacts on the Fates of Individual Nanotechnology Companies.

Starting a nanotechnology company is an enormous project. Founders must devote their lives to the task. Decisions made in the early days of a company will influence the fate of a company and in Chapter 10, we highlighted some of these critical issues. In many cases, the most important issue involved in launching a company is negotiating the terms and conditions of license agreements with universities or government labs. Another matter that warrants serious consideration is naming the company. As illustrated by recent stock activity of companies with the term "nano" in their names, unsophisticated investors may be willing to pour money into companies simply because they sound catchy. Founders must also decide whether the company will be organized as a C Corp or a limited liability company (LLC). While an LLC could provide initial investors with tax benefits, the C Corp could be more suitable for a variety of reasons. Setting up a framework for corporate governance and advisory boards are key to running the company in a smooth and effective manner, and installing a sensible stock options plan is necessary to ensure founders a fair return and attract skilled employees. Founders must decide where to locate the company; being located near nanotech hubs will enable companies to access the human capital necessary to grow.

Perhaps the most important early-stage task for entrepreneurs is drafting the business plan. We devoted all of Chapter 11 to nanotech business plan writing. The primary message in this chapter is that a thorough assessment of the most attractive markets is essential, and entrepreneurs must realistically describe how they plan to reach their target markets. Because the current climate for funding early-stage nanotech companies is one of cautious optimism, overly optimistic entrepreneurs will turn investors off. While investors are excited about the potential of nanotech, the unique array of technological, regulatory, and legal considerations makes nanotech different from any industry in their experience. The business plan will be the starting point for addressing these concerns.

2. Intellectual Property Is Critical to Commercialization of Nanotechnology.

Investors, managers, and lawyers must understand legal issues involving nanotech patents and must carefully monitor patent issuances, licenses, and litigation. Indeed, at the company's inception, intellectual property issues are a primary consideration. In most cases, licensing intellectual property from universities or government labs is necessary to launch the company. When a company spins out of another company, the start-up must navigate through complex trade secret and contract law issues. Once a company has acquired basic intellectual property (IP) rights to protect its core technology, it must then begin to traverse through the fragmented and chaotic IP landscape. In several different places in the book, we have highlighted the current obsession with patenting in nanotechnology. As discussed in detail in Chapter 13, a start-up must develop strategies for when it finds itself infringing patents held by others and when it discovers others infringing its patents. Most nanotech start-ups should focus on building their IP portfolios by licensing or filing as many patents as possible. In prosecuting, licensing, and litigating patents, companies must understand legal issues affecting the scope and validity of patent claims.

3. Nanotech Companies Should Be Prepared for Multiple Stages of Financing.

Developing nanotechnology tools, materials, devices, and systems is an extremely expensive process. Early signs suggest that the path to commercialization will be slow and painful. Nanotech research often involves expensive capital equipment and highly qualified employees who can command high salaries. Additionally, different products based on nanotechnology could be subject to costly regulations.

As such, companies must prepare for multiple stages of financing. In Chapter 12, we explored the different financing mechanisms available to nanotech start-ups. To get the company off the ground, founders should rely on friends, family, and angel investors. Companies should then apply for federal grants through SBIR and STTR programs at different agencies, DARPA, or the Advanced Technology Program. Although nearly all nanotech companies will require venture capital, managers should delay it as long as possible to increase valuation and minimize dilution.

Venture capitalists are extremely excited about the market potential of nanotechnology. However, many investors do not have a sophisticated understanding of the scientific or engineering principles involved. When seeking venture financing, CEOs should be resourceful and prepared for a long and exhaustive process. Further, when accepting venture financing, founders and managers must clearly understand the terms and conditions of the deal.

4. Nanotech Companies Should Pursue Partnerships, Cooperation, and Consolidation.

In light of enormous costs and significant business and technology challenges, nanotech companies must keep their eyes and ears open to partnerships, cooperation, and consolidation. Specifically, nanotech companies should explore both vertical and horizontal relationships.

In Chapter 14, we analyzed vertical relationships between nanotech start-ups and large corporations. Because nanotechnology is primarily an enabling technology for a number of different industries, corporate partnering will play a central role in the commercialization of nanotechnology. Partnering with a large corporation enables the start-up to leverage market, technical, and regulatory expertise, obtain additional capital, and enhance its credibility. There are several different types of collaborations that can take place between a start-up and a large company. Agreements can include equity investment, joint technology development, manufacturing rights, distribution rights, and licensing provisions. The success of a partnership is dependent on its design and implementation. Factors likely to increase the chances of a partnership succeeding are clarity and precision, "win-win" arrangements, strong personal relationships, and effective communication channels and procedures. While a good partnership can pave the road to commercial success, a poor deal can cripple a fledgling start-up.

In Chapter 15, we made the case for stronger horizontal relationships between nanotechnology start-ups. A survey of the hundreds of companies engaged in nanoscale research reveals that there are large numbers of companies competing to develop similar technologies. For example, we identified a couple dozen companies focused on developing similar technologies based on carbon nanotubes. Fragmentation could inhibit the commercialization of nanotechnology. Start-ups should consider horizontal mergers with other companies engaged in similar R&D projects. In certain circumstances, consolidation could eliminate destructive patent disputes, put an end to inefficient, duplicative research, and increase the likelihood of developing commercial products within a shorter time horizon. Mergers between nanotechnology start-ups are complicated by a variety of legal and business issues. Nevertheless, skilled and determined managers, lawyers, and investors should be able to negotiate deals.

In addition to consolidation, companies engaged in nanoscale research would benefit from standardization. Standards are currently needed for consistent characterization of nanomaterials and will be required for the future interoperability of devices and systems. As demonstrated by the by the IEEE's launch of the Nanotechnology Standards Initiative in the fall of 2003, some progress is already being made in this direction. The IEEE and the nanotechnology community must be aware of several complicated legal and business issues that could emerge.

5. The Road to Payoff Will Be Long and Arduous, but the Rewards for Nanotech Companies Will Match Those of the Most Significant New Technology Areas in History.

At the time of this writing, a "nano euphoria" seems to be beginning to grip the capital markets. Yet, it is still unclear how nanotech companies will fare in the public markets. The stock rallies of companies with the word "nano" in their names in 2004 illustrate the willingness of some investors to bet on nanotechnology. However, due to the backlash to egregious misbehavior during the Internet boom, the world is unlikely to witness a wave of nanotech IPOs similar to the wave of Internet IPOs that characterized the end of the 1990s, but the possibility still remains.

In addition, because developing a nanotech company requires so much investment and is regulated so closely, reaping the benefits of the technology will be a long and arduous process. We believe that, in order to go public, most nanotech companies will need to offer promising technology, defensible IP, real revenue, and clear potential for strong earnings. Thus, rather than an IPO explosion, we anticipate that the majority of nanotech activity in financial markets during the next few years will take place in private equity markets.

Despite the long road faced by nanotech companies, we believe that the payoffs will, at least in some cases, be great. Investors are paying close attention to "nano" companies, and there is every reason to believe that the truly outstanding companies to emerge from this sector will be rewarded handsomely in both public and private equity markets. Though it will take time and effort, nanotech companies that develop profitable products and avoid believing their own hype will be limited only to the long-term historical contribution that the technology is able to make to society.

Notes

Introduction

1. Richard Smalley, June 22, 1999, at http://www.house.gov/science/smalley_062299.htm.
2. Philip G. Collins and Phaedon Avouris, *Nanotubes For Electronics*, 283 SCI. AM. 62, 69 (Dec. 2000).
3. National Science and Technology Council, Committee on Technology, Subcommittee on Nanoscale Science, Engineering, and Technology, *National Nanotechnology Initiative: The Initiative and Its Implementation Plan*, July 2000, at http://www.nano.gov/nni2.htm#page=15.
4. Mihail C. Roco, and William S. Bainbridge, NANOSCALE SCI. & EDUC. TASK FORCE, SOCIETAL IMPLICATIONS OF NANOSCIENCE AND NANOTECHNOLOGY, Mar. 2001, *available at* http://www.wtec.org/loyola/nano/societalimpact/nanosi.pdf.
5. T. Kalil (White House), *National Nanotechnology Initiative*, Sept. 28, 2000.

Chapter 1

1. ALBERT EINSTEIN. THE EXPANDED QUOTABLE EINSTEIN (Princeton University Press, June 2000).
2. Gary Stix, *Little Big Science*, 285 SCI. AM. 32, 34 (Sept. 2001) (quoting Steven Block).
3. *Id*.
4. *See Foresight FAQ General Nanotechnology Information*, at http://www.foresight.org/NanoRev/FIFAQ1.html#FAQ1.
5. NATIONAL NANOTECHNOLOGY INITIATIVE ("NNI"), *What is Nanotechnology?*, at http://www.nano.gov/html/facts/whatIsNano.html (last visited Mar. 5, 2004).
6. Richard P. Feynman, *There's Plenty of Room at the Bottom*, at http://www.zyvex.com/nanotech/feynman.html. ("Consider the possibility that we, too, can make a thing very small which does what we want—that we can manufacture an object that maneuvers at that level!")
7. In the original Feynman interpretation, the top-down approach means to build generations of successively smaller machines, until the nanometer scale became accessible. *See id*.
8. *See generally* George M. Whitesides, et al., *Unconventional Methods For Fabricating and Patterning Nanostructures*, 99 CHEM. REV. 1823 (1999).

9. This process involves placing a desired pattern of chromium film on a glass plate. A beam of ultraviolet light shines through the chromium mask and is focused on a photosensitive coating of organic polymer on the surface of a silicon wafer. When the parts of the organic polymer coating struck by light are removed, silicon is displayed in the form of the original pattern. *See* George M. Whitesides and J. Christopher Love, *The Art of Building Small*, 285 SCI. AM 39, 40 (Sept. 2001).
10. This method involves writing the circuitry pattern on a thin polymer film with a beam of electrons. *Id.* at 41.
11. We describe carbon nanotubes as inorganic nanostructures, because they are not compounds.
12. Richard Smalley and Robert F. Curl, *Fullerenes*, 265 SCI. AM. 54 (Oct. 1991).
13. Phillip G. Collins and Phaedon Avouris, *Nanotubes For Electronics*, 283 SCI. AM. 62, 63 (Dec. 2000).
14. *Id.*
15. Armchair nanotubes are always metallic, whereas the zigzag and chiral tubes can be either metallic or semiconducting. The electronic band gap determines the conductivity of a substance. The electronic band gap is the energy needed to knock electrons from valence states to conducting states. Because nanotubes have different band gaps depending on the chiral twist of the nanotube, they have different conductivities. *See* Collins and Avouris, *supra* note 13, at 63–65.
16. *Id.* at 69.
17. *Id.*
18. *Id.*
19. *Id.*
20. *Id.*
21. *See, e.g.*, M.H. Huang, et al., *Room-Temperature Ultraviolet Nanowire Nanolasers*, 292 SCI. 1897 (June 8, 2001).
22. *See generally* Mark Reed, *Quantum Dots*, 268 SCI. AM. 118 (Jan. 1993).
23. *See, e.g.*, Mei Li, et al., *Coupled Synthesis and Self-Assembly of Nanoparticles To Give Structures With Controlled Organization*, 402 NATURE 393 (Nov. 25, 1999); John C. Hulteen, et al., *Nanosphere Lithography: Size-Tunable Silver Nanoparticle and Surface Cluster Arrays*, 103 J. PHYSICAL CHEM. B 3854 (May 13, 1999).
24. *See, e.g.*, Kimberly Hamad-Schifferli, et al., *Remote Electronic Control of DNA Hybridization Through Inductive Coupling To An Attached Metal Nano-crystal Antenna*, 415 NATURE 152 (2002); Mao, et al., *A Nanomechanical Device Based on the B–Z Transition of DNA*, 397 NATURE 144 (Jan. 1999).
25. N.C. Seeman, *DNA nanotechnology: Novel DNA Constructions*, 27 ANNU. REV. BIOPHYS. BIOMOL. STRUCT. 225–248 (1998).
26. W.M. Shih, J.D. Quispe and G.F. Joyce, *A 1.7-kilobase Single-Stranded DNA That Folds Into A Nanoscale Octahedron*, 247 NATURE 618–21 (Feb. 2004).
27. L. Wang, A. Brock, B. Herberich, and P.G. Schultz, *Expanding the Genetic Code of Escherichia Coli*, 292 SCIENCE 498–500 (2001); V. Döring, H.D. Mootz, L.A. Nangle, T.L. Hendrickson, V. de Crécy-Lagard, P. Schimmel and P. Marlière, *Enlarging The Amino Acid Set of Escherichia Coli By Infiltration of the Valine Coding Pathway*, 292 SCIENCE 501–504 (2001).
28. *See* Jonathan Trent, *Nanotechnology R&D*, at http://ipt.arc.nasa. gov/trent.html.
29. *See* Chad A. Mirkin, et al., *Protein Nanoarrays Generated By Dip-Pen Nanolithography*, 295 SCI. 1702 (March 1, 2002).
30. Amy Higgins and Sherri Koucky, *Smaller Electronics Through Proteins*, 75 MACHINE DESIGN 47 (Feb. 20, 2003).
31. Q. Wang, et al., *Icosahedral Virus Particles as Addressable Nanoscale Building Blocks*, 41 CHEM. INT. ED. 459–462 (2002).

32. *See generally* Harm-Anton Klok and Sebastian Lecommandoux, *Supramolecular Materials via Block Copolymer Self-Assembly*, 13 ADVANCED MATERIALS 1217 (2001).
33. Donald A. Tomalia, *Dendrimer Molecules*, 272 SCI. AM. 62, 62 (May 1995).
34. The importance of this field is evidenced by the National Science Foundation's launch of the Network for Computational Nanotechnology in September 2002. The purpose of the Network is to facilitate collaboration between experimentalists, theorists, and computational scientists and develop high-performance software packages that are available to the scientific community.
35. For example, computational methods such as quantum MD and Monte Carlo are being used to build complex models. *See* Deepak Srivasta, Madhu Menon, and Kyeongjae Cho, *Computational Nanotechnology With Carbon Nanotubes and Fullerenes*, COMPUTING IN SCIENCE & ENGINEERING (Jul./Aug. 2001). Some of these models are available online. *See, e.g.,* "Nanohub" at http://nanohub.purdue.edu (registration required).
36. For a more detailed discussion of the STM microscope, see Robert Service, *Atom-Scale Research Gets Real: Outlook For Nanotechnology*, 290 SCI. 1523, 1526 (Nov. 24, 2000).
37. For a more detailed discussion of the AFM microscope, see Carol Wright-Smith and Christopher M. Smith, *Atomic Force Microscopy*, 15 THE SCIENTIST 23, 23 (Jan. 22, 2001).
38. *See* Robert F. Service, *Nanoscientists Look to the Future; Meeting: IEEE Nano 2001; Nanotweezers and Movable Platform Research*, 294 SCI. 1448, 1448 (Nov. 16, 2001).
39. M.J. Lang and S.M. Block, *Resource Letter: LBOT-1: Laser-Based Optical Tweezers*, 71 AM. J. PHYS. 201 (2003).
40. J.M. Garces and M.C. Cornell, *Impact of Nanotechnology on the Chemical and Automotive Industries*, in SOCIETAL IMPLICATIONS OF NANOSCIENCE AND NANOTECHNOLOGY 55 (2001).
41. Andrew Wood and Alex Scott, *Nanomaterials: A Big Market Potential*, CHEM. WEEK, Oct. 16, 2002, at LEXIS-NEXIS, News Library.
42. Jennifer L. Schenker, *It's The N_Generation; Nanotechnology, Which Offers Super-Small Solutions to Some Very Big Problems, May Be Coming of Age*, TIME INT'L, July 29, 2002, at 35.
43. For a more detailed discussion of the use of nanomaterials in coatings and films, see *Harnessing Innovation; A Manufacturer's Guide to Nanotechnology*, INDUSTRY WEEK, Dec. 2002.
44. A. P. Alivisatos, *Less Is More In Medicine*, 285 SCI. AM. 67, 70 (Sept. 2001).
45. For a good discussion of why Moore's Law will reach its limit, see Steve Jurvetson, *Transcending Moore's Law With Molecular Electronics and Nanotechnology*, 1 NANOTECH. L.&B. 70 (2004). Of course, if a fundamental limit is hit and no alternative technology is available, current architectures will simply have to be optimized.
46. Collins and Avouris, *supra* note 13, at 65.
47. Mark A. Reed and James M. Tour, *Computing With Molecules*, 282 SCI. AM. 87, 87 (June 2000).
48. Charles Lieber, et al., *Gallium Nitride Nanowire Nanodevices*, 2 NANO LETT. 101 (2002).
49. Zhen Yao, et al., *Carbon Nanotube Intramolecular Junctions*, 402 NATURE 273 (1999).
50. Yu-Chih Tseng, Peiqi Xuan, Ali Javey, Ryan Malloy, Qian Wang, Jeffrey Bokor, and Hongjie Dai, *Monolithic Integration of Carbon Nanotube Devices with Silicon MOS Technology*, 4 NANO LETT. 123–127 (2004).
51. Robust computer architectures like these have been demonstrated in the past and have attracted theoretical interest lately. *See* J. Heath, et. al, *A Defect-Tolerant Computer Architecture: Opportunities for Nanotechnology*, 280 SCIENCE 1716

(June 1998); Seth Goldstein, et. al, *Reconfigurable Computing and Electronic Nanotechnology*, available at http://www2.cs.cmu.edu/~phoenix/papers/asap03.pdf.
52. Collins and Avouris, *supra* note 13, at 68–69.
53. Reed and Tour, *supra* note 47, at 86.
54. Thomas Rueckes, et al., *Carbon Nanotube Based Nonvolatile Random Access Memory For Molecular Computing*, 289 SCI. 94–97 (2000).
55. Zhiming Liu, et al, *Molecular Memories That Survive Silicon Device Processing and Real-World Operation*, 302 SCI. 1543–1545 (2003).
56. The technology uses thousands of "nano sharp tips" to punch data onto a thin plastic film. Benjamin Pimentel, *New IBM Storage Technology Can Pack a Trillion Bits Onto Chip*, S.F. CHRONICLE, June 11, 2002, at B3.
57. T.A. Taton, et al., *Device Physics: Defective Promise In Photonics*, 416 NATURE 685–686 (2002).
58. S.A. Maier, et al., *Plasmonics—A Route To Nanoscale Optical Devices*, 13 ADVANCED MATERIALS 1501–1505 (2001).
59. There are several technical and economic issues associated with the exploitation of vast hydrocarbon reserves such as gas hydrates, oil shale and tar sands that could make their extraction unfeasible. *See generally* Charles Hall, et al., *Hydrocarbons and the Evolution of Human Culture*, 426 NATURE 318 (Nov. 20, 2003).
60. Michael Grätzel, *Photoelectrochemical Cells*, 414 NATURE 338 (2001).
61. Michael McGehee, STANFORD UNIV., *Organic and Polymeric Photovoltaic Cells*, Presented at NSF Organic 2003, available at http://www.mrc.utexas.edu/NSFWorkshop/Presentations/mcgehee1.pdf.
62. L. Schlapbach and A. Züttel, *Hydrogen-storage Materials For Mobile Applications*, 414 NATURE 353–358 (2001).
63. N.L. Rosi, et al., *Hydrogen Storage in Microporous Metal-Organic Frameworks*, 300 SCI. 1127–1129 (2003).
64. J.M. Tarascon and M. Armand, *Issues and Challenges Facing Rechargeable Lithium Batteries*, 414 NATURE 359–367 (2001).
65. D. Larbalestier, *High-Tc Superconducting Materials For Electric Power Applications*, 414 NATURE 368–377 (2001).
66. *See generally* Celia M. Henry, *Drug Delivery*, 80 CHEMICAL AND ENGINEERING NEWS 39 (Aug. 22, 2002).
67. Alan Dove, *Proteomics: Translating Genomics into Products?*, 17 NATURE BIOTECHNOLOGY 233–235 (1999).
68. The nano-needle is a 50 nm-diameter silver-coated optical fiber carrying a helium-cadmium laser beam. Monoclonal antibodies that attach and bind to BPT (a product of a chemical reaction between cellular DNA and a cancer-causing pollutant) are attached to the tip of the nano-needle, which is inserted into a tissue sample. The laser light causes the antibody-BPT complex to fluoresce, and the fluorescent light travels up the fiber to an optical detector. According to its inventors, the device can detect damaged DNA with carcinogens attached, in addition to recognizing the presence of carcinogens even when there is no damage to the DNA. *See* Allen Menezes, et al., *Within a Nanometer of Your Life*, 123 MECH. ENG. 54, 56 (Aug. 2001).
69. For example, researchers from Scottish Universities of Glasgow, Edinburgh, and Strathclyde are working on project "Robodoc", a capsule that will travel through the body finding and diagnosing illnesses. *See generally* David Roman, *Lab-on-a-pill May Replace Invasive Endoscopy*, ELECTRONIC ENGINEERING TIMES, Nov. 3, 2003, available at http://www.eetimes.com/story/OEG20031031S0060 ("The fully encapsulated lab-on-a-pill includes sensors, system-on-chip circuitry, wireless communications capability and an on-board battery with a life of about 24

hours, though the 35 mm long pill is expected to complete its work over a 12-hour period.").
70. *See* Allen J. Menezes, et al., *Within A Nanometer of Your Life*, ME, Aug. 2001, *available at* http://www.memagazine.org/backissues/aug01/features/nmeter/nmeter.html.
71. Tissue reconstruction efforts have involved the use of a scaffold to grow tissue. However, researchers have encountered difficulty in directing organized tissue growth. By dotting the surface of the scaffold with nano-pores, researchers believe that they will be able to "shape and weave" the tissue. Research involving a "smart bandage" composed of biodegradable plastic with nano-sized grooves that aids in healing severed tendons illustrates this concept. Traditionally, operations on severed tendons have been highly ineffective, because the regenerating natural tissue sheathing attaches itself to the tendon, precluding mobility. The nano pores on the bandage allow macrophages, which are the cells that attach the tendon to the rest of the tissue, to grow into the grooves of the bandage separated from the sheathing. Researchers at Glasgow University believe that this technique will allow researchers to direct the growth of any pattern of tissue. *See id.* at 57.
72. Nanomedical products can facilitate the integration of synthetic materials into the body. For example, researchers are experimenting with a biomaterial containing nano-sized pores to treat spinal disc deterioration. When the material is fused with vertebrae, the porous structure allows the bone to gradually infiltrate into and throughout the device. The natural and artificial materials join together resulting in spinal reconstruction and less nerve compression and pain than traditional therapies. *See id.* at 55.
73. *The Smaller the Better*, THE ECON., June 23, 2001, *available at* LEXIS, News Library.
74. V. Vogel, *Societal Impacts of Nanotechnology in Education and Medicine*, in SOCIETAL IMPLICATIONS OF NANOSCIENCE AND NANOTECHNOLOGY, 143, 146 (Mihail C. Roco & William Sims Bainbridge eds., March 2001).
75. Chappell Brown, *NASA Ramps Nanotech. To Explore Space*, ELECTRONIC ENGINEERING TIMES, May 5, 2003, at 51.
76. *See* Mindy Rittner, *National Security Needs Drive Nanomaterials Development*, NANOPARTICLE NEWS, Dec. 2002, *available at* LEXIS, News Library.
77. A number of envisioned "building large" applications involve small devices; since their general purpose seems to involve a meaningful interaction with macroscopic scales—by assembling objects, restoring health, and so forth—we include them under this heading.
78. *See generally* K. ERIC DREXLER, ENGINES OF CREATION (Anchor Press 1986); K. ERIC DREXLER, ET. AL., UNBOUNDING THE FUTURE (William Morrow and Co. 1991).
79. Menezes, *supra* note 68, at 58 ("Biomedical nanotechnology will make it possible to build nanorobots having cellular dimensions with the ability to eliminate infections, unclog arteries, and a range of other applications. . . . Who can say? Biomedical nanotechnology's future may one day eliminate old age, or at least its symptons."); Robert Freitas, *What Would Be The Biggest Benefit To Be Gained For Human Society From Nanomedicine?*, (1998), at http://www.foresight.org/Nanomedicine/ NanoMedFAQ.html#FAQ19.
80. *See generally*, Ralph Merkle, *Molecular Manufacturing: Adding Positional Control to Chemical Synthesis*, (Sept./Oct. 1993), *available at* http://www.zyvex.com/nanotech/CDAarticle.html.
81. A crude, first-order approximation to the atomic diameter of a carbon atom is 0.22 nm. Atomic volume is roughly 10^{-23} cm3 per atom.
82. A simple assembly of n atoms is a chain with $n-1$ bonds. Given 10^{24} bonds (10^{24} -1 is for all purposes the same as 10^{24}), a billion assemblers working simultane-

ously at a frequency of 1Ghz will perform 10^{18} total operations per second. Therefore, it will take 10^6 s (~11.6 days) to complete the task.
83. There are various estimates for the age of the Universe. One such estimate places it between 13 and 14 billion years. *See* Hansen, et al., *The White Dwarf Cooling Sequence of the Globular Cluster Messier 4*, (May 6, 2002), *available at* http://arxiv.org/abs/astro-ph/0205087. The manufacturing operation described would take 1018 s, about 32 billion years.
84. Drexler and others have advanced ideas of self-replication and exponential assembly. *See generally* SKIDMORE ET AL., EXPONENTIAL ASSEMBLY (2004), *available at* http://www.zyvex.com/ Research/exponentialGS.html; Ralph Merkle, SELF REPLICATION AND NANOTECHNOLOGY, online at http://www.zyvex.com/nanotech/selfRep.html.
85. *See generally* K. Eric Drexler, NANOSYSTEMS MOLECULAR MACHINERY MANUFACTURING AND COMPUTATION (John Wiley & Sons, 1998).
86. In addition to mechanical control, Drexler's ideas include the fact that mechanosynthesis be carried out in a high-vacuum environment. Biological molecules work immersed in water, which is fundamental to their synthetic abilities.
87. Richard E. Smalley, *Of Chemistry, Love and Nanobots*, SCI. AM. 76, 77 (Sept. 2001).
88. R. Baum, *Nanotechnology—Drexler and Smalley Make The Case For And Against 'Molecular Assemblers'*, 81 CHEMICAL AND ENGINEERING NEWS 37–42 (December 1, 2003).

Chapter 2

1. Unnamed speaker, presentation given at Stanford University, Fall 2002.
2. http://www.nanoinvestornews.com/modules.php?name=Company_Profiles. The site can be accessed for free after registering.
3. The conference, which focused on developing a Northern California Nanotechnology Initiative, was held on Jan. 30, 2003 in San Francisco by the NanoBusiness Alliance.

Chapter 3

1. Bill Joy, *Why The Future Doesn't Need Us*, WIRED, Apr. 2000, at http://www.wired.com/wired/archive/8.04/joy.html."Spiritual machines" was a concept popularized by futurologist and computer scientist Ray Kurzweill.
2. This is the so-called "grey goo" problem.
3. MICHAEL CRICHTON, PREY (2002).
4. *See generally* IRVIN C. BUPP & JEAN CLAUDE DERIAN, LIGHT WATER: HOW THE NUCLEAR DREAM DISSOLVED (1978).
5. Harold P. Green, *The Law-Science Interface in Public Policy Decisionmaking*, 51 OHIO ST. L.J. 375, 388 (1990).
6. Dan L. Burk and Barbara A. Boczar, *Biotechnology and Tort Liability: A Strategic Industry At Risk*, 55 U. PITT. L. REV. 791, 837–8 (Spring 1994).
7. *See, e.g.*, Jane Kay, *Chefs Shun Fish With Altered DNA*, S.F. CHRON., Sept. 19, 2002, at A4 ("[C]elebrity chefs nationwide are joining grocery stores and seafood distributors to boycott biotech fish.").
8. William Charles Lucas, *Book Review: From Alchemy to IPO: The Business of Biotechnology*, 8 WID. L. SYMP. J. 153, 154 (2001) ("[T]he public backlash over "Frankenstein" (i.e. genetically modified) foods in Europe have had a significant impact on the bioagricultural industry. There are very real public concerns over the potential harm of biotechnology. The maladroit management of such sensi-

tivities on Capitol Hill, or in the media, can upset a business plan as assuredly as a science hurdle.").

9. *See* Robert A. Freitas, Jr., FORESIGHT INST., *Some Limits to Global Ecophogy by Biovorous Replicators, with Public Policy Recommendations*, Apr. 2000, at http://www.foresight.org/NanoRev/ Ecophagy.html.

10. Michael Dertouzos, *Kurzweil vs. Dertouzos*, 104 TECH. REV. 80 (Jan. 1, 2001) ("[T]ransistors and the systems made with them are used by people. And that's where exponential change stops! Has word-processing software, running on millions of transistors, empowered humans to contribute better writings than Socrates, Descartes, or Lao Tzu? . . . We have no basis to assert that machine intelligence will or will not be achieved.").

11. *See, e.g.*, Glenn H. Reynolds, Nat. Space Soc., *Space, Nanotechnology, and Techno-Worries*, in Roadmap for the Settlement of Space 7, *available at* http://www.nss.org/community/roadmap/Aapdfs/Reynolds.pdf. ("It is not obvious, however, that intelligence has much to do with world domination. It may be that, like James Branch Cabell's eponymous protagonist Jurge, superintelligent machines would find that 'cleverness was not on top of things, and never had been.' While scientists and computer experts would tend to regard superior intelligence as the sin qua non of world domination, that view should be dispelled by a glance at the headlines.").

12. RAY KURZWEIL, THE AGE OF SPIRITUAL MACHINES: WHEN COMPUTERS EXCEED HUMAN INTELLIGENCE (1999).

13. *See, e.g.*, K. ERIC DREXLER, ENGINES OF CREATION (Annchor Press 1986); K. ERIC DREXLER, ET. AL., UNBOUNDING THE FUTURE (William Morrow and Co. 1991).

14. Raymond Kurzweil, *Kurzweil v. Dertouzos*, 104 TECH. REVIEW 80 (Jan. 1, 2001).

15. For example, researchers in countries that have banned research on embryos have moved to other countries permitting such research. *See* Andy Coghlan, *Highly Cultured*, NEW SCIENTIST, Aug. 19, 2000.

16. For example, the NPT norm against the pursuit of nuclear weapons has been violated many times. The list of states that have broken or are thought to have broken the norm includes Argentina, Brazil, India, Iran, Iraq, Israel, North Korea, Pakistan, South Africa, South Korea and Taiwan. Kathleen C. Bailey, *The Comprehensive Test Ban Treaty: The Costs Outweigh The Benefits*, POLICY ANALYSIS, Jan. 15, 1999, at 23, *available at* http://www.cato.org/pubs/pas/pa330.pdf. The Biological Weapons Convention has also failed to prevent development of biological weapons. Dave Kopel and Glenn Reynolds, *Another Bad Treaty*, NAT. REV., Sept. 6, 2001 ("But rather than hope that the United Nations will produce a better protocol, the United States ought to realize that the Biological Weapons Convention is a proven failure—having already induced the creation of massive stockpiles of sophisticated biowar agents by the Soviet Union . . ."). *See also* COLIN GRAY, HOUSE OF CARDS 107-108 (1992) ("Germany in the 1920s and early 1930s, Japan in the 1930s, and the Soviet Union in the 1970s, all took advantage of what amounted to arms control 'cover' to press for unilateral military improvements.").

17. Glenn H. Reynolds, *supra* note 11, at 7.

18. *Id.*

19. As Glenn Reynolds has argued, a "regime that banned only the construction of, for example, 'assembler' devices would leave unregulated huge amounts of research that would be readily translatable into such devices." GLENN HARLAN REYNOLDS, FORWARD TO THE FUTURE: NANOTECHNOLOGY AND REGULATORY POLICY (Pacific Res. Inst. Policy Paper) (Nov. 2002).

20. *See* Scott Pace, FORESIGHT INST., *Military Implications of Nanotechnology*, Foresight Update No. 6, *available at* http://www.foresight.org/Updates/

Update06/Update06.2.html ("How might nanotechnology contribute to U.S. military power at these different levels of conflict? In peacetime or crisis, nanocomputers may allow more capable surveillance of potential aggressors. The flood of data from worldwide sensors could be culled more efficiently to look for truly threatening activities. In low-intensity warfare, intelligent sensors and barrier systems could isolate or channel guerrilla movements depending on the local terrain. In conventional theatre war, nanotechnology may lead to small, cheap, highly lethal anti-tank weapons.") ("President Reagan's goal of making nuclear weapons 'impotent and obsolete' could be reached not by space-based defenses, but by terrestrial nanoweapons making nuclear weapons irrelevant.... These potential military applications would allow the United States a greater range of options in deciding how to respond to aggression.... Nanoweapons could deter war by threatening unacceptable damage to an aggressor, as with today's strategic nuclear weapons, or by denying any plausible achievement of an aggressor's objectives, as is the potential with space-based missile defenses.").

21. The First Amendment provides in pertinent part that "Congress shall pass no law abridging . . . the freedom of speech." U.S. CONST. amend. I.
22. Although the United States Supreme Court has never ruled on a case involving protection of scientific speech, the Court has stated that "in the area of freedom of speech and press the courts must always remain sensitive to any infringement on genuinely serious literary, artistic, political, or scientific expression." Miller v. California, 413 U.S. 15, 22–23 (1973). The only case in which scientific expression was an issue involved restricting publication of an article that described how to build an atom bomb. *See* United States v. Progressive, Inc., 467 F. Supp. 990 (W.D. Wis. 1979). However, the case was dismissed as moot by the Seventh Circuit when the article was published by another magazine. *See* JUDITH AREEN ET AL., LAW, SCIENCE, AND MEDICINE 320 (2d ed. 1996).
23. Spence v. Washington, 418 U.S. 405, 409 (1974).
24. *See* Barnes v. Glen Theatre, 501 U.S. 560, 567 (1991).
25. 491 U.S. 397 (1989).
26. *Id.* at 405.
27. *Spence*, 418 U.S. at 410-11 (1974).
28. *Barnes*, 501 U.S. 560 (1991) (holding that nude dancing was expressive conduct even though the dancers were motivated primarily by the pursuit of profit).
29. "A scientific assertion is one that can be tested and, if the results of the test so indicate, be rejected as false." *See* Areen, *supra* note 22.
30. Richard Delgado et al., *Can Science Be Inopportune? Constitutional Validity of Governmental Restrictions on Race-IQ Research*, 31 UCLA L. REV. 128, 160 (1983) (concluding that there is a fundamental right to engage in IQ research); Roy G. Spece, Jr. & Jennifer Weinzierl, *First Amendment Protection of Experimentation: A Critical Review and Tentative Synthesis / Reconstruction of the Literature*, 8 S. CAL. IINTERDISC. L.J. 185, 213–19 (1998) (arguing that experimentation is protected by the First Amendment because (1) it is an integral part of a systematic process—the scientific method; and (2) experimentation is uniquely and powerfully facilitative of highly valued thought); John A Robertson, *The Scientist's Right to Research: A Constitutional Analysis*, 51 S. CAL. L. REV. 1203, 1217–18 (1977); SUBCOMM. ON SCIENCE, RESEARCH AND TECHNOLOGY OF THE HOUSE COMM. ON SCIENCE AND TECHNOLOGY, 95TH CONG., 2d Sess., SCIENCE POLICY IMPLICATIONS OF DNA RECOMBINANT MOLECULE RESEARCH 60 (Comm. Print 1978) (statement of T. Emerson, Professor, Yale Law School) (arguing that because scientific inquiry cannot proceed without scientific experimentation, some experiments are necessarily expression).

31. *See generally* RICHARD RORTY, CONTINGENCY, IRONY, AND SOLIDARITY (1989).
32. 391 U.S. 367 (1968).
33. *Id.* at 377.
34. Nanotechnology Research and Development Act of 2003, H.R. 766, 108th Cong. (2003), *available at* http://www.house.gov/science/hearings/full03/may01/hr766.pdf.
35. For a more thorough discussion of nanoethics centers, *see* Edward Etzkorn & Susan Hackwood, *Chapter 6: Social and Ethical Impacts of Nanotechnology*, in NANOSCIENCE AND NANOTECHNOLOGY: OPPORTUNITIES AND CHALLENGES IN CALIFORNIA 106 (2004).
36. Victoria Griffith, *Big Risks on A Microscopic Scale: Technology: Fears About the Potential Dangers of Nanotechnology Are Threatening Its Wider Use*, FIN. TIMES, Sept. 25, 2002, at LEXIS, Nexis Library (quoting Charles Lieber, a Harvard professor and leading researcher in nanoelectronics).
37. The North Carolina Citizens Technology Forum is a National Science Foundation Project to explore ways of improving public involvement in discussions of technology policy. *See* http://www.ncsu.edu/chass/communication/ciss/sponsored.html.
38. *See* Tom Kalil, *Next Steps For The NNI*, 1 NANOTECH. L&B 55 (2004).
39. The Guidelines classify experiments based on their risks and require physical and biological safeguards commensurate with the risk. Physical safeguards include required laboratory installations, procedures, and equipment. Biological safeguards include production of experimental organisms that require special supplements (that are not available outside of the laboratory) to survive. *See* NIH Guidelines For Research Involving Recombinant DNA Molecules, 48 Fed. Reg. 24,555, 24,557 (June 1, 1983).
40. Richard A. Merrill and Bryan J. Rose, *FDA Regulation of Human Cloning: Usurpation or Statesmanship*, 15 HARV. J. LAW & TECH. 85, 117 (Fall 2001).
41. *Foresight Guidelines on Molecular Manufacturing*, FORESIGHT INST., June 4, 2000, at http://www.foresight.org/guidelines/current.html.
42. *Id.*
43. *Id.*

Chapter 4

1. *Nanotechnology: Hearing Before Science, Technology, and Space Subcomm. on Senate Comm. on Commerce, Science, and Transportation*, 107th Cong. (2002) (statement of F. Mark Modzelewski, Exec. Dir. of NanoBusiness Alliance), FED. DOCUMENT CLEARING HOUSE CONG. TESTIMONY, at LEXIS, Nexis Library.
2. *See* EROSION, TECHNOLOGY, & CONCENTRATION GROUP, *No Small Matter! Nanotech. Particles Penetrate Living Cells and Accumulate in Animal Organs*, 76 ETC Communique 1, 1, May/Jun. 2002, at http://www.etcgroup.org/documents/Comm_NanoMat_July02.pdf. [hereinafter ETC Report]. The ETC Report is vague in the course of action that it recommends. Although the Report does not explicitly state that the EPA should act, its reference to "heads of state" taking action can be understood as a call for agency action. *See id.*
3. Barnaby J. Feder, *As Uses Grow, Tiny Materials' Safety Is Hard To Pin Down*, N.Y. TIMES, Nov. 3, 2003, at C1.
4. *Id.*
5. *Id.*
6. *Id.*
7. *Id.*

8. David M. Ewalt, *The Next (Not So Big) Thing—Nanotechnology Sounds Like Science Fiction—But the Promise Is Real*, INFO. WEEK, May 13, 2002, at 42.
9. The area of the nanotube image Is 2.2 x 1.5 micrometers, while that of the asbestos image is approximately 140 x 100 micrometers. The area for the nanotube image is available at http://www.phys.unt.edu/stm/images.htm. The area of the asbestos image is measured from the indicated scale bar.
10. Email from Nora Savage, Ph.D., Environmental Engineer, EPA, ORD, NCER, to author, Sept. 20, 2002, on file with author (noting that the EPA is "attempting to fund extramural research (to academic institutions) on that issue."). *See also* Environmental Futures Research in Nanoscale Science, Engineering and Technology Science To Achieve Results (STAR) Program, NATIONAL CENTER FOR ENVIRONTMENTAL RESEARCH, *available at* http://es.epa.gov/ncer/rfa/current/02nanotech.html#SUMMARY OF PROGRAM.
11. For example, in March 2002, the EPA held a meeting called: "Nanotechnology: Environmental Friend or Foe?"
12. Doug Brown, *U.S. Regulators Want To Know Whether Nanotech Can Pollute*, SMALL TIMES, Mar. 8, 2002, at http://www.smalltimes.com/document_display.cfm?document_id=3231.
13. *See generally* Toxic Substance and Control Act [hereinafter TSCA] §§ 2–412, 15 U.S.C. §§ 2601–2692 (2002).
14. The notice must, "insofar as is known or reasonably ascertainable," include the intended uses of the chemical, the amount in which it will be manufactured, data concerning health and environmental effects, and estimated safe exposure levels for humans and the environment. 15 U.S.C. §§ 2604(d)(1)(A), 2607(a)(2) (2002). The manufacturer may also be required to submit test data for EPA evaluation with the PMN. 15 U.S.C. § 2604(b) (2002).
15. 15 U.S.C. § 2602(9) (2002).
16. For example, carbon, zinc oxide, and titanium dioxide are listed on the TSCA Inventory: Carbon, 7440-44-0; Zinc oxide, 1314-13-2; Titanium dioxide, 13463-67-7. (This information was obtained from an extract prepared by Cornell University, at http://msds.pdc.cornell.edu/tsca.) (authorization of website required).
17. Under section 5(a)(1)(B) of TSCA, persons must submit a significant new use notice to EPA at least 90 days before they manufacture, import, or process a substance for a significant new use. 15 U.S.C. § 2605(a)(1)(B) (2002). Persons subject to a SNUR must follow the same rules and procedures as persons who are required by section 5(a)(1)(A) of TSCA to submit a PMN. 15 U.S.C. §2605(a)(1)(A) (2002).
18. EPA considers the following factors in deciding whether to issue a SNUR: (1) the projected volume of manufacturing and processing of a chemical substance; (2) the extent to which a use changes the type or form of exposure of human beings or the environment to a chemical substance; (3) the extent to which a use increases the magnitude and duration of exposure to human beings or the environment to a chemical substance; and (4) the reasonably anticipated manner and methods of manufacturing, processing, distribution in commerce, and disposal of a chemical substance. 15 U.S.C. § 2604(a)(2) (2002).
19. TSCA 8(e), 15 U.S.C. § 2607(e) (2002).
20. This conclusion is based on a survey conducted by emailing a dozen companies that produce different types of nanomaterials. Although most did not respond to the email questions, the few that did stated that they did not need to seek regulatory review because the chemicals had already been approved in their bulk forms. The email responses are on file with the author.

21. Nanophase sells zinc oxide nanoparticles to companies such as BASF for use in sunscreen.
22. In Fall 2002, Mitsui & Co. opened the world's first large-scale nanotube plant to produce 10 tons of multi-walled nanotubes per month. Carbon Nanotechnologies was established in 2000 to produce single-walled nanotubes. Both NEC Corp. and Mitsubishi plan on beginning mass production of carbon nanotubes in 2004.
23. R. FLAGAN & D.S. GINLEY, NAT'L SCI. & TECH. COUNCIL, *Nanoscale Processes in the Environment*, in NANOTECHNOLOGY RESEARCH DIRECTIONS: IWGN WORKSHOP REPORT: VISION FOR NANOTECHNOLOGY RESEARCH AND DEVELOPMENT, (M.C. Roco, S. Williams & P. Alivisatos eds., Sep. 1999), *available at* http://www.wtec.org/loyola/nano/IWGN.Research.Directions/chapter10.pdf ("[P]art of the difficulty in assessing the impact of nanoscale materials on biological systems is finding analytical techniques suitable for monitoring both their presence and their impact.").
24. Professor Jeffrey Rachlinski argues that negative public perceptions increased liability for industries emitting air and water pollution in the 1970s and tobacco manufacturers in the 1990s. "[S]ocial perceptions of the nature of the underlying activity had changed, leading to an atmosphere more conducive to liability for industry." Jeffrey J. Rachlinksi, *Regulating in Foresight Versus Judging Liability in Hindsight: The Case of Tobacco*, 33 GA. L. REV. 813, 815 (1999). He concludes: "Changes in public attitudes could put many products, including tobacco, firearms, and pharmaceuticals, at risk for punitive damages in the future, even though the manufacturers' present conduct would not attract such a sanction." *Id.* at 844.
25. Dan L. Burk & Barbara A. Boczar, *Biotechnology and Tort Liability: A Strategic Industry At Risk*, 55 U. PITT. L. REV. 791, 838 (Spring 1994).
26. Erin K.L. Mahaney, *Assessing the Fitness of Novel Scientific Evidence in the Post-Daubert Era: Pesticide Exposure Cases As A Paradigm For Determining Admissibility*, 26 ENVTL. L. 1161, 1182 (1996) ("Because there are few epidemiological studies available, animal studies constitute the primary source of information regarding the carcinogenic, teratogenic, or other disease-inducing properties of pesticides.").
27. Mark Dickie, *Environmental Toxicology and Health Risk Assessment in the United States: Economic and Policy Issues*, in MANAGING POLLUTION 43–45 (Clive L. Splash & Sandra McNally eds., 2001) ("[D]ose response assessment differs according to assumptions about the presence of a threshold and the slope and curvature of the function. For non-cancer effects, thresholds usually are assumed. Dose-response function may be linear or non-linear. . . . For cancer, the US EPA assumes linear dose-response functions unless no gene mutations occur and there is conclusive evidence of non-linearity.").
28. Because exposure levels are extremely difficult to determine, risk assessors must often make arbitrary determinations. *Id.* at 45 ("Human exposures are undoubtedly highly variable. . . . Data on actual human exposures are expensive to collect and often are unavailable.").
29. Characterization of the risk depends on whether the chemical has threshold effects. For chemicals with threshold effects, the reference dose is compared to the estimated exposure dose (EED). If the RfD exceeds the EED, the risk is minimal. For chemicals without threshold effects, risk is characterized as a probability. The slope of the dose-response function at low doses is multiplied by the EED. *Id.* at 46.
30. *See, e.g.*, K. Donaldson et al., *Ultrafine (Nanometre) Particle Mediated Lung Injury*, 29 J. AEROSOL SCI. 553 (1998); D. Hohr, et al., *Hydrophobic Coating of Ultrafine*

Titanium Dioxide Reduces The Acute Inflammatory Response After Instillation in The Rat, INT'L. J. HYG. ENVIRON. HEALTH (2001); A. Churg, et al., *Comparison of the Uptake of Fine and Ultrafine TiO$_2$ in a Tracheal Explant System,* 274 AM. J. PHYSIOL. L81 (1998); H.P. Dy & D.R. Chen, *Nanometer Particles: A New Frontier For Multidisciplinary Research,* 28 J. AEROSOL. SCI. 539 (1997); W.G. Kuschner, et al., *Human Pulmonary Responses To Experimental Inhalation of High Concentration Fine and Ultrafine Magnesium Oxide Particles,* 105 ENVIRON. HEALTH PERSPECT. 1234 (1997); W.G. Kuschner, et. al., *Early Pulmonary Cytokine Responses To Zinc Oxide Fume Inhalation,* 75 ENVTL. RES. 7 (1997); W.G. Kuschner, et al., *Pulmonary Responses to Purified Zinc Oxide Fume,* 43 J. INVESTIGATIVE MED. 371 (1995); K.P. Lee, et al., *Pulmonary Responses of Rats to Titanium Dioxide By Inhalation From Two Years,* 79 TOXICOL. APPL. PHARMACOL. 179 (1985); A. Nemmar, et al., *Passage of Intratracheally Instilled Ultrafine Particles Into The Systemic Circulation of the Hamster,* 164 AM. J. RESPIR. CRIT. CARE MED. 1665 (2001); A Nemmar, et al., *Effect of Ultrafine Particles On Experimental Thrombus Formation in Hamster Model,* 60 TOXICOL. SCI. 191 (2001); G. Oberdorster, et al., *Association of Acute Air Pollution and Acute Mortality: Involvement of Ultrafines?,* 7 INHAL. TOXICOL. 111 (1994); G. Oberdorster, et al., *Pulmonary Effects of Inhaled Ultrafine Particles,* 74 INT. ARCH. OCCUP. ENVIRON. HEALTH 1 (2001); A. Peters, et al., *Respiratory Effects Are Associated With The Number of Ultrafine Particles,* 155 AM. J. RESPIR. CRIT. CARE MED. 1376 (1997); V. Stone, et al., *Increased Calcium Influx in A Monocytic Cell Line On Exposure To Ultrafine Carbon Black,* 15 EUR. RESPIR. J. 297 (1997); P.J. Anderson, et al., *Respiratory Tract Deposition of Ultrafine Particles In Subjects With Obstructive or Restrictive Lung Disease,* 97 CHEST 1115 (1990); H.C. Yeh, *In Vivo Deposition of Inhaled Ultrafine Particles in the Respiratory Tract of Rhesus Monkeys,* 27 AEROSOL SCI. TECH. 465; Q. Zhang, et. al., *Differences in the Extent of Inflammation Caused By Intratracheal Exposure to Three Ultrafine Metals: Role of Free Radicals,* 53 J. TOXICOL. ENVTL. HEALTH 423 (1998); Zhang, et al., *Toxicity of Ultrafine Nickel Particles in Lungs After Intratracheal Instillation,* 40 OCCUP. HEALTH 171 (1998).
31. H-Erich Wichmann, et al., *Daily Mortality and Fine and Ultrafine Particles in Erfurt Germany. Part I: Role of Particle Number and Particle Mass,* HEALTH EFFECTS INSTITUTE RESEARCH REPORT No. 98 (Cambridge, MA: HEI 2000).
32. Air particulate measurements were obtained at one site near a road. *Id.* at 8. Although the sample size was statistically small (only 200,000 people live in Erfurt), Erfurt was chosen because geography allowed testing of exposure at a single site. *Id.* at 9. The statistical methods used were Poisson regression with a generalized additive model to smooth time trends, weather, and other variables. The regression procedure included polynomial distributed lags to measure the timing of the effects. A lag is the assumed time period between exposure and effect. Thus, pollutant levels were examined on the day of death (lag 0), the day prior to death (lag 1), two days prior to death (lag 3), four days prior to death (lag 4), and five days prior to death (lag 5). *Id* at 18–20. A time-series approach was used to analyze whether day-to-day changes in fine and ultrafine particles corresponded to day-to-day changes in death. The study lasted for 3.5 years.
33. The study involved exposing human volunteers to 33 mg m-3 zinc oxide particles with an average primary particle size of 8–40 nm for 30 minutes. The volunteers experienced large increases in the bronchoalveolar lavage (BAL) concentrations of the pro-inflammatory cytokines tumor necrosis factor (TNF) and interleukin (IL)-8 after termination of exposure. W.G. Kuschner, A. D'Alessandro, H. Wong, and P.D. Blanc, *Early Pulmonary Cytokine Responses To Zinc Oxide Fume Inhalation,* 75 ENVTL. RES. 7–11 (1997).
34. One study exposed two groups of rats to an equal airborne mass concentration of fine and ultrafine TiO$_2$ particles. Researchers found that there was much

more bronchoalveolar inflammation in the rats exposed to the ultrafine particles. J. Ferin, G. Oberdorster, & D.P. Penney, *Pulmonary Retention of Ultrafine and Fine Particles in Rats*, 6 CELL MOL. BIOL. 535–542 (1992). Another study instilled ultrafine titanium dioxide particles directly into the trachea of mice. Gunter Oberdorster, et al., *Acute Pulmonary Effects of Ultrafine Particles In Rats and Mice*, HEALTH EFFECTS INSTITUTE (Aug. 2000).

35. A study involving the instillation of cabosil amorphous silica approximately seven nm in size for two days to 12 weeks showed that there was no inflammation. S.A. Murphy, K.A. Berube, F.D. Pooley & R.J. Richards, *The Response of Lung Epithelium to Well Characterized Fine Particles*, 62 LIFE SCI. 1789–1799 (1998).

36. When subjects were exposed to magnesium oxide particles where 28% of the particles were less than 100 nm in diameter, there was no effect on the cytokine levels in the lavage. W.G. Kuschner, H. Wong, A. D'Alessandro, P. Quinlin & P.D. Blanc, *Human Pulmonary Responses To Experimental Inhalation of High Concentration Fine and Ultrafine Magnesium Oxide Particles*, 105 ENVTL. HEALTH PERSPECT. 1234–1237 (1997).

37. In one study, rats were exposed to carbon black particles, which were approximately 14 nm in size and carbon black particles that were 260 nm in size at 1mg m^{-3} for seven hours in rats. Although the larger carbon black particles had no impact on the rats, the nano-sized particles caused several inflammatory effects. Further, risk factors associated with cardiovascular disease were detected in the rats exposed to the nano-sized particles. Researchers concluded that the study marks a "clear link between low exposure to ultrafine particles . . . and physiological changes relevant to pulmonary inflammation, heart attacks and strokes[.]" Ken Donaldson, Vicki Stone, and William MacNee, *The Toxicology of Ultrafine Particles*, in PARTICULATE MATTER: PROPERTIES AND EFFECTS UPON HEALTH 121 (R.L. Maynard & C.V. Howard eds. 1999) [hereinafter Donaldson, Stone & MacNee, *Toxicology*]. Other studies of carbon black have similar results. *See* K.E. Driscoll, et al., *Pulmonary Inflammatory, Chemokine, and Mutagenic Responses in Rats After Subchronic Inhalation of Carbon Black*, TOXICOL. APPL. PHARMACOL. 136 (1996); X.Y. Li, et al., *Free Radical and Pro-Inflammatory Activity of Particulate Air Pollution In Vivo and In Vitro*, 51 THORAX 1216–1222 (1996).

38. Gunter Oberdorster, et al., *Acute Pulmonary Effects of Ultrafine Particles in Rats and Mice*, HEALTH EFFECTS INSTITUTE, Synopsis of Research Report 96 (Aug. 2000). *See also* email from Joe Mauderly (toxicologist) to author (Nov. 11, 2002) (on file with author) ("There have been a few studies of experimental exposures of humans to arc-generated carbon ultrafines, but with little evidence of effects in normals. . . Animal studies done here and elsewhere using similar particles have not suggested that ultrafine carbon is a particular concern unless 'decorated' with organics and metals, as in combustion soot.").

39. A team of researchers in Warsaw carried out experiments to explore whether carbon nanotubes act in lung tissue the way asbestos does. The test group of pathogen free guinea pigs was given an intratracheal instillation of 25 mg of carbon nanotubes in soot, and the control group was given soot without carbon nanotubes. The pulmonary functions of pigs in the two groups did not differ. Further, autopsies did not reveal significant differences in the animals' inflammatory reactions. Therefore, the researchers concluded that "working with soot containing carbon nanotubes is unlikely to be associated with any health risk." *See* A. Huczko, et. al., *Physiological Testing of Carbon Nanotubes: Are They Asbestos-Like?*, 9 FULLERENE NANOTUBES AND CARBON NANOSTRUCTURES 251 (April 15, 2001) (hereinafter "Warsaw Study").

40. Toxicologist Chui-wing Lam at NASA instilled fine-particle suspensions of single-walled nanotubes into the trachea of mice, and seven or 90 days later, evaluated lung tissue for pathological changes. A different group of mice were exposed to carbon black or quartz for comparison. All three different preparations caused microscopic nodules called granulomas, which can potentially lead to lung lesions. The results were presented at the ACS National Meeting in April 2003. *See* Ron Dagani, *Nanomaterials: Safe or Unsafe*, 81 CHEM. & ENG. NEWS 30 (April 28, 2003).
41. Toxicologist David B. Warheit carried out this study. Nanotube/soot mixtures were instilled into the tracheas of rats. Lungs were examined after one day, one week, one month, and three months. Particles of quartz or high-purity iron were also instilled in other rats for comparison. 15% of the rats receiving a high dose of nanotubes (5mg per kg of body weight) died within 24 hours due to suffocation. Researchers also observed granulomas, but their numbers were not correlated with nanotube dose. Additionally, the number of granulomas remained the same or declined during the final two-month observation period of the study. While the surviving rats showed no sustained inflammatory response, exposure to quartz particles resulted in a sustained and dose-dependent inflammatory response. *Id.*
42. Dagani, *supra* note 40.
43. Email from Joe Mauderly, toxicologist, to author (Nov. 11, 2002) (on file with author) ("Epidemiology has tried to compare effects from environmental exposures to fine and ultrafine particles, but with mixed success and mixed results. . . The epidemiology is far from clear, in large part because personal exposure is never known with any accuracy.").
44. Wichmann, et al., *supra* note 31.
45. Mark Eliot Shere, *The Myth Of Meaningful Environmental Risk Assessment*, 19 HARV. ENVTL. L. REV. 409, 438 (Winter 1995) (citing Identification, Classification, and Regulation of Potential Occupational Carcinogens, 45 Fed. Reg. 5002, 5193 (1980)).
46. *Id.* at 439.
47. *Id.*
48. Gerald W. Boston, *A Mass-Exposure Model of Toxic Causation: the Content of Scientific Proof and the Regulatory Experience*, 18 COLUM. J. ENVTL. L. 181, 225-26 (1993). ("There are a number of reasons why the reliability of animal data to establish causation in humans can be questioned. The major reason is that species-related differences exist in the metabolism and disposition of chemicals in general. Differences in the balance between bioactivation and detoxification mechanisms and in the dose of the chemical actually delivered to the target organ represent two common problems concerning the extrapolation of the animal response as a means of predicting the outcome of similar human exposures. Moreover, known species differences in biological processes important in chemical carcinogenesis exist, including differences in the rate of DNA repair, the background incidence of tumors, and the anatomy and biochemical functioning of affected organs or systems.").
49. Shere, *supra* note 45, at 439 ("As a National Cancer Institute researcher explained, 'Now, you would have never predicted that from animal tests . . . . And that is very unfortunate.'").
50. Dr. Selikoff of the Mount Sinai School of Medicine testified that "[f]or years I was told that asbestos was not a powerful carcinogen, because the animals did not show cancer with it. Animal studies began with asbestos in 1930, and the first cancers were produced only in 1962." *Id.* at 439.
51. Donaldson, Stone, and MacNee, *Toxicology*, *supra* note 37, at 119.
52. *Id.* ("[A] sufficiently high exposure to any airborne particle, even a 'low-toxicity

particle' that is non-pathogenic at plausible levels of human exposure, will result in false-positive pathological outcome.").
53. Dickie, *supra* note 27, at 37-8.
54. Interview with Dr. Phil Sayre (Associate Director, Risk Assessment EPA), Jan. 2004 [hereinafter Sayre Interview].
55. Email from Joe Mauderly, toxicologist, to author (Nov. 11, 2002) (on file with author) ("What we know is that, at least at some dose, a portion of deposited ultrafine poorly-soluble particles can get into respiratory surface epithelium, be transported into blood and lymph, and go to other organs. . . .[W]e know that it is at least probable that you might have an ultrafine, solid particle in your big toe that you inhaled, but we don't know whether you should care!"). Dickie, *supra* note 27, at 38 ("Yet data obtained from one route of exposure may not indicate even qualitatively the toxicity associated with other routes, owing to pharmacokinetic differences at different portals of entry. Moreover, there is no generally accepted method for quantitative extrapolation from one route to another.").
56. Sigmund F. Zakrzekwski, ENVIRONMENTAL TOXICOLOGY 19 (Oxford Univ. Press 2002) (noting that the biological effect of a chemical is related to its dose). *See also* Dickie, *supra* note 27, at 33 ("Toxic responses differ in their relationships to dose.").
57. Paul A. Baron, et al., *Evaluation of Aerosol Release During The Handling of Unrefined Single Walled Carbon Nanotube Material*, NIOSH DART-02-191 REV. 1.1 (April 2003).
58. Sayre Interview, *supra* note 54.
59. Dickie, *supra* note 27, at 39.
60. *Id.*
61. Email from Joe Mauderly, toxicologist, to author (Nov. 11, 2002) (on file with author).
62. *Societal Implications of Nanotechnology: Hearing Before the House Science Comm.*, 108th Cong. (April 9, 2003) (Statement of Dr. Vicki L. Colvin, Director Center for Biological and Environmental Nanotechnology (CBEN), and Associate Professor of Chemistry, Rice University, Houston, Texas), FED. DOC CLEARING HOUSE CONG. TESTIMONY, *available at*, LEXIS, Nexis Library.
63. Email conversation between Kathy Jo Wetter (ETC) and author (on file with author).
64. Jim Krane, *Environmentalists Fearful of Nanotechnology*, CHATTANOOGA TIMES, Sept. 10, 2002, at E1.
65. Stephen Charest, *Bayesian Approaches to the Precautionary Principle*, 12 DUKE ENV.L. & POL'Y 265, 365 (Spring 2002).
66. Mark Geistfeld, *Reconciling Cost-Benefit Analysis With The Principle That Safety Matters More Than Money*, 76 N.Y.U.L. REV. 114 (2001). For example, the Rio Declaration on Environment and Development states: "Where there are threats of serious or irreversible damage, lack of full scientific certainty shall not be used as a reason for postponing cost-effective measures to prevent environmental degradation." *The Rio Declaration on Environment and Development*, UN CONFERENCE ON ENVIRONMENT AND DEVELOPMENT, U.N. Doc. A/CONF. 151/5/Rev.1 (1992), *reprinted in* 31 I.L.M. 874 (1992).
67. AARON WILDAVSKY, BUT IS IT TRUE? 428 (1995).
68. *See generally* Frank B. Cross, *When Environmental Regulations Kill*, 22 ECOLOGY L.Q. 729, 730 (1995) [hereinafter Cross, *Regulations*]; Frank B. Cross, *A Syncretic Perspective on Environmental Protection and Economic Growth*, 2 KAN. J. L. & PUB. POL'Y 53 (1992) (highlighting the direct relationship between economic growth and environmental quality); Gene M. Grossman & Alan B. Krueger, *Economic*

Growth and the Environment, Q.J. Econ. 353 (1995) (concluding that there is a positive relationship between national income and measures of environmental quality in developed nations); Douglas Holtz-Eakin & Thomas M. Selden, *Stoking the Fires: CO[2] Emissions and Economic Growth*, 57 J. Pub. Econ. 85 (1995) (finding that higher-income countries have lower levels of carbon dioxide emissions as per capita gross domestic product increases).

69. Ralph L. Keeney, *Mortality Risks Induced By Economic Expenditures*, 10 Risk Analysis 147, 148 (1990).
70. *See* Cross, *Regulations*, *supra* note 68, at 742.
71. Cost-benefit analysis has been defined as a determination of "whether the reduction in risk of material health impairment is significant in light of the costs of attaining that reduction." Natural Res. Defense Council v. EPA, 824 F.2d 1146, 1159 n.6 (D.C. Cir. 1987) (quoting American Textile Mfrs. Inst. v. Donovan, 452 U.S. 490, 506 (1981)).
72. Attempting to prove this would be difficult. Two different methods might be used. One method would measure the costs in terms of lives lost. First, a dollar figure would have to be derived for the economic impact of regulations, and then this figure would have to be translated into lives lost. Several studies have attempted to estimate the number of lives lost as the result of a certain amount of economic loss. Second, an estimate would have to be made regarding the number of lives that would be directly saved as the result of nanotechnology products. Estimating the number of lives that would be saved as the result of a new drug or technology to reduce air pollution would be nearly impossible. Third, there would have to be a projection regarding the number of lives that would be lost as the result of the worst environmental harms possible from nanomaterials. Finally, the number of lives lost as the result of regulation would be compared to the number of lives lost without regulation. A second method would measure the costs in terms of dollars lost. First, a dollar figure would have to be derived for the economic impact of regulations. Second, there would have to be a dollar estimate for the number of lives directly saved as the result of nanotechnology products. Finally, the number of lives that would be lost assuming the worst environmental consequences would have to be converted to a dollar amount.
73. Richard L. Revesz, *Environmental Regulation, Cost-Benefit Analysis, and the Discounting of Human Lives*, 99 Colum. L. Rev. 941, 943 (May 1999) ("The use of cost-benefit analysis has become commonplace in environmental and other health-and-safety regulation.").
74. Executive Order 12,866 mandates that agencies perform cost-benefit analysis for all major regulations. Exec. Order No. 12,866, 3 C.F.R. 638 (1993), *reprinted in* 5 U.S.C. § 601 (1994). It should be noted that this mandate does not supersede statutory provisions. Numerous federal statutes require CBA. *See* Am. Textile Mfrs. Inst. v. Donovan, 452 U.S. 490, 510–11 & 510 n. 30 (1981). Ten states require CBA of all proposed agency rules, and seven states require CBA of selected rules. Robert W. Hahn, State And Regulatory Reform: A Comparative Analysis 3 (AEI-Brookings Joint Ctr. for Regulatory Studies, Working Paper 98-3, 1998), at http://www.aei.brook.edu/publications/working/working_98_03.pdf.
75. 15 USCS § 2605 (2002).
76. 40 C.F.R. § 720.75(d).
77. First, EPA must determine that: (1) there is insufficient information "to permit a reasoned evaluation of the health and environmental effects of a chemical substance"; and (2) "the manufacture, processing, distribution in commerce, use, or disposal of such substance, or any combination of such activities, may present an unreasonable risk of injury to health or the environment, or such substance is

or will be produced in substantial quantities, and such substance either enters or may reasonably be anticipated to enter the environment in substantial quantities or there is or may be significant or substantial human exposure to the substance". 15 U.S.C. § 2604(e)(1)(A) (2002). Second, EPA must issue a proposed order 45 days before the expiration of the notification period. 15 U.S.C. § 2604(e)(1)(B) (2002). The order, which takes effect on the expiration of the notification period, can "prohibit or limit the manufacture, processing, distribution in commerce, use, or disposal of such substance or prohibit or limit any combination of such activities." 15 USCS § 2604(e)(1)(A) (2002). A manufacturer may avoid a section 5(e) order by responding with the requisite specificity to those aspects of the order it deems objectionable. If the manufacturer objects, "the proposed order shall not take effect." 15 U.S.C. § 2604(e)(1)(C) (2002). EPA may then apply to a United States district court for an injunction that would force compliance with the proposed order. 15 U.S.C. § 2604(e)(2) (A)(i) (2002).
78. 15 U.S.C. § 2604(e)(2)(B).
79. Approximately 80% of chemicals in use today have not been adequately tested. The National Academy of Sciences concluded that of the tens of thousands of commercially useful chemicals, only a few have been subjected to toxicity testing and most have scarcely been tested at all. NATIONAL ACADEMY OF SCIENCES, TOXICITY TESTING: STRATEGIES TO DETERMINE NEEDS AND PRIORITIES 92–99 (1984). *See also* Andrew Hanan, *Pushing the Environmental Regulatory Focus a Step Back: Controlling the Introduction of New Chemicals Under the Toxic Substances Control Act*, 18 AM. J. L. & MED. 395, 409 (1992) ("TSCA directs EPA to regulate new chemicals, but overlooks the lack of toxicity information. This information deficit results from an insufficient amount of chemical testing."); John S. Applegate, *The Perils of Unreasonable Risk: Information, Regulatory Policy and Toxic Substances Control*, 91 COLUM. L. REV. 261, 265–66 (March 1991). Applegate explains as follows:

> The prediction of excess deaths for regulatory purposes is, as we shall see, an extremely information-intensive undertaking for which sufficient data is rarely available. The result is pervasive uncertainty in the regulatory process. . . .
> The legal effects of uncertainty are at least as troubling when the agency has the burden of proof. If it is the agency's obligation to justify its actions, uncertainty undermines the factual support it needs to withstand challenges from affected industries. Subjecting agency action to an intensive standard of judicial review, for example, increases the likelihood that a court will find the agency's justification wanting. *Id.*

80. GAO Report, *Chemical Risk Assessment*, (Aug. 2001), at 14–15.
81. Only 7% of chemicals have basic tests for minimum understanding of toxicity. *Id.*
82. EPA officials sort through divergent data, rely on data from similarly structured chemicals and animal tests, and make dose-response extrapolations, route-to-route extrapolations, and predictions of exposure assessment. *See* Hanan, *supra* note 79, at 409 ("This lack of toxicity information forces EPA to employ its own methods of analysis. Due to the large number of PMNs received each year and the costs of testing, EPA is incapable of undertaking a thorough, large-scale testing program. Consequently, EPA uses a simple, relatively unrefined method to analyze these substances. The end result is an immense void of information with regard to the health hazards of chemicals that are approved for commercial use.") (internal citations and footnotes omitted). Evidence does suggest, however, that these estimation techniques are quite accurate. *See id.*

83. For example, a GAO study found that EPA had issued regulations to control "only nine chemicals in almost 18 years" and had imposed controls on new chemicals pending the development of sufficient information "for a small percentage of chemicals." GAO Report, *Toxic Substances Control Act—Legislative Changes Could Make the Act More Effective*, Oct. 26, 1994, at LEXIS, Nexis Library. *See also* Hanan, *supra* note 79 (arguing that a lack of information regarding the majority of chemical substances and an overly strict standard of judicial review mean that most chemicals are approved).
84. 15 U.S.C. § 2601(c) directs EPA to "carry out this chapter in a reasonable and prudent manner" and "consider the environmental, economic, and social impact of any action" taken by it. *See also* 15 U.S.C. § 2601(b)(3) (2002) (declaring policy that "authority over chemical substances . . . be exercised . . . as not to impede unduly or create unnecessary economic barriers to technological innovation."); Chem. Mfrs. Ass'n v. EPA, 899 F.2d 344, 347–8 (5th Cir. 1990) ("Congress also plainly intended the EPA to consider the economic impact of *any* actions taken by it under TSCA.") (emphasis in original); Corrosion Proof Fittings v. EPA, 947 F.2d 1201 (5th Cir. 1991) ("Congress did not enact TSCA as a zero-risk statute. The EPA, rather, was required to consider both alternatives to a ban and the costs of any proposed actions and to 'carry out this chapter in a reasonable and prudent manner [after considering] the environmental, economic, and social impact of any action.'") (quoting 15 U.S.C. § 2601(c) (2002)).
85. First, the EPA must find that there is "insufficient data and experience upon which the effects of" the manufacturing or processing of the chemical "on the health or the environment can reasonably be determined or predicted," and that testing of the chemical "with respect to such effects is necessary to develop such data." *Chem. Mfrs. Ass'n.*, 899 F.2d 344 (5th Cir. 1990) (citing TSCA § 4(a)). Second, EPA must find that the chemical's manufacturing or processing "may present an unreasonable risk of injury to health or the environment" or that the chemical "is or will be produced in substantial quantities, and (I) it enters or may reasonably be anticipated to enter the environment in substantial quantities or (II) there is or may be significant or substantial human exposure to such substance or mixture, . . ." 15 U.S.C. § 2603 (a)(1)(A)(i), (a)(1)(B)(i)(I)(II) (2002).
86. In order to restrict a chemical, EPA must prove that that there is a "reasonable basis to conclude that the manufacture, processing, distribution in commerce, use, or disposal of a chemical substance or mixture, or that any combination of such activities, presents or will present an unreasonable risk of injury to health or the environment." Corrosion Proof Fittings, et al., v. EPA, 947 F.2d 1201, 1214 (5th Cir. 1991).
87. For a comprehensive list of studies, see Lee S. Siegel, Note, *As the Asbestos Crumbles: A Look at New Evidentiary Issues In Asbestos-Related Property Damage Litigations*, 20 Hofstra L. Rev. 1139, n. 132 (Summer 1992).
88. 40 C.F.R. § 763.160.
89. *Id.*
90. *Corrosion Proof Fittings*, 947 F.2d at 1230.
91. *Id.*
92. The six categories of asbestos-containing products that continue to be banned under TSCA include the following: corrugated paper, rollboard, commercial paper, specialty paper, flooring felt, and new uses of the substance (those not existing on July 12, 1989). *Six Product Categories Remain Banned, Restrictions on Eight Others Lifted by EPA*, Daily Env't Rep. (BNA), Nov. 5, 1993, at A-6.

93. General Agreement on Tariffs and Trade, Oct. 30, 1947, 61 Stat. A3, 55 U.N.T.S. 194 [hereinafter GATT]. The GATT was amended as part of the Uruguay Round in 1994. *See* General Agreements on Tariffs and Trade 1994, April 15, 1994, Marraeksh Agreement Establishing the World Trade Organization, Apr. 15, 1994 [hereinafter WTO Agreement], Annex 1A, Legal Instruments—Results of the Uruguay Round vol. 1 (1994). For U.S. implementing legislation, see Uruguay Round Agreements Act, Pub. L. No. 103-465, 108 Stat. 1809 (1994). Relevant GATT provisions are article 1 (mandating non-discrimination against imported goods on the basis of their national origin), article 3 (mandating non-discrimination between foreign and domestic goods), and article 11 (prohibiting quantitative restrictions of imports or exports).
94. Agreement on the Application of Sanitary and Phytosanitary Measures, Apr. 15, 1993, WTO Agreement, Annex 1A, Legal Instruments—Results of the Uruguay Round vol. 1 (1994) [hereinafter SPS Agreement]. Under this agreement, a member state has the right to take necessary sanitary measures (i.e. measures protecting human or animal health) or phytosanitary measures (i.e. measures protecting plant life or health) within its territory that affect trade, but only if they are based on "scientific principles, . . . and not maintain without sufficient scientific evidence." *Id.* art 2(2). Further, members must "take into account the objective of minimizing negative trade effects," *id.* art. 5(4), and must "avoid arbitrary or unjustifiable distinctions in the levels it considers to be appropriate in different situations, if such distinctions result in discrimination or a disguised restriction on international trade." *Id.* art. 5(5).
95. Agreement on Technical Barriers to Trade, Apr. 15, 1994, WTO Agreement, Annex 1A, Legal Instruments—Results of the Uruguay Round vol. 1 (1994) [hereinafter TBT Agreement].
96. *See* WTO: EC Measures Concerning Meat and Meat Products (Hormones)—AB-1997-4, WT/DS26/AB/R & WT/DS48/AB/R, Jan. 16, 1998, *available at* http://www.wtp.org/wto/ddf/ep/public.html. *See generally* Sean D. Murphy, *Biotechnology and International Law*, 42 Harv. Int'l L.J. 47, 80-83 (2001); David A. Wirth, *International Decisions*, 92 Am. J. Int'l L. 755 (1998); Dave E. McNiel, *The First Case Under The WTO's Sanitary and Phytosanitary Agreement: The European Union's Hormone Ban*, 39 Va. J. Int'l L. 89 (1998).
97. *See* SPS Agreement, *supra* note 94, at art. 3(2).
98. Murphy, *supra* note 96, at 85.
99. 15 U.S.C. § 2604(a)(1)(B) (1994). EPA has implemented an electronic filing system for companies with reporting responsibilities under TSCA §§ 4, 5, 8d, 8e and 12b. *See Filing Chemical Reports Electronically Should Be Possible Soon*, 188 Daily Env't Rep. (BNA), Sept. 29, 1999, at A-9.
100. Jessica Gorman, *Taming High-Tech Particles*, Science News, March 30, 2002, at http://www.ruf.rice.edu/~cben/NanoEnviHealth.shtml. *See also* Vicki Colvin, *Responsible Nanotechnology: Looking Beyond the Good News*, EurekAlert!, Nov. 2002, at http://www.eurekalert.org/context.php?context=nano&who=essays&essaydate=1102 (describing the research taking place at CBEN).
101. Gorman, *supra* note 100.
102. *Id.*
103. *See* Victoria Griffith, *Big Risks on a Microscopic Scale: Technology: Fears About The Potential Dangers of Nanotechnology Are Threatening Its Wider Use*, Fin. Times, Sept. 25, 2002, at LEXIS, Nexis Library.
104. *See Research Opportunities: Impacts of Manufactured Nanomaterials On Human Health and the Environment*, at http://es.epa.gov/ncer/rfa/current/2003_nano.html.

Chapter 5

1. Raj Bawa, *Nanotechnology Patenting In The U.S.*, 1 NANOTECH. L&B 31 (2004).
2. U.S. CONST., art. I, § 8, cl. 8.
3. Fritz Machlup identified four different positions on which advocates of patent protection have rested their case: (1) the "natural law" thesis; (2) the "reward-by-monopoly" thesis; (3) the "monopoly-profit-incentive" thesis; and (4) the "exchange for secrets" thesis. Fritz Machlup, *An Economic Review of the Patent System*, Study No. 15 of the Subcomm. on Patents, Trademarks, and Copyrights of the Committee on the Judiciary United States Sen., 85th Cong. 21 (1958). The "natural-law" thesis holds that people have natural property rights in their own ideas. Society is morally obligated to recognize and protect this property right. *Id.* The "reward-by-monopoly" thesis maintains that justice requires that people receive reward for their services in proportion to their usefulness to society. *Id.* The "monopoly-profit-incentive thesis" assumes that inventions and/or their exploitation will not be obtained in sufficient measure if inventors and capitalists can hope only for such profits as the competitive exploitation of all technical knowledge will permit. *Id.* Finally, the "exchange for secrets" thesis portrays the issuance of a patent as a bargain between inventor and society. *Id.* The inventor surrenders the possession of secrecy knowledge in exchange for the protection of a temporary exclusivity in its industrial use. *Id.* Machlup concluded that none of the theories justifies the existence of the patent system. "[I]f we did not have a patent system, it would be irresponsible, on the basis of our present knowledge of its economic consequences, to recommend instituting one. But since we have had a patent system for a long time, it would be irresponsible, on the basis of our present knowledge, to recommend abolishing it." *Id.* at 80.
4. In 1958, Fritz Machlup reported to the Senate that the patent system did not have a substantial effect on innovation. *See id*. Economist Adam Jaffe recently concluded that the "value of patent rights might still be too small relative to overall costs and returns to have a measurable impact on innovative behavior." Adam Jaffe, *The U.S. Patent System in Transition: Policy Innovation and the Innovation Process*, (NBER Working Paper Series, August 1999), 46. *See also Patently Absurd*, THE ECON., June 21, 2001, at LEXIS, Nexis Library ("Do firms become more innovative when they increase their patenting activity? Studies of the most patent-conscious business of all—the semiconductor industry—suggest they do not.").
5. 35 U.S.C. § 102(b) (2002).
6. 35 U.S.C. § 112 (2002). There are three requirements within section 112. First, the application must disclose the best mode for practicing the invention. Second, the enablement doctrine requires the inventor to provide sufficient information to enable a person skilled in the relevant art to make and use the claimed invention without "undue experimentation." CHISUM, ET. AL., PRINCIPLES OF PATENT LAW: CASES AND MATERIALS 162 (2001). The purpose of the doctrine is to provide the public with the benefits of a technological disclosure and to define the scope of patent rights. *Id.* at 172. Third, the written description requirement is "broader than to merely explain how to 'make and use'." The application must also "convey with reasonable clarity to those skilled in the art that, as of the filing date sought, he or she was in possession of the invention." Vas-Cath Inc. v. Mahurkar, 935 F.2d 1555, 1563-64 (Fed.Cir. 1991). In practice, difficulties encountered in attempting to distinguish the two requirements have generated judicial confusion and uncertainty. *Compare In re Wilder*, 736 F.2d 1516, 1520 (Fed.Cir.1984), *cert. denied*, 469 U.S. 1209 (1985) ("The written description requirement is found in 35 U.S.C. § 112 and is sepa-

rate from the enablement requirement of that provision.") *with* Kennecott Corp. v. Kyocera Intr'l, Inc., 835 F.2d 1419, 1421 (Fed.Cir. 1987), *cert. denied*, 486 U.S. 1008 (1988) (The purpose of the [written] description requirement [of § 112] is to state what is needed to fulfill the enablement criteria. These requirements may be viewed separately, but they are intertwined."). *See also* Mark D. Janis, *On Courts Herding Cats: Contending with the "Written Description" Requirement (and Other Unruly Patent Disclosure Doctrines)*, 2 WASH. U. J. LAW & POL'Y 53, 62-3 (2000) (asserting that "neither the Federal Circuit nor the C.C.P.A. has ever articulated a persuasive rationale for distinguishing the written description requirement from the enablement requirement."); Kevin S. Rhoades, *The Section 112 "Description Requirement" - A Misbegotten Provision Confirmed*, 74 J. PAT. & TRADEMARK OFF. SOC'Y 869 (1992) (arguing that written description requirement is not needed in light of the enablement requirement). The Federal Circuit recently confirmed that the written description requirement is separate and distinct from the enablement requirement in 35 U.S.C. § 112, ¶ 2. Univ. of Rochester v. G.D. Searle & Co., Slip No. 03-1304 (Fed. Cir. Feb. 13, 2004), *available at* 2004 U.S. App. LEXIS 2458.
7. Under section 101, the invention must fall within at least one of four classes of statutory subject matter to be patentable: processes, machines, manufactures, or compositions of matter. 35 U.S.C. § 101 (2002).
8. 35 U.S.C. § 101 (2002). Under the PTO's 2001 utility guidelines, upstream research that does not have a "specific, substantial, and credible utility" is excluded from patentability. Utility Examination Guidelines, 66 Fed. Reg. 1092, 1098 (Jan. 5, 2001).
9. 35 U.S.C. § 102(a) (2002). A patent claim is anticipated if each element of the claim in issue is found, either expressly or under principles of inherency, in a single prior art reference. Minnesota Mining & Mfg. Co. v. Johnson & Johnson Orthopaedics, Inc., 976 F.2d 1559, 1565 (Fed. Cir. 1992).
10. The nonobvious requirement has been described as the "most significant obstacle that a patent applicant faces." CHISUM, *supra* note 6, at 514.
11. The suggestion may come (1) "from the references themselves"; (2) "from knowledge of those skilled in the art that certain references, or disclosures in the references, are known to be of special interest or importance in the particular field"; and (3) "from the nature of the problem to be solved, leading inventors to look to references relating to possible solutions to that problem." Pro-Mold & Tool Company v. Great Lakes Plastics, Inc., 75 F.3d 1568, 1573 (Fed. Cir. 1996). Thus, the suggestion does not have to be explicit. *See In re* Nilssen, 851 F.2d 1401, 1403 (Fed. Cir. 1988)("[F]or the purpose of combining references, those references need not explicitly suggest combining teachings, much less specific references.").
12. *In re* Dow Chemical Co., 837 F.2d 469, 473 (Fed. Cir. 1988) ("The consistent criterion for determination of obviousness is whether the prior art would have suggested to one of ordinary skill in the art that this process should be carried out and would have a reasonable likelihood of success, viewed in the light of the prior art."). Although the obviousness determination is a legal conclusion, the court must make factual inquiries including: the scope and content of prior art; differences between the prior art and the claims at issue; and the level of ordinary skill in the art. Graham v. John Deere Co., 383 U.S. 1, 17 (1966). Courts have repeatedly emphasized that "obvious to try" is not the standard and that the use of "hindsight" is prohibited in making obviousness determinations. *See, e.g., In re* O'Farrell, 853 F.2d 894, 903 (Fed. Cir. 1988) ("The admonition that 'obvious to try' is not the standard under § 103 has been directed mainly at two kinds of errors. In some cases, what would have been 'obvious to try' would have been to vary all parameters or try each of numerous possible choices until

one arrived at a successful result, where the prior art gave either no indication of which parameters were critical or no direction as to which of many possible choices is likely to be successful. In other words, what was 'obvious to try' was to explore a new technology or general approach that seemed to be a promising field of experimentation, where the prior art gave only general guidance as to the particular form of the claimed invention or how to achieve it.").

13. A patent gives its owner the "right to exclude others from making, using, offering for sale, or selling" the invention "throughout the United States or importing the invention into the United States . . . ." 35 U.S.C. § 154(a)(1)-(2) (2002).
14. Texas Instruments, Inc. v. United States Int'l Trade Comm'n, 805 F.2d 1558, 1562 (Fed. Cir. 1986).
15. Warner-Jenkinson Co. v. Hilton Davis Chem. Co., 520 U.S. 17, 29 (1997).
16. Graver Tank & Mfg. Co. v. Linde Air Prod. Co., 339 U.S. 605, 608 (1950).
17. 35 U.S.C. § 283 authorizes injunctive relief. Under 35 U.S.C. § 284, the claimant may receive damages adequate to compensate for infringement, but no less than a reasonable royalty. Courts will evaluate several factors in determining if there is willful infringement. *See* Georgia-Pacific Corp. v. United States Plywood Champion Papers Inc. Corp., 446 F.2d 295 (2d. Cir. 1971). Damages for lost profits have been allowed under 284. Rite-Hite Corp. v. Kelley Co., 56 F.3d 1538, 1549 (Fed. Cir. 1995) (en banc). Attorney's fees and treble damages are also granted under 284 when there is willful infringement. *See id.* at 1560 ("The court may increase the damages up to three times the amount found or assessed.") (quoting 35 U.S.C. § 284 (1988)).
18. U.S. Patent Statistics, Calendar Years 1963-2001, at http://www.uspto.gov/web/offices/ac/ido/oeip/taf/us_stat.pdf.
19. *Id.*
20. *PTO Plan to Relieve Patent Backlog Faces Political Hurdles*, NAT'L J's CONGRESS DAILY, June 20, 2003, at LEXIS, Nexis Library.
21. First, the restraints on patenting have gradually eroded. In 1980, the Supreme Court waived a green flag to patentees in biotechnology by declaring that "anything under the sun [made by man] is patentable", including biological organisms. Diamond v. Chakrabarty, 447 U.S. 303, 309 (1980). The Court further opened the floodgates to patentees in the software industry by obliterating the traditional notion that algorithms could not be patented. Courts now uphold algorithm-containing inventions as long they produce a tangible, useful result. *In re* Alappat, 33 F.3d 1526 (Fed. Cir. 1994). In the *State Street* decision, the Supreme Court held that there is no bar to patenting business methods. State St. Bank & Trust Co. v. Signature Fin. Group, Inc., 149 F.3d 1368 (Fed. Cir. 1998), *cert. denied*, 525 U.S. 1093 (1999). As courts chip away at the barriers to obtaining patents and validate more patents than ever before, the PTO has demonstrated an increased willingness to issue patents. *See* John R. Allison & Mark A. Lemley, *Empirical Evidence on the Validity of Litigated Patents*, 26 AIPLA Q.J. 185 (1998) (percentage of patents held valid rose from 35% in the 1970s to 54% in the early 1990s); Robert P. Merges, *Commercial Success and Patent Standards: Economic Perspectives on Innovation*, 76 CAL. L. REV. 803 (1988) (comparing pre-and post-Federal Circuit era statistics); Robert P. Merges, *As Many As Six Impossible Patents Before Breakfast: Property Rights For Business Concepts and Patent System Reform*, 14 BERKELEY TECH. L.J. 577 (1999) (arguing that PTO has become more lenient in reviewing applications); Cecil D. Quillen, Jr. & Ogden H. Webster, *Continuing Patent Applications and Performance of the U.S. Patent and Trademark Office*, 11 FED. CIR. B.J. 1, 12 (2001-2002) (concluding that the PTO may approve as many as 97% of the applications placed before it). In addition to legal changes, technological trends may also be contributing to the patent

frenzy. The "high-tech" revolution has ushered in industries that rely on laboratory innovations rather than large, capital-intensive investments. *See* John R. Allison and Mark A. Lemley, *The Growing Complexity of the United States Patents System*, 82 B.U.L. Rev. 77, 78 (Feb. 2002). Further, firms have recognized that patents can serve as revenue-generating tools as well as strategic bargaining chips. Finally, the quest to patent may simply be another manifestation of a culture that is consumed by property rights. *Id.* at 79 ("[T]he increasing attention paid to intellectual property in the media and popular press has led people to pay more attention to patents.").

22. William Smith, *Patent This!*, 87 A.B.A.J. 49 (March 2001) (quoting Charles P. Baker, Chair of the ABA Intellectual Property Law Section).
23. *See* Allison and Lemley, *supra* note 21, at 77.
24. Hartman, et al., U.S. Patent No. 5,960,411, *Method and System For Placing A Purchase Order Via a Communications Network*, (issued Sept. 28, 1999).
25. John R. Thomas, *Patent System Reform: The Responsibility of the Rulemaker: Comparative Approaches to Patent Administration*, 17 Berkeley Tech. L.J. 727, 728 (Spring 2002).
26. The Strategic Plan increases prosecution and maintenance fees, utilizes foreign prosecution strategies regarding searching and handling of applications, improves the reviewable record, changes patent rules, and increases the quality of examiners. Steven J. Moore and James W. Jakobsen, *The 21st Century Strategic Plan: An Overhaul of the USPTO or A Wrench in American Innovation?*, Intellectual Prop. Today, Sept. 2002, at LEXIS, Nexis Library.
27. Under the Patent and Trademark Authorization Act of 2002, Congress authorized the PTO to receive appropriations for fiscal years 2003 through 2008 "in amounts equal to those fees collected by the agency in each such fiscal year." *See* Robert O. Lindefield, *IP Reforms Become A Reality With Signing of Bill*, 227 Legal Intelligencer 110, at 5, Dec. 5, 2002, *available at* LEXIS, Nexis Library. However, this provision does not guarantee increased resources for the PTO as the accompanying Conference Report makes clear that "[i]f enacted, . . . this full-funding authorization would still be subject to appropriations." *Id.* The Act also directed the PTO to create and implement a computer-based filing and processing system to enable patent practitioners and examiners alike to harness the full power of the Internet and computers to "send all communications electronically" and allow the PTO to "process, maintain, and search electronically the contents and history of each application." *Id.*
28. Bawa, *supra* note 1. It should be noted that PTO is working to develop a cross-reference art collection for nanotechnology patents. Cross-reference art collections serve as a tool for examiners to search across different classes. The nanotechnology art collection will include documents from the IPC Class B82B, additional documents obtained through key word searches and screening of relevant U.S. classifications, and additional IPC groups and subclasses.
29. Tennent, U.S. Patent No. 4,663,230, *Carbon Fibrils, Method For Producing Same and Compositions Containing Same* (issued May 5, 1987) (claiming "an essentially cylindrical discrete carbon fibril characterized—by a substantially constant diameter between—about 3.5 and about 70 nanometers, length greater than—about 102 times the diameter, an outer region of multiple—essentially continuous layers of ordered carbon atoms and—a distinct inner core region, each of the layers and core—disposed substantially concentrically about the cylindrical—axis of the fibril.").
30. *See, e.g.*, Candace Stuart, *Nanotech Is Old Tech To Hyperion, Churning Out Fibrils Since the 80s*, Small Times, Sept. 16, 2002, at http://www.smalltimes.com/document_display.cfm?document_id=4622 ("Tennent developed a process for cat-

alytically growing nanotubes using hydrocarbon feedstocks in 1982."); Tennent, et al., U.S. Patent No. 5,165,909, *Carbon Fibils and Method For Producing Same* (issued Nov. 24, 1992) ("Multiwalled carbon nanotubes of a morphology similar to the catalytically grown fibrils described above have been grown in a high temperature carbon arc... It is now generally accepted that these arc-grown nanofibers have the same morphology as the earlier catalytically grown fibrils of Tennent.").
31. Bethune, et. al., U.S. Patent No. 5,424,054, *Carbon Fibers and Method for Their Production* (issued June 13, 1995).
32. *See, e.g.*, Moy, et al., U.S. Patent No. 6,221,330, *Process For Producing Single Wall Nanotubes Using Unsupported Metal Catalysts* (issued April 24, 2001) ("Pat. No. 5,424,054 to Bethune et al. describes a process for producing single-walled carbon nanotubes by contacting carbon vapor with cobalt catalyst.").
33. Dai et al., U.S. Patent No. 6,346,189, *Carbon Nanotube Structures Made Using Catalyst Islands* (issued Feb. 12, 2002).
34. Brandes, et al., U.S. Patent No. 6,445,006, *Microelectronic and Microelectromechanical Devices Comprising Carbon Nanotube Components, and Methods of Making Same* (issued Sept. 3, 2002).
35. *Id.*
36. Smalley, et al., U.S. Patent No. 6,683,783, *Carbon Fibers Formed From Single-Wall Carbon Nanotubes* (issued Jan. 27, 2004).
37. Goldstein, U.S. Patent No. 6,268,041, *Narrow Size Distribution Silicon and Germanium Nanocrystals* (issued July 31, 2001).
38. Alivisatos, U.S. Patent No. 5,505,928, *Preparation of III-V Semiconductor Nanocrystals* (issued April 9, 1996).
39. Bawendi, et al., U.S. Patent No. 6,322,901, *Highly Luminescent Color-selective Nano-crystalline Materials* (issued Nov. 27, 2001).
40. Weiss, et al., U.S. Patent No. 5,990,479, *Organo Luminescent Semiconductor Nanocrystal Probes For Biological Applications and Process For Making and Using Such Probes* (issued Nov. 23, 1999); *see also* Weiss, et al., U.S. Patent No. 6,207,392, *Semiconductor Nanocrystal Probes For Biological Applications and Process For Making and Using Such Probes* (issued March 27, 2001).
41. *See* Rebecca Eisenberg, *Public Research and Private Development: Patents and Technology Transfer in Government-Sponsored Research*, 82 VA. L. REV. 1663, 1710 (1996) ("The Cohen-Boyer patents have been widely licensed to biotechnology firms and pharmaceutical firms on terms that have been set low enough that they have generated few complaints from industry and have probably not created a significant impediment to commercial development."); Patent and Trademark Office, *Patent Pools: A Solution to the Problem of Access in Biotechnology Patents*, Dec. 5, 2000, at 3 [hereinafter *Patent Pools*] ("[T]wo of the most profitable patents in the biotechnology area are those of Cohen and Boyer, which are owned by Stanford University. These patents cover the fundamental technology used throughout molecular biology, including recombinant DNA research. By minimizing licensing fees and extending non-exclusive licenses, potential infringers were inclined to obtain licenses and the technology was therefore broadly distributed. The dominance of these patents did not inhibit further development but instead spurred further innovation while providing profits to the patent owner.").
42. For example, PTO is working with the NSF to develop a nanotechnology patent activity report. Developing the report has been difficult due to difficulties in defining what a nanotechnology patent is. Database searches to determine the number of the nanotechnology patents can lead to numbers anywhere from 1100 to 17,000—depending on the search terms used.

43. Economist Carl Shapiro has used this term to describe industries where large numbers of patents must be licensed to commercialize a product. *See* Carl Shapiro, *Navigating the Patent Thicket: Cross Licenses, Patent Pools, and Standard-Setting*, in 1 INNOVATION POLICY AND THE ECONOMY (Adam Jaffe, Joshua Lerner, and Scott Stern, eds., 2001).
44. Rodriguez, et al., U.S. Patent No. 6,159,538, *Method for Introducing Hydrogen into Layered Nanostructures*, (issued Dec. 12, 2000); Maeland, et al., U.S. Patent No. Patent 6,290,753, *Hydrogen Storage in Carbon Material*, (issued Sept. 18, 2001); Okazaki, et al., U.S. Patent No. 6,481,217, *Gas Storage Method and System, and Gas Occluding Material*, (issued Nov. 19, 2002); Chen, et al., U.S. Patent No. 6,471,936, *Method of Reversibly Storing H2 and H2 Storage System Based on Metal-Doper Carbon-Based Materials*, (issued Oct. 29, 2002); Schutz, U.S. Patent No. 6,541,974, *Device for Storing a Gaseous Medium in a Storage Container* (issued April 1, 2003); Pratt, et al., U.S. Patent No. 6,584,825, *Method and Apparatus for Determining the Amount of Hydrogen in a Vessel*, (issued July 1, 2003); Fleckner, et al., U.S. Patent No. 6,589,682, *Fuel Cells Incorporating Nanotubes in Fuel Feed*, (issued July 8, 2003); Wolfe, U.S. Patent No. 6,591,617, *Method and Apparatus for Hydrogen Storage and Retrieval*, (issued July 15, 2003); Cooper, et al., U.S. Patent No. 6,596,055, *Hydrogen Storage Using Carbon-Metal Hybrid Compositions*, (issued July 22, 2003); Schmitman, U.S. Patent No. 6,610,193, *System and Method for the Production and Use of Hydrogen on Board a Marine Vessel*, (issued August 26, 2003); Norley, et al., U.S. Patent No. 6,613,252, *Molding of Materials from Graphite Particles*, (issued September 2, 2003); Hussain, et al., U.S. Patent No. 6,634,321, *Systems and Method for Storing Hydrogen*, (issued October 21, 2003); Zagaja, et al., U.S. Patent No. 6,659,049, *Hydrogen Generation Apparatus for Internal Combustion Engines and Method thereof*, (issued December 9, 2003); Bradley, et al., U.S. No. Patent 6,672,077, *Hydrogen Storage in Nanostructure with Physisorption*, (issued January 6, 2004); Emori, et al., U.S. Patent No. 6,680,600, *Power Supply Unit, Distributed Power Supply System and Electric Vehicle Loaded therewith*, (issued January 20, 2004).
45. NANOBUSINESS ALLIANCE, *Quantum Dot and Semiconductor Nanocrystal Survey*, 2002 (The survey was prepared by Foley & Lardner and can be purchased by contacting the Alliance).
46. There were 18 general patents directed towards methods and processes of making quantum dots, as well as the quantum dots themselves. The survey identified 75 patents on building generic semiconductor devices using quantum dots. There were 27 patents covering the different light absorbing and emitting characteristics of semiconductor nanocrystals. There were also 11 patents on optical switches, optical spatial modulators, optical fibers, optical filters, and general optically responsive mediums. According to the study, 46 patents covered a variety of different electronic devices including switches, modulators, task lighting, flat panel display lighting, LEDs, and radiation detectors. Finally, 39 patents dealt with memory devices.
47. Interview with Ken Barovsky (Quantum Dot Corp.), November 2002.
48. Interview with Larry Bock (Nanosys Inc.), September 2003.
49. J. Steven Rutt, FOLEY & LARDNER, *Dendrimers and Nanotechnology: A Patent Explosion*, Apr. 29, 2002, *available at* http://www.foleylardner.com/FILES/tbl_s31Publications/FileUpload137/840/rutt_dendrimer.pdf.
50. Mike McGehee, Prof. Materials Science, Stanford University, March 2003.
51. A careful analysis of the semiconductor industry reveals that the patent litigation that does take place is primarily the result of dominant firms attempting to "keep outsiders from entering." John H. Barton, *Antitrust Treatment of Oligopolies With Mutually Blocking Patent Portfolios*, 3 ABA: ANTITRUST L.J. 69, 854-55 (2002).

52. Arti Rai, *Fostering Cumulative Innovation in the Biopharmaceutical Industry: The Role of Patents and Antitrust*, 16 BERKELEY TECH. L.J. 813, 834 (Spring 2001).
53. Michael Heller & Rebecca S. Eisenberg, *Can Patents Deter Innovation? The Anticommons in Biomedical Research*, 280 SCI. 698, 701 (May 1998) (noting that acrimonious licensing negotiations are particularly likely in biotechnology research because the negotiating parties are often scientists who may overestimate the value of their work).
54. *See* Suzanne Scotchmer, *Standing on the Shoulders of Giants: Cumulative Research and the Patent Law*, 5 J. ECON. PERSPECT. 29, 32 (1991) ("A second inventor who cannot market the next generation product without a license has a very weak bargaining position. If the second inventor does not get all the surplus being bargained over, he will earn only a fraction of the new product's market value and presumably only a fraction of its social value, and this fraction may be less than the cost of developing it. Hence the incentive for an outside firm to develop second generation products can be too weak."); *see also* Rai, *supra* note 52, at 833 (arguing that it is impossible to divide the surplus ex post in a manner that provides adequate incentives for both the initial inventor and the improver).
55. *See* Michael S. Greenfield, Note, *Recombinant DNA Technology: A Science Struggling With The Patent Law*, 44 STAN. L. REV. 1051, 1073-74 (1992) (explaining the hold-out problem in the context of biotechnology).
56. *Id.*
57. Interview with Craig Zolan, CEO Uventures, March 2003. Mr. Zolan recounted stories told by managers of start-up companies attempting to obtain a handful of licenses from different entities. Since pricing is so subjective, each entity would demand absurdly high amounts of equity in the company in exchange for the license. Managers found it difficult to explain to patent holders the dilemma of having to pay a substantial price for a number of different licenses.
58. Experts estimate that the cost of negotiating a license is approximately $50,000 per license. Mark A. Lemley, *Rational Ignorance At The Patent Office*, 95 Nw. U. L. REV. 1495, 1507 (Summer 2001).
59. *See* Rai, *supra* note 52, at 832 (noting that a study conducted in 1997 and 1998 by the NIH Working Group on Research Tools concluded that the high transaction costs associated with licensing negotiations over research tools blocked the licensing of many research tools). *See also* Merges & Nelson, *On The Complex Economics of Patent Scope*, 90 COLUM. L. REV. 839, 874 (1990) ("A substantial literature documents the steep transaction costs of technology licensing, and there is indirect evidence that these costs increase when major innovations are transferred. Moreover, various studies have indicated that transaction costs tend to be very high if licenses are tailored to particular licensees.").
60. Informal "cross-licensing" arrangements can be observed in the semiconductor, consumer electronics, and chemical industries.
61. John H. Barton, *Antitrust Treatment of Oligopolies With Mutually Blocking Patent Portfolios*, 3 ABA: ANTITRUST L.J. 69 , 854 (2002).
62. *Id.*
63. Robert P. Merges, *Contracting Into Liability Rules: Intellectual Property Rights and Collective Rights Organizations*, 84 CALIF. L. REV. 1293, 1353 (Oct. 1996).
64. *Id.* at 1340.
65. *Id.*
66. *Id.*
67. *See Pooling of Patents: Hearings on H.R. 4523 Before the House Comm. on Patents*, 74th Cong. 1140, 1144-45 (1935) ("In all of the following major industries

which the committee has included within the scope of its activities some form of patent consolidation are in use in an attempt to circumvent the existence of patent deadlocks and overlapping inventions: automobile, agricultural machinery, aviation, building equipment and supplies, chemicals, communications, electrical-equipment industries, food industries, glass, machinery and machine equipment, mining, munitions, oil, office equipment and machinery, radio, railroad equipment, rubber, steel, scientific instruments, utilities.").

68. *See* Merges, *supra* note 63, at 1342.
69. *See* Harry T. Dykman, *Patent Licensing within the Manufacturer's Aircraft Association (MAA)*, 46 J. PAT. OFF. SOC'Y 646, 648 (1964).
70. The patent pool was formed in 1997 by the Trustees of Columbia University, Fujitsu Limited, General Instrument Corp., Lucent Technologies Inc., Matushita Electric Industrial Co., Ltd., Mitsubishi Electric Corp., Philiips Electronics N.V. (Philips), Scientific-Atlanta, Inc., and Sony Corp. *See* Letter from Joel I. Klein, Assistant Attorney General, Department of Justice, Antitrust Division, to Gerrard R. Beeney, Esq. (June 26, 1997), at http://www.usdoj.gov/atr/public/busreview/1170.htm.
71. This pool was formed in 1999 by Toshiba Corporation, Hitachi, Ltd., Matsushita Electric Industrial Co., Ltd., Mitsubishi Electric Corp., Time Warner, Inc., and Victor Company of Japan, Ltd. *See* Letter from Joel I. Klein, Assistant Attorney General, Department of Justice, Antitrust Division, to Carey R. Ramos, Esq. (June 10, 1999), at http://www.usdoj.gov/atr/public/busreview/2121.htm.
72. Rai, *supra* note 52, at 844.
73. Heller and Eisenberg, *supra* note 53, at 700 ("Large corporations with substantial legal departments may have considerably greater resources for negotiating licenses on a case-by-case basis than the public sector institutions or small start-up firms."); Donna M. Gitter, *International Conflict Over Patenting Human DNA Sequences in the United States and the European Union: An Argument For Compulsory Licensing And A Fair-Use Exemption*, 76 N.Y.U.L. REV. 1623, 1681 (Dec. 2001) ("There are a lot of small, hungry companies out there whose only asset is intellectual property. It's less likely that broad cross-licensing agreements can happen. If you have too many people owning small, overlapping slices of the same pie, there could be a breakdown.").
74. Shapiro, *supra* note 43, at 8. ("The hold-up problem is worst in industries where hundreds if not thousands of patents, some already issued, others pending, can potentially read on a given product. In these industries, the danger that a manufacturer will 'step on a land mine' is all too real. The result will be that some companies avoid the minefield altogether, i.e. refrain from introducing certain products for fear of hold-up.").
75. *See* Heller and Eisenberg, *supra* note 53.
76. *Id. See also* Alexander K. Haas, *Intellectual Property B. Patent 3. Patentability b) Genome Data: The Wellcome Trust's Disclosure of Gene Sequence Data into the Public Domain & the Potential for Proprietary Rights in the Human Genome*, 16 BERKELEY TECH. L.J. 145, 160-61 (2001) ("Licensing of patent rights may not solve the problem. In the genomics field, the need to get licenses from many patent holders to produce a downstrean product leads to high transaction costs and raises a real danger of an anticommons tragedy."); James Bradshaw, *Gene Patent Policy: Does Issuing Gene Patents Accord With the Purposes of the U.S. Patent System*, 37 WILLIAMETTE L. REV. 637, 657-59 (Fall 2001) ("Gene patents, however, have the potential to interfere with the process of transforming scientific discoveries into commercial products... This situation involves the prolifera-

tion of patents in upstream discoveries stifling downstream innovations, such as product development.").
77. "On the margin, the costs of a patent system will harm small inventors more than large; negotiating a patent system is easier for IBM than for the garage inventor." LAWRENCE LESSIG, THE FUTURE OF IDEAS 209 (2001).
78. Reid G. Adler, *Controlling the Applications of Biotechnology: A Critical Analysis of the Proposed Moratorium on Animal Patenting*, 1 HARV. J. LAW & TECH. 1, 16 (Spring 1998) ("Because much of the present commercial development of biotechnology is performed by small start-up ventures, companies may depend heavily on patent protection to justify the major research and development investments necessary to undertake difficult technological challenges."); *Hearings Before the Subcomm. on Regulation and Business Opportunities of the House Comm. on Small Business*, 100th Cong., 2d Sess. (Mar. 29, 1988) (Statement of Subcomm. Chair, Ron Wyden (D, Oregon)) ("[T]he 'patent approval process can shape—or warp—the future of an entire fledgling industry' since 'patents are the financial and legal backbone of any biotech firm.'"); KENNETH J. BURCHFIEL, BIOTECHNOLOGY AND THE FEDERAL CIRCUIT 476 (1995) ("[P]redictability is an essential requirement for industrial research and investment in new technology.").
79. Lisa A. Small, *Offensive and Defensive Insurance Coverage For Patent Infringement, Litigation: Who Will Pay*, 16 CARDOZO ARTS & ENT. L.J. 707, 708 ("For small, start-up high-tech companies, protecting their intellectual property from the illegal exploitation of larger competitors can lead to financial ruin. 'Anyone who has been involved in an intellectual property case as counsel or as a party will tell you it is an expensive experience.' In particular, patent suits are among the most cumbersome and expensive cases in the legal system because they typically involve complicated technology requiring a tremendous amount of research for litigation and extensive expert testimony."); FRED WARSHOFSKY, THE PATENT WARS: THE BATTLE TO OWN THE WORLD'S TECHNOLOGY (1994) ("For the highly innovative and usually underfunded companies that make up much of the biotech industry, the mere threat of patent litigation is enough to force them to shut down a production or shutter the business itself.").
80. Carolina Braunschweig, *Nano Nonsense*, 1120 VENTURE CAPITAL JOURNAL 18, 20 (Jan. 2003).
81. *See* Yusing Ko, *An Economic Analysis of Biotechnology Patent Protection*, 102 YALE L.J. 777, 800 (1992) ("The traditional phamarceutical companies, despite their superior innovative resources, lag far behind the small start-up companies in contributing to biotechnological innovations.").
82. Mark A. Lemley, *supra* note 58, at n. 2.
83. *Id. See also* MARK A. LEMLEY AT AL., SOFTWARE AND INTERNET LAW 334-34 (2000); Julie E. Cohen & Mark A. Lemley, *Patent Scope and Innovation in the Software Industry*, 89 CAL. L. REV. 1, 42-45 (2001).
84. *Id.* at 56 (concluding that the issuance of broad claims in the software industry is "much more likely to hinder innovation than to foster it"); Pamela Samuelson, et al., *Toward a Third Intellectual Property Paradigm: A Manifesto Concerning the Legal Protection of Computer Programs*, 94 COLUM. L. REV. 2308, 2422 (Dec. 1994) (noting that the "issuance of a large number of questionable patents for software-related ideas may impede competitive development and follow-on innovation in the software industry"); David A. Burton, *Software Developers Want Changes in Patent and Copyright Law*, 2 MICH. TELECOMM. & TECH. L. REV. 87, 87 (1996) (describing the results of a poll of computer programmers that concluded that most programmers believe that software patents impede development).

85. *See, e.g., Patently Absurd*, THE ECON., June 21, 2001, at LEXIS, Nexis Library.
86. State Street Bank & Trust Co. v. Signature Fin. Group, Inc., 149 F.3d 1368 (Fed. Cir. 1998), *cert. denied*, 525 U.S. 1093 (1999).
87. Lemley, *supra* note 58, at n. 2 (Summer 2001); *see also* John Schwartz, *Online Patents to Face Tighter Review*, WASH. POST, Mar. 30, 2000, at E1.
88. For a more detailed explanation of the review guidelines, see Seth H. Ostrow and Silvana M. Merlino, *PTO, Congress Seek To Improve 'Business Method' Patents*, 17 E-COMMERCE 4, April 2001, at LEXIS, Nexis Library.
89. Raymond Van Dyke, *E Wars—Episode One: The Patent Menace*, 6 COMP. L. REV. & TECH. J 1, 9 (Fall 2001).
90. Behfar Bastani & Dennis Fernandez, *Intellectual Property Rights in Nanotechnology*, INTELL. PROP TODAY, Aug. 2002.
91. Kelly Kordzik, *Small New World*, NAT. L. J., Dec. 16, 2002, at LEXIS, Nexis Library.
92. The Atlantic Nano Forum holds monthly technology seminars to sharpen skills of the examiners and improve perception that the PTO is sufficiently trained to handle nanotechnology.
93. For a review of the issues discussed at the meeting, see PTO Customer Partnership Meeting, 1 NANOTECH. L & B 133 (2004).
94. *See, e.g.*, Joseph A. Yosick, *Note: Compulsory Patent Licensing For Efficient Use of Inventions*, 2001 U. ILL. L. REV. 1275 (2001).
95. The Supreme Court has stated that "compulsory licensing is a rarity in our patent system." Dawson Chemical Co. v. Rohm & Haas Co., 448 U.S. 176, 215 (1980). *See also* Cont'l Paper Bag Co. v. Eastern Paper Bag Co., 210 U.S. 405 (1908) (noting that "it is the privilege of any owner of property to use or not use it, without question of motive."). Several proposals have been advanced to amend U.S. patent law to require compulsory licensing under certain circumstances. For example, the Hart Bill was proposed in 1973 to permit compulsory licensing of patents related to "public health, safety, or protection of the environment." *See* Jason Mirabito, *Compulsory Patent Licensing For the United States: A Current Proposal*, 57 J. PAT. OFF. SOC'Y 404, 432 (1975). In 1999, the Affordable Prescription Drugs Act was proposed to require compulsory licensing of patents relating to human health under certain conditions, including if "the patented material is priced higher than may be reasonably expected." H.R. 2927, 106th Cong. (1999). However, these bills have failed due to strong political opposition. Yosick, *supra* note 94, at 1278.
96. 35 U.S.C. § 271(d)(4) (1994).
97. The Atomic Energy Act provides for the licensing of patents "useful in the production or utilization of special nuclear material or atomic energy." 42 U.S.C. § 2183(c) (2002).
98. The Clean Air Act requires licensing if the invention is needed to comply with emission requirements, there is no reasonable alternative available to meet the requirements, and the unavailability of the use of the invention would result in a "lessening of competition or [the] tendency to create a monopoly." 42 U.S.C. § 7608(B)(2).
99. *In re* Indep. Serv. Orgs. Antitrust Litig., 203 F.3d 1322, 1327 (Fed. Cir. 2000).
100. *See generally* Leroy Whitaker, *Compulsory Licensing - Another Nail in the Coffin*, 2 AM. INTELL. PROP. L. ASS'N. Q.J. 155 (1974).
101. *Patent Pools, supra* note 41, at 8 ("By creating a patent pool of these basic patents, businesses can easily obtain all the necessary licenses required to practice that particular technology concurrently from a single entity.").

102. *Id.* at 8-9 ("[P]atent pools can reduce or eliminate the need for litigation over patent rights because such disputes can be easily settled, or avoided, through the creation of a patent pool. A reduction in patent litigation would save business time and money, and also avoid the uncertainty of patent rights caused by litigation.").
103. *Id.* at 8.
104. Merges, *supra* note 63, at 1346 (noting that aircraft patent pool was "lauded far and wide as a success"). Michael A. Heller, *The Boundaries of Private Property*, 108 YALE L.J. 1163 (April 1999) ("In the automobile, aircraft, and synthetic rubber industries, patent pools have emerged, sometimes with the help of government, when licenses under multiple patent rights were thought necessary to develop important new products."); Dykman, *Patent Licensing within the Manufacturer's Aircraft Association (MAA)*, 46 J. PAT. OFF. SOC'Y 646 (1964) (describing government intervention to form industry licensing pool, because "no one would license the other under anything like a reasonable basis"). *See also* Gen. Tire & Rubber Co. v. Firestone Tire & Rubber Co., 489 F.2d 1105, 1107–08 (6th Cir. 1973) (describing a pool in the area of synthetic rubber research formed at the request of the U.S. government).
105. Patent pools have been criticized as being anticompetitive for several reasons. First, the pooling of such patents creates a barrier to competitive alternatives to the technology. *Patent Pools, supra* note 41, at 10 ("This argument is based on the assumption that while certain patents may be considered to be legally blocking, such patents actually cover competitive alternatives to a certain technology, and that the pooling of these patents will expand monopoly pricing."); *see also* Steven C. Carlson, Note, *Patent Pools and the Antitrust Dilemma*, 16 YALE J. ON REG. 359, 373 (1999). Second, because pools shield invalid patents, they force the public to pay royalties on technology that would become part of the public domain if there was litigation. *Id.* at 386–87. Third, pools encourage collusion and price fixing. *Id.* at 387. Due to these concerns, patent pools must pass administrative and judicial antitrust scrutiny.
106. In reviewing the legitimacy of patent pooling agreements, the DOJ and FTC make two determinations: (1) "whether the proposed licensing program is likely to integrate complementary patent rights"; and (2) "if so, whether the resulting competitive benefits are likely to be outweighed by competitive harm posed by other aspects of the program." *Patent Pools, supra* note 41, at 7; *see also* Letter from Joel I. Klein, Assistant Attorney General, Department of Justice, Antitrust Division, to Carey R. Ramos, Esq. (June 10, 1999), at http://www.usdoj.gov/atr/public/busreview/2121.htm. The Justice Department recently maintained that it will approve a pooling scheme if the following conditions are satisfied: (1) the patents in the pool are valid and not expired; (2) there is no aggregation of competitive technologies and setting a single price for them; (3) an independent expert is used to determine whether a patent is essential to complement technologies in the pool; (4) the pool does not disadvantage competitors in downstream product markets; and (5) the pool participants must not collude on prices outside the scope of the pool. *Id.*
107. Merges, *supra* note 63, at 1357.
108. *Id.* at 1356.
109. *See id.* ("In several cases where the government was concerned that technology useful to the military was not being developed because of a logjam of conflicting property rights, the lurking threat of the eminent domain power contributed to the formation of patent pools.").
110. Email from Lita Nelson, Director of technology licensing at MIT, to author, (Winter 2003) (on file with authors).

111. For example, the pool on aircraft was only created after Congress threatened to pass legislation mandating licensing. *See* Merges, *supra* note 63, at 1356-57.

Chapter 6

1. Jane Henney, *Jane Henny Delivers Remarks at the National Press Club*, FED DOCUMENT CLEARING HOUSE POL. TRANSCRIPTS, Dec. 12, 2000, at LEXIS, Nexis Library.
2. *The Food and Drug Administration: An Overview*, CTR. FOR FOOD SAFETY & APPLIED NUTRITION (CFSAN Pub. No. BG 99-2), June 11, 1999, at http://www.cfsan.fda.gov/fdaoview.html.
3. The President's budget request for fiscal year 2004 totals $1.7 billion, including $1.4 billion in budget authority and $300 million in user fees. *See Fiscal Year 2004: Hearing Before the Subcomm. on Agriculture of the House Comm. On Appropriations*, 108th Cong. (2003) (statement of Mark McClellan, Commissioner, Food and Drug Administration), at LEXIS, Nexis Library.
4. FDA Modernization Act, Pub. L. No. 105-115, 111 Stat. 2296 (codified at 21 U.S.C. §§ 301 et seq. (1994)) [hereinafter FDAMA]. The most significant initiative to enhance the efficiency of drug regulation was the reauthorization of "user fees" on drug manufacturers. FDAMA § 101—107 (1994). The legislation also attempted to increase patient access to experimental drugs through the fast-track process and increase the similarities between regulations on drugs and biologics and streamline the approval process for clinical research on drugs and biologics. *Id.* at § 112, § 123(f) ("The Secretary of Health and Human Services shall take measures to minimize differences in the review and approval of products required to have approved biologics license applications under section 351 of the Public Health Service Act (42 U.S.C. 262) and products required to have approved new drug applications under section 505(b)(1) of the Federal Food, Drug, and Cosmetic Act (21 U.S.C. 355(b)(1))"). In an attempt to streamline the approval process for clinical research on drugs and biologics, FDAMA enables clinical investigations to begin 30 days after the manufacturer provides the FDA with a submission containing required information. *Id.* at § 117. A manufacturer can request to meet with the FDA to collaborate in designing clinical trials for NDA's and BLA's. *Id.* at § 119. The legislation also intended to increase the use of scientific advisory panels and simplify the approval process for drug and biological manufacturing changes. *Id.* at § 120, § 116. In addressing regulation of medical devices, Congress authorized the FDA to allow third party review of low-risk 510(k)'s and put a higher priority on FDA review of life-saving devices. *Id.* at § 210, § 202. The act directed the FDA to institute "early meetings" with product sponsors. *Id.* at § 201, § 205. Finally, the act mandated that the FDA consider the "least burdensome appropriate means of evaluating device effectiveness that would have a reasonable likelihood of resulting in approval." *Id.* at § 205. The legislation also included several broad policies intended to improve the overall effectiveness of the FDA. First, the legislation introduced a process for dispute resolution when a scientific controversy arises between the manufacturer and the agency. *Id.* at § 404. Second, the legislation placed a strong emphasis on consultation and cooperation with domestic and international entities. The Act notes that the FDA should "participate...with representatives in other countries to reduce the burden of regulation, harmonize regulatory requirements, and achieve appropriate reciprocal arrangements;" and "carry out [its mission] in consultation with experts . . . and in cooperation with consumers, users, manufacturers, importers, packers, distributors, and retailers of regulated products."

Id. at § 406. Under § 411, the FDA was directed to implement programs and policies to enhance collaboration with the NIH and other science-based Federal agencies, and under § 415, the FDA was encouraged to enter into contracts with organizations or individuals to enhance product evaluation. The agency was also directed to establish a system for recognizing national and international standards in product review. *Id.* at § 410.

5. *See generally FDA's Drug Review and Approval Times*, FDA /CTR. FOR DRUG EVALUATION & RESEARCH, at http://www.fda.gov/cder/reports/reviewtimes/default.htm, (last updated Jul. 30, 2001) (describing new procedures reducing review times).
6. The act included a number of provisions such as allowing establishment inspection by accredited persons (third-parties) under carefully prescribed conditions and new regulatory requirements for reprocessed single-use devices. For a good summary of the law, see *Summary of the Medical Device User Fee and Modernization Act of 2002*, FDA/CTR. FOR DEVICES & RADIOLOGICAL HEALTH, Nov. 7, 2002, at http://www.fda.gov/cdrh/mdufma/mdufmasummary.html.
7. Michael Brooks, *Thanks But No Thanks*, NEW SCIENTIST, Oct. 6, 2001, at 33. ("[I]t still isn't clear whether anyone actually wants to be a nanomedicine guinea pig. It's all very well to dream up and develop cell-repair machines, but what if the prospect of rampaging nanorobots or unexpected immune reactions means that no one is prepared to have the technology implanted?. . . The fate of trial subjects in similar endeavors might make saying no to nanomedicine not seem like such an extreme reaction.").
8. Bogdan Dziurzynski, *FDA Regulatory Review and Approval Process: A Delphi Inquiry*, 51 FOOD AND DRUG L.J. 143, 144 (1996) (noting that manufacturer's "inability to establish a consistent track record of success further complicates the ability of companies to raise capital").
9. *Id.* at 145 (noting that "FDA reviewers do all they can to avoid being publicly criticized for a purported lack of regulatory oversight" and that this has caused the agency to become "one of the most conservative government health protection agencies in the world").
10. *See id.* at 144 ("Trade press reports on the complexities of drug development and the failure of some biotechnology companies to successfully navigate the regulatory waters jaded analysts and made investors apprehensive.").
11. LEWIN GROUP, OUTLOOK FOR MEDICAL TECHNOLOGY INNOVATION: WILL PATIENTS GET THE CARE THEY NEED? REPORT 4: THE IMPACT OF REGULATION AND MARKET DYNAMICS ON INNOVATION 11 (4th Report 2001) (noting that "[t]o the extent that new technology raises this form of regulatory uncertainty or otherwise challenges the readiness of the agency to manage regulation in a timely and predictable manner, companies may be discouraged from attempting to develop some of these more innovative technologies").
12. *FDA's Growing Responsibilities For the Year 2001 and Beyond*, DEPT. OF HEALTH & HUMAN SERV., FDA, at http://www.fda.gov/oc/opacom/budgetbro/budgetbro.html (last updated Jun. 13, 2001) ("This peace of mind is an important contribution to the special quality of life, confidence, and vitality that makes the United States the envy of the world—and it is a part of FDA's proud tradition.").
13. *Id.*
14. Elizabeth Jacobson, 2000 FDA Science Board Meeting, Nov. 17, 2000, 28. (further explaining: "In April, NASA and NCI announced a Memo of Understanding to develop nano explorers, their term, for the human body in the form of injectable nanorobots or nanorobots that will roam the body to detect,

diagnose, and treat disease.... These little nano bots would be biosensors, and probably drug use delivery systems as well.").
15. CDRH has noted that it will be challenged to resolve complex issues connected with emerging technological developments such as nanotechnology. *See Better Health Care With Quality Medical Devices: FDA on the Cutting Edge of Device Technology*, FDA (Pub. No. FS 01-5), Feb. 2002, at http://www.fda.gov/opacom/factsheets/justthefacts/5cdrh.html. CBER acknowledges nanotechnology could be the delivery vehicle for gene therapy in the near future. *See Human Gene Therapy and the Role of the Food and Drug Administration*, FDA/CTR. FOR BIOLOGICS EVALUATION & RESEARCH, Sept. 2000, at http://www.fda.gov/cber/infosheets/genezn.htm.
16. However, fierce public opposition could force the agency to take ethical and social issues into account in regulating nanomedicine. With regard to cloning, the FDA has asserted jurisdiction and argued that it will attempt to prohibit all cloning based on safety and ethical considerations. *See* Bernard Schwetz, *Remarks of the Acting Principal Deputy Commissioner of Food and Drugs*, 56 FOOD DRUG L.J. 123, 127 (2001) ("FDA considers the use of cloning technology to clone a human being as a serious public health issue. There are many unresolved safety concerns with this technology . . . FDA has the authority to regulate human cloning technology, and no investigators have the approval to use the technology at this time. As recognized by the National Bioethics Advisory Commission there are, of course, broader social and ethical implications of using cloning technology to clone a human being. As Dr. Kathyrn Zoon recently testified before a congressional committee, 'FDA is opposed unequivocally to the cloning of human beings because of moral, ethical, and scientific issues.'"). Indeed, the FDA may be able to creatively employ statutory tools to prohibit nanomedical research. *See* Frederick Degnan, *Emerging Technologies and Their Implications: Where Policy, Science, and Law Intersect*, 53 FOOD & DRUG L.J. 594 (1998) ("Agencies can be adept at imposing such requirements, even under statutory provisions that do not call specifically for such requirements. Courts generally uphold such agency actions in deference to the overriding public-health mission that these agencies are charged to carry out.... Fortunately, there are any number of precedential examples where a forward-looking agency policy has relied on a statutory provision and interpreted it in light of new scientific or technological developments, and, in effect, changed if not revolutionized how the agency has regulated a given area.").
17. The 1906 Food and Drugs Act empowered the FDA to regulate drugs that were adulterated (unsanitary or unsafe) or misbranded. The 1906 Act limited the effectiveness of FDA regulation in two significant ways: the FDA could not regulate drugs before they were sold, and the FDA had to prove that the seller knew its claims were false. The 1938 Federal Drug and Cosmetic Act enabled the FDA to require pre-market notification. The manufacturer could market a drug within 180 days of notifying the FDA unless the FDA challenged its safety. The tragic births associated with Thalidomide in the 1950s resulted in the Congress enacting amendments in 1962. The amendments required the FDA to confirm the effectiveness and safety of the drug before marketing could take place. The FDA was also given greater authority in designing and supervising clinical trials. *See generally* PETER BARTON HUTT & RICHARD A. MERRILL, FOOD AND DRUG LAW: CASES AND MATERIALS 8-9 (2d ed. 1992); Richard A. Merrill, *The Architecture of Government Regulation of Medical Products*, 82 VA. L. REV. 1753 (1996).
18. 21 U.S.C. § 321(g)(1) (2001).
19. *See* JAMES T. O'REILLY, FOOD AND DRUG ADMINISTRATION § 13.11 (2d ed. 1995).

20. *Id.*
21. *Id.*
22. 21 U.S.C. §§ 379g-379h (1994).
23. Prior to 1938, medical devices were not subject to federal regulation. The 1938 Federal Drug and Cosmetic Act authorized the FDA to regulate medical devices. However, regulatory authority was limited to adulterated or misbranded products, and the FDA was primarily focused on drugs. In 1976, Congress amended the act to substantially increase FDA regulation of medical devices. *See* Merrill, *supra* note 17, at 1800-1812.
24. 21 U.S.C.S. § 321(h) (2001).
25. "General controls" include: (1) regulations against adulteration or misbranding; (2) regulations requiring establishment regulation and product listing; (3) premarket notification; and (4) compliance with good manufacturing practices. *See* O'REILLY, *supra* note 19, at § 18.05.
26. "Special controls" include: (1) some performance standards; (2) requirements for patient registries; and (3) post-market surveillance of the device. *See* 21 U.S.C. 360c(a)(1)(B)(1994); O'REILLY, *supra* note 19, at § 18.06.
27. The FDA can treat as "substantially equivalent" devices that have the same intended use and the same technology as an existing device or differ in design or technology from a predicate device but perform the same function and are shown to be as safe and effective. O'REILLY, *supra* note 19, at § 18.07.
28. Ch. 288, 37 Stat. 309 (1912) (codified at 42 U.S.C. §§ 201 et seq. (1994)).
29. FDAMA § 123(d)(I) (codified at 21 U.S.C. §§ 301 et seq. (1994)).
30. Lorrie Harrison McNeill, *CBER Update—Spring 2001*, UPDATE 2001, Spring 2001, at http://www.fdli.org/pubs/Update/2001/Issue4/McNeill/print.html.
31. O'REILLY, *supra* note 19, at § 13.22.
32. Safe Medical Devices Act of 1990, Pub. L. No. 101-629, § 16, 104 Stat. 4511, 4526 (codified as amended at 21 U.S.C. § 353 (1994).
33. 21 C.F.R. Part 3, Subpart A, § 3.2 (e).
34. FDAMA § 416.
35. United States v. Midwest Pharmaceuticals, Inc., 633 F. Supp. 316, 326 n. 1 (D. Neb. 1986), *aff'd in pertinent part*, 825 F.2d 1238 (8th Cir. 1987). *See also* Jay M. Zitter, *What is A 'Device' Within Meaning of Federal Food, Drug, and Cosmetic Act*, 129 ALR Fed 343 (2001); S. Rep. No. 361, 74th Cong., 1st Sess. 4 (1935).
36. *See* Action on Smoking & Health v. Harris, 655 F.2d 236 (D.C. Cir. 1980); Biotics Research Corp. v. Heckler, 710 F.2d 1375 (9th Cir. 1983); Weinberger v. Bentex Pharmaceuticals, Inc., 412 U.S. 645, 654 (1973).
37. FDAMA § 123(f) ("The Secretary of Health and Human Services shall take measures to minimize differences in the review and approval of products required to have approved biologics license applications under section 351 of the Public Health Service Act (42 U.S.C. 262) and products required to have approved new drug applications under section 505(b)(1) of the Federal Food, Drug, and Cosmetic Act (21 U.S.C. 355(b)(1))").
38. For example, CDRH must review a PMA within 180 days and a 510(k) within 90 days. CDER has 360 days to review an NDA for standard drugs and 180 days to review an NDA for priority drugs. LEWIN GROUP, *supra* note 11, at 30-31.
39. Michael Baram, et al., *Regulatory and Liability Considerations*, 6 B.U. J. SCI. & TECH. L. 5, (Spring 2000) (noting that "medical device regulation traditionally was deemed to be a shorter and easier route to market than regulation as a pharmaceutical product or a biological product").
40. The Centers employ different evidence standards. CDER places more emphasis on methodological aspects such as randomized controlled trials than CBER and CDRH. LEWIN GROUP, *supra* note 11, at 30.

41. FDA Intercenter Agreement between the Center for Drug Evaluation and Research and the Center for Biologics Evaluation and Research (Oct. 31, 1991), *reprinted in* FDLI, 3 FOOD & DRUG L REP. No. 2, Feb. 1992, at 29.
42. FDA Intercenter Agreement between the Center for Drug Evaluation and Research and the Center for Devices and Radiological Health (Oct. 31, 1991), *reprinted in* FDLI, 3 FOOD & DRUG L REP. No. 2, Feb. 1992, at 44.
43. FDA Intercenter Agreement between the Center Biologics Evaluation and Research and the Center for Devices and Radiological Health (Oct. 31, 1991), *reprinted in* FDLI, 3 FOOD & DRUG L REP. No. 2, Feb. 1992, at 17.
44. LEWIN GROUP, *supra* note 11, at 29.
45. *Id.* at 30.
46. Memorandum from Jerome Davis, CBER, to Director, Division of Emergency and Investigational Operations, Tissue Products Regulated by CBER and CDRH (Dec. 17, 1998).
47. *Id.*
48. FDA Intercenter Agreement between the Center for Drug Evaluation and Research and the Center for Devices and Radiological Health, *supra* note 42.
49. *Id.*
50. Email, from CSO/Jurisdiction & Device Status Expert, to the author (Jan. 17, 2002) (copy on file with author).
51. *See* Robert Freitas, *Say Ah! Nanorobots the Size of Bacteria Might One Day Roam People's Bodies, Rooting Out Disease Organisms and Repairing Damaged Tissue*, THE SCIENCES, July / Aug. 2000, at 26, 29.
52. CBER is responsible for regulating the following classes of products: (f) protein, peptide, or carbohydrate products produced by cell culture excepting antibiotics, hormones, products listed in A.3 above, and products previously derived from human or animal tissue and regulated as approved drugs. FDA Intercenter Agreement between the Center for Drug Evaluation and Research and the Center for Biologics Evaluation and Research, *supra* note 41.
53. *See* Ralph J. Fessenden & Joan S. Fessenden, ORGANIC CHEMISTRY 991-3 (4th ed. 1990); *see also* Frederick A. Fielder & Glenn H. Reynolds, *Legal Problems of Nanotechnology: An Overview*, 3 S. CAL. INTERDIS. L.J. 593, 609 (Winter 1994).
54. 21 U.S.C. § 355(j)(2)(iv)(2000). Generally, the proposed product must be identical in active ingredient, dosage form, strength, route of administration, labeling, quality, performance characteristics, and intended use, among other things, to a previously approved product.
55. Bioavailability is defined as the "rate and extent to which the active ingredient or therapeutic ingredient is absorbed from a drug and becomes available at the site of drug action." 21 U.S.C. § 355 (j)(7)(1988). *See* 21 C.F.R. 320.1 (1993) (giving same definition of bioavailability).
56. *See* Elizabeth H. Dickinson, *Striking the Right Balance Between Innovation and Drug Price Competition: Understanding the Hatch-Waxman Act: FDA's Role in Making Exclusivity Determinations*, 54 FOOD DRUG L.J. 195, 195-196 (1999).
57. 21 C.F.R. § 314.54 (2004).
58. 21 C.F.R. § 814.39 (2004).
59. Jonathan Kahan, *Public Workshop: Innovative Systems For Delivery of Drugs and Biologics: Scientific, Clinical, and Regulatory Challenges*, in REGULATORY & LEGAL CHALLENGES FOR DEVELOPERS OF DRUG DELIVERY DEVICES, Jul. 8, 2003, at 288, *available at* http://www.fda.gov/ohrms/dockets/dockets/03n0203/03n-0203-tr 00001-vol.2.DOC.
60. *Id.*
61. *See also* 21 C.F.R. 892.1000 (1989).

62. *FDA Guidance Concerning Demonstration of Comparability of Human Biological Products, Including Therapeutic Biotechnology-Derived Products*, FDA/Ctr. For Biologics Evaluation and Research (CBER) & Ctr. For Drug Evaluation and Research (CDER), 1996, *available at* http://www.fda.gov/cber/guidelines.htm and http://www.fda.gov/cber/gdlns/comptest.txt.
63. *Id.*
64. In 2003, MedMarket Diligence released a report examining the status of technologies, applications and markets for tissue engineering. The report reviews therapeutic technologies under various stages of development at over 150 companies. *See Tissue Engineering, Cell Therapy and Transplantation: Products, Technologies, & Market Opportunities*, 2001-2013, MEDMARKET DILIGENCE Rep. No. S505, Aug. 2003, at http://www.mediligence.com/rpt-s505.htm.
65. *Societal Implications of Nanotechnology: Hearing Before the House Science Comm.*, 108th Cong. (April 9, 2003) (Statement of Dr. Vicki Colvin, Director Center For Biological and Environmental Nanotechnology (CBEN), and Associate Professor of Chemistry, Rice University, Houston, Texas), *available at* LEXIS, Nexis Library.
66. Michael Brooks, *Thanks But No Thanks*, NEW SCIENTIST, Oct. 6, 2001, at 33.
67. Robert A. Freitas Jr., FORESIGHT INST., *What Could Go Wrong During a Nanomedical Procedure*, in NANOTECH FAQ, 1998, at http://www.foresight.org/Nanomedicine/NanoMedFAQ.html#FAQ18.
68. LEWIN GROUP, *supra* note 11, at 26 (citing the FDA FY 2001 Congressional Budget Request).
69. Henney, *supra* note 1.
70. Kathryn Zoon, *The Impact of New Biotechnology on the Regulation of Drugs and Biologics*, 41 FOOD DRUG COSM. L.J. 429, 430 (Oct. 1986).
71. David T. Bonk, *FDA Regulation of Biotechnology*, 43 FOOD DRUG COSM. L.J. 67, 77 (Jan. 1988).
72. The first "Points To Consider" concerned Interferon, products derived from r-DNA technology, monoclonal antibody products, and cell lines used to produce biologicals. *See* Zoon, *supra* note 70, at 431.
73. For example, in 1991, the FDA published a "Points To Consider" document on human somatic cell and gene therapies. In 1995, the FDA published a "Points to Consider" document on therapeutic products derived from transgenic animals. Martha J. Carter, *The Ability of Current Biologics Law to Accommodate Emerging Technologies*, 51 FOOD DRUG L.J. 375, 376 (1996).
74. *Id.*
75. 55 Fed. Reg. 12, 284 (1990).
76. 55 Fed. Reg. 12,284 (1990).
77. Sandra H. Cuttler, *The Food and Drug Administration's Regulation of Genetically Engineered Human Drugs*, 1 THE J. PHARM. & LAW 191, 210 (1992); *see also* David Hanson, *Pharmaceutical Industry Optimistic About Improvements at FDA*, CHEM & ENG. NEWS, Jan. 27, 1992, at 28-29.
78. *FDA Dumps Office of Biotechnology: Miller to Stanford as Visiting Scholar*, BIOTECHNOLOGY NEWSWATCH, Jan. 17, 1994, at 13 ("Abolishing the office is 'no signal that biotechnology is less important', an FDA spokesman said. 'The agency looks at that office as one established when biotechnology was an emerging technology,' he said. It is no longer 'emerging', the spokesman pointed out. These days, most decisions regarding biotechnology are being made at the Center for Biologics Evaluation and Research rather than at the commissioner's level, he said.").
79. Larry Thompson, *Science at FDA: The Key to Making the Right Decision*, FDA CONSUMER MAGAZINE, March-April 2000, at http://www.fda.gov/fdac/features/2000/200_sci.html.

80. Jane Henney, *Science and the FDA*, Speech delivered at 6th Annual FDA Science Forum (Feb. 14, 2000) (prepared remarks *available at* http://www.fda.gov/oc/speeches/2000/scienceforum.html).
81. LEWIN GROUP, *supra* note 11, at 28.
82. In a speech made in February 2000, Henney highlighted FDA personnel traveling to Merk's manufacturing site to learn about developments in barrier isolation technology as an example of joint training. Other examples of joint training session topics include: new ELISA technologies in food inspections, microarray technology, nucleic acid amplification testing, and new trends in sterilization. *See* Gary Dykstra, *Scientific Training Outside the Boundaries: FDA & Industry Partnerships for Emerging Technology Training*, Speech delivered at 6th Annual FDA Science Forum (Feb. 15, 2001) (abstract *available at* http://vm.cfsan.fda.gov/~frf/forum01/abst01sp.html).
83. Examples of CRADA's include: (1) CDER and MULTICASE working together to develop software strategies for predicting drug toxicities; (2) CDER and Boehringer-Ingelheim Pharmaceuticals developing a model of carcinogenic potential of chemicals. *See* Bernard Schwetz, Susan A. Homire, and James T. MacGregor, DEPT. OF HEALTH & HUMAN SERVS., FDA, *Science at FDA: Improving the Scientific Basis of Regulation Through Collaboration With 'Stakeholders'*, at http://www.fda.gov/oc/oha/fdascience.htm.
84. For example, JIFSAN, a partnership between the FDA and the University of Maryland, was designed to explore risk assessment and the Food Safety Initiative. There is a similar agreement involving food safety issues between the FDA and the Illinois Institute of Technology Research Institute. *Id.*
85. For example, FDA is working with the Department of Health and Human Services, NIH, CDC, the Department of Defense, USDA, and EPA. FDA and EPA are collaborating to research endocrine disruptors. *Id.*
86. For example, PQRI is a nonprofit foundation formed under the umbrella of the American Association of Pharmaceutical Scientists. Its purpose is to facilitate FDA, university, and industry collaboration to address critical issues in pharmaceutical product quality. *Id.*
87. For example, the ICH(2) conference between regulatory bodies and global industry organizations resulted in a worldwide set of uniform recommendations for approval of new drugs. *See* Schwetz, Homire, & MacGregor, *supra* note 83.
88. OFF. OF SCI. & TECH., *Annual Report: Fiscal Year 2000* [hereinafter OST Annual Report], at http://www.fda.gov/CDRH/ANNUAL/FY2000/OST/OST-ANNUALREPORT2000.HTML.
89. *See Testimony: Agriculture, Rural Development and Related Agencies Before Senate Appropriations Comm.*, 107th Cong. (May 10, 2001) (statement of Bernard Schwetz, Acting Principal Deputy Commissioner, FDA), at LEXIS, Nexis Library.
90. *Id.* ("The project's accomplishments include the development of methods for early disease detection, the identification of new therapeutic targets and the discovery of new biomarkers for drug-induced patient toxicity. This bench-to-bedside model has resulted in a first-of-its kind clinical trial that incorporates a 'proteomic portrait' of the disease in human tissue that could lead to customized, patient tailored therapeutics. Currently, this research has identified over 150 proteins that are aberrantly expressed in human prostate, lung, breast, ovary, esophageal, and colon cancer.").
91. Other examples of cutting edge laboratory research conducted by FDA scientists include research into the mechanisms by which organ replacement technology interacts with the body, the testing procedures available for evaluating potential adverse effects of biomaterials on the immune system, a standardized screening

assay for measurement of mutation induction in the p53 gene for studying cancer risk associated with technologies, tissue engineering, computational modeling, and genetic testing. *See* OST Annual Report, *supra* note 88.
92. The FDA has taken numerous steps to adequately prepare for the regulation of genetic testing devices in the near future. OST scientists have served as members of scientific advisory committees for other FDA Centers reviewing genetic devices, have taught courses on biocompatibility to update review staff, and have been involved in laboratory research projects. Information sessions, including presentations by developers of genetic and genomic technologies, have been organized by ODE's Division of Clinical Laboratory Devices. A Genomics / Proteomics working group has been formed to develop priorities for action related to FDA readiness in assessing new genetic technologies. *See* OST Annual Report, *supra* note 88.
93. The Tissue Action Plan was formalized in March 1998 to develop on a timely basis the policies, regulations, and guidance needed to implement the FDA's February 1997 Proposed Approach to the Regulation of Cellular and Tissue Based Therapies. *See CBER Update*, UPDATE 2001, Spring 2001, at http://www.fdli.org/pubs/Update/2001/Issue4/McNeill/print.html. *See also* Darin Weber, *Hearing Before Biological Response Modifiers Advisory Committee, Dept. of Health & Human Servs. Food & Drug. Admin. Ctr. for Drug Evaluation and Research*, CONG. HEARING TRANSCRIPTS, Mar. 20, 2000, at LEXIS, Nexis Library ("FDA's regulatory framework for cell and tissue-based products wasn't formulated overnight. It has been evolving for most of the last century and is continuing to evolve today[.]").
94. The FDA has spent a great deal of time and resources recruiting and training staff to regulate gene therapy. *See Human Gene Therapy and the Role of The Food and Drug Administration*, FDA/CTR. FOR BIOLOGICS EVALUATION & RESEARCH, Sept. 2000, at http://www.fda.gov/cber/infosheets/genezn.htm. In 2000, the FDA announced that it would increase its inspections of gene therapy studies. It also announced the Gene Therapy Clinical Trial Monitoring Plan, under which the sponsors of gene therapy trials must routinely submit monitoring plans to the FDA. *See* Edward Korwek and Mark D. Learn, *Biologics Update*, UPDATE 2001, at http://www.fdli.org/pubs/Update/2001/Issue2/Issue2/Korwek_Learn/print.html.
95. *Implementation of FDA Modernization, Hearing Before House Commerce Committee*, 105th Cong. (Oct. 7, 1998) (Statement of Joseph C. Scodari, President and Chief Operating Officer, Biotechnology Industry Organization), FED. DOC. CLEARING HOUSE CONG. TESTIMONY, Oct. 7, 1998, at LEXIS, Nexis Library ("Our experience and those of other BIO member companies points to numerous examples where both clinical development and complex manufacturing issues were speedily resolved because of the scientific expertise within the Center for Biologics Evaluation and Research (CBER)."); *see also* LEWIN GROUP, *supra* note 11, at 12 ("Regulation of the medical device industry by FDA has improved in recent years. . . . Improvements in FDA regulation are attributable to such main factors as the agency's reengineering efforts, collaboration with industry, and a commitment to the implementation of the FDA Modernization Act of 1997 (FDAMA).").
96. Interview with Donald Marlowe, Food and Drug Administration, January 2004.
97. *See Tufts Study: Staff Departures Will Slow FDA Operations*, Jan. 21, 2004, at http://www.fdanews.com/pub/pcr/4_2/fda/20723-1.html (subscription required).
98. LEWIN GROUP, *supra* note 11, at 26.
99. *FDA's Growing Responsibilities for the Year 2001 and Beyond*, DEPT. OF HEALTH & HUMAN SERVS., FDA, at http://www.fda.gov/oc/opacom/budgetbro/budgetbro.html (last updated Jun. 13, 2001).

100. ROBERT FREITAS, NANOMEDICINE 31, (1999) (quoting Gergory M. Fahy in *Molecular Nanotechnology and its Possible Pharmaceutical Implications*, 2020 VISIONS: HEALTH CARE INFORMATION STANDARDS AND TECHNOLOGIES (U.S. Pharmacopoeial Convention, Inc., 1993), at 152-159).

Chapter 7

1. DANIEL RATNER & MARK RATNER, NANOTECHNOLOGY AND HOMELAND SECURITY: NEW WEAPONS FOR NEW WARS (2004) (excerpt taken from foreword).
2. *Id.* For an excellent review of the book, see Drew Harris, *Shrinking the Battlefield: A Review of Nanotechnology and Homeland Security*, 1 NANOTECH. L&B 116 (2004) (reviewing DANIEL RATNER & MARK RATNER, NANOTECHNOLOGY AND HOMELAND SECURITY: NEW WEAPONS FOR NEW WARS (2003)).
3. *See, e.g.*, Trade Promotion Coordinating Comm., Toward a National Export Strategy: Report to the United States Congress 29, 54 (1993).
4. *See, e.g.*, John B. Reynolds, et al., *Export Controls and Economic Sanctions*, 37 INT'L L. 263 (Summer 2003) ("The national focus on combating terrorism dominated the development of economic sanctions and export controls in 2002. . . . Federal responses to terrorism and homeland security will continue to shape export controls and sanctions policies in 2003.").
5. *See* Christopher Corr, *The Wall Still Stands! Complying with Export Controls on Technology Transfers in the Post-Cold War, Post-9/11 Era*, 25 HOUS. J. INT'L L. 441, n.8 (2003).
6. Donald W. Smith, *Defense of Export Controls Enforcement Actions*, in COPING WITH U.S. EXPORT CONTROLS 1999, at 745, 751 (PLI/Com. L. & Practice Course, Handbook Series No. 798, 1999).
7. United States v. Gregg, 829 F.2d 1430 (8th Cir. 1987).
8. *Id.* at 1436.
9. *See, e.g.*, United States v. Frade, 709 F.2d 1387 (11th Cir. 1983); United States v. Hernandez, 662 F.2d 289 (5th Cir. 1981); United States v. Ortiz de Zevallos, 748 F. Supp. 1569 (S.D. Fla. 1990), *rev'd sub nom.* United States v. Macko, 994 F.2d 1526 (11th Cir. 1993); United States v. Fuentes-Coba, 738 F.2d 1191 (11th Cir. 1984).
10. Ronald J. Sievert, *Urgent Message to Congress—Nuclear Triggers to Libya, Missile Guidance to China, Air Defense to Iraq, Arms Supplier to the World: Has the Time Finally Arrived to Overhaul the U.S. Export Control Regime?—The Case For Immediate Reform of Our Outdated, Ineffective, and Self-Defeating Export Control System*, 37 TEX. INT'L L.J. 89, 92 (2002).
11. H.R. Rep. No. 102-37, at 15 (1991).
12. *See generally* Corr, *supra* note 5.
13. EAR was promulgated under the Export Administration Act (EAA) as supplanted by the International Emergency Economic Powers Act (IEEPA). Although the EAA expired in 1994, the provisions of the EAR are still in place through Executive Order. 15 C.F.R. § 730.2 (2004); International Emergency Economic Powers Act (IEEPA), 50 U.S.C. §§ 1701-1706 (2000).
14. *See Introduction to Commerce Department Export Controls*, BUREAU OF INDUSTRY AND SECURITY: DEPT. OF COM., at http://www.bxa.doc.gov/Licensing/exportingbasics.htm (last updated May 8, 2003).
15. 15 C.F.R. § 734.2 (2002).
16. 15 C.F.R. § 734.3 (2004).
17. *Id.*
18. 15 C.F.R. § 774 (2004).

19. 15 C.F.R. § 734.3(c) (2004).
20. An exporter does not need to submit a license application unless: the action is prohibited by a denial order; the technology is knowingly exported to an end-user or end-use that is prohibited by part 744; the manufacturer seeks to export or reexport to an embargoed destination; the export supports proliferation activities; the technology is exported through or transit through Albania, Armenia, Azerbaijan, Belarus, Buglaria, Cambodia, Cuba, Estonia, Georgia, Kazakhstan, Kyrgyzstan, Laos, Latvia, Lithuania, Mongolia, North Korea, Russia, Tajikistan, Turkmenistan, Ukraine, Uzbekistan, or Vietnam. *See* 15 C.F.R. § 736.2(b)(4-10) (2004).
21. 15 C.F.R. § 734.7 (2002).
22. 15 C.F.R. § 736 (2002).
23. 15 C.F.R. § 734.2(b)(3) (2002).
24. 15 C.F.R. § 734.2(b)(ii) (2002).
25. 15 C.F.R. § 734, Supp. No. 2 (2004).
26. 15 C.F.R. § 738 (2003).
27. 15 C.F.R. § 740 (2003).
28. Exec. Order No. 12,981, 60 Fed. Reg. 62,981, 62,981 (Dec. 5, 1995).
29. Corr, *supra* note 5, at 469.
30. 15 C.F.R. § 764.3 (2003). Civil and administration sanctions include fines and "denial orders." Denial orders bar exports by the sanctioned person—and related parties—as well as the supply to such person of items that have been exported from the United States.
31. *See Unequal Justice on Tech Transfer Enforcement*, EXPORT PRACTITIONER, Feb. 2002, at 6 ("In light of the [post 911] paradigm shift ... [BIS] might be out for blood.").
32. Nanocrystalline materials are defined as materials having a crystal grain size of 50 nm or less, as determined by X-ray diffraction. 15 C.F.R. § 774, 1C003 (2004).
33. 15 C.F.R. § 774, 1C011 (2004).
34. 15 C.F.R. § 744.17 (2004).
35. *See, e.g.*, 15 C.F.R. § 774, 2B007 (2004).
36. International Traffic in Arms Regulations, 22 C.F.R. § 120-30 (2001).
37. 22 C.F.R. § 121.1 (2004).
38. 22 C.F.R. § 120.4 (2004).
39. (1) Firearms, Close Assault Weapons and Combat Shotguns; (2) Guns and Armament; (3) Ammunition / Ordinance; (4) Launch Vehicles, Guided Missiles, Ballistic Missiles, Rockets, Torpedoes, Bombs and Mines; (5) Explosives and Energetic Materials, Propellants, Incendiary Agents and Their Constituents; (6) Vessels of War and Special Naval Equipment; (7) Tanks and Military Vehicles; (8) Aircraft and Associated Equipment; (9) Military Training Equipment; (10) Protective Personnel Equipment; (11) Military Electronics; (12) Fire Control, Range Finders, Optical and Guidance and Control Equipment; (13) Auxiliary Military Equipment; (14) Toxicological Agents, Including Chemical Agents, Biological Agents, and Associated Equipment; (15) Spacecraft Systems and Associated Equipment; (16) Nuclear Weapons, Design and Testing Related Items; (17) Classified Articles, Technical Data and Defense Services Not Otherwise Enumerated; (18) Directed Energy Weapons; (19) Submersible Vessels, Oceanographic and Associated Equipment; and (20) Miscellaneous Articles.
40. Registration is not required for: (1) Officers and employees of the United State Government acting in an official capacity; (2) Persons whose pertinent business activity is confined to the production of unclassified technical data only; (3)

Persons all of whose manufacturing and export activities are licensed under the Atomic Energy Act of 1954, as amended; or (4) Persons who engage only in the fabrication of articles for experimental or scientific purpose, including research and development. 22 C.F.R. § 122.1 (2004).
41. 22 C.F.R. § 123.1 (1993).
42. 22 C.F.R. § 127.3 (1993).
43. Corr, *supra* note 5, at 463. OFAC normally defines U.S. persons to include U.S. citizens, permanent U.S. residents, and entities organized under U.S. jurisdiction, including foreign branches of U.S. companies. *Id.*
44. OFAC publishes a list of "Specially Designated Nationals." *See* 31 C.F.R. ch. V., app. A.
45. Corr, *supra* note 5, at 462 ("These OFAC controls often leave businesses that operate internationally in the uncomfortable position of having no clear answer as to whether a proposed transaction is permissible, forcing them to either forgo business opportunities or accept potentially unmanageable risks.").
46. Executive Order No. 12,829, 58 Fed. Reg. 3479, 3479 (Jan. 6, 1993). Although the National Security Council has ultimate authority over NISP, the Secretary of Defense is charged with the task of implementing the Program rules and requirements.
47. The rules and guidelines governing such acquisitions are set out in the National Industrial Security Operating Manual (NISPOM).
48. 31 C.F.R. § 800.101 (2002).
49. 31 C.F.R. § 800.401 (2003).
50. *See* Corr, *supra* note 5, at 498.
51. 10 C.F.R. § 810.3 (2000).
52. 10 C.F.R. § 110.8 (2000).
53. R. Aylan Broadbent, *U.S. Export Controls on Dual-Use Goods and Technologies: Is the High Tech Industry Suffering?*, Currents: Int'l Trade L.J. 49, 52 (1999).
54. In 1996, the Clinton administration launched a plan prohibiting exports and re-exports of encryption products without key "recovery" features. *See* Encryption Items Transferred from the U.S. Munitions List to the Commerce Control List, 61 Fed. Reg. 68,572, 68,574 (Dec. 30, 1996). "Neither private industry, nor the government agencies actually imposing the controls, appeared to comprehend the details of how a key recovery system should or could be implemented. . . . The software industry strongly opposed the plan, stating that it would impede the competitiveness of the U.S. industry in this sector by favoring competitors in countries not subject to similar restrictions." Jonathon Westreich, *Regulatory Controls on U.S. Exports of Weapons and Weapons Technology: The Failure to Enforce the Arms Export Control Act*, 7 Admin. L.J. U. 463, 485 (1993). In response to criticism of the policy, the Clinton administration attempted to liberalize the regulations. *See* Press Release, Office of the Press Secretary, The White House, *Administration Updates Encryption Policy* (Sept. 16, 1998), *available at* http://www.cdt.org/crypto/admin/whousepress091698.html. The revised plan was "criticized as overly restrictive and unrealistic." Westreich, *supra*, at 486.
55. Corr, *supra* note 5, at 484.
56. *See* Westreich, *supra* note 54, at 466 n.111021 (1993).
57. *See* Sievert, *supra* note 10, at 91.
58. *See id.* at 104.
59. Indeed, a GAO study concluded that an effort to place authority for export controls in a single agency would fail due to political resistance from both Commerce and Customs. *See generally* House Subcomm. On Econ. Policy, Trade & The Env't, Comm. On Foreign Affairs, GAO report to the Chairman, Export Controls: Actions Needed To Improve Enforcement (1993).

Chapter 8

1. LINDA COHEN AND ROGER NOLL, THE TECHNOLOGY PORK BARREL 1 (Brookings Inst. 1991)
2. Cal. NanoSystems Inst., at http://www.cnsi.ucla.edu/mainpage.html. *See also* Kenneth Weiss, *Davis Awards Science Funds; Education: The Governor Selects Three Institutes, Including One Based at UCLA, to Receive $300 Million,* L.A. TIMES, Dec. 8, 2000, *available at* LEXIS, Nexis Library.
3. THE NANOTECHNOLOGY INSTITUTE, at http://www.sep.benfranklin.org/resources/nanotech.html. *See also Nanotech Hubs Spread All Over,* UPI, July 11, 2001, at LEXIS, Nexis Library UPI file.
4. TEXAS NANOTECHNOLOGY INITIATIVE, at http://www.texasnano.org/. *See also Texas Nanotechnology Initiative Elects Board of Directors: Consortium Is Dedicated to Positioning Texas As Leader,* PR NEWSWIRE, Jan. 18, 2002, at LEXIS, Nexis Library (quoting the President of the Texas Nanotechnology Initiative: "Texas has already begun establishing a nanotechnology community with roots in Austin, Dallas and Houston. TNI will focus on growing the academic, corporate, governmental, and investment infrastructure necessary to make Texas a hotbed for nanotechnology.").
5. *Nanotechnology: Hearing Before Science, Technology, and Space Subcomm. of Senate Comm. On Commerce, Science, and Transportation,* 107th Cong. (2002) (statement of F. Mark Modzelewski, Exec. Dir. of Nanobusiness Alliance), FED. DOCUMENT CLEARING HOUSE CONG. TESTIMONY, at LEXIS, Nexis Library.
6. Lerwen Liu, *Asia Pacific Nanotechnology R&D and Commercialization Efforts,* 1 NANOTECH. L&B 104 (2004).
7. Yumura Motoo, *Carbon Nanotubes, Materials For Nanotechnology,* J. OF JAPANESE TRADE & INDUSTRY, Dec. 3, 2001, at LEXIS, Nexis Library. *See also* Watanabe Makoto, *A Future Society Built By Nanotechnology,* J. OF JAPANESE TRADE & INDUSTRY, Dec. 3, 2001, at LEXIS, Nexis Library.
8. *Trouble in NanoLand,* THE ECON., Dec. 7, 2002, at LEXIS, Nexis Library.
9. *See Institute Profile,* NAT'L INSTITUTE FOR NANOTECHNOLOGY, at http://www.nint.ca/nav01.cfm?nav01=12888.
10. COHEN AND NOLL, *supra* note 1, at 2.
11. *Id.* at 18 ("[N]ot all the returns to investment in R&D can be appropriated by the innovator, and both capital markets and large organizations do not accurately evaluate the kinds of risks involved in pursuing commercial R&D"). *See also* Robert A. Book, *Public Research Funding and Private Innovation: The Case of the Pharmaceutical Industry,* Sept. 5, 2002, at 4, *available at* http://home.uchicago.edu/~rbook/; Kenneth J. Arrow, *Economic Welfare and the Allocation of Resources For Invention,* in THE RATE AND DIRECTION OF INVENTIVE ACTIVITY, NBER Conference Report (1962), 609-625; Richard R. Nelson, *The Simple Economics of Basic Scientific Research,* 67 J. POL. ECON. 297—306 (June 1959).
12. COHEN AND NOLL, *supra* note 1, at 2.
13. *Id.* at 26.
14. *See, e.g.,* Nathan A. Adams, *Monkey See, Monkey Do: Imitating Japan's Industrial Policy in the United States,* 31 TEX. INT'L L.J. 527 (Summer 1996).
15. *See generally* COHEN AND NOLL, *supra* note 1, at 37-52.
16. *Id.* at 365.
17. *Id.* at 53-76.
18. *Id.* at 6.
19. NAT'L SCI. & TECH. COUNCIL, at http://www.ostp.gov/NSTC/html/NSTC_Home.html.

20. COMMITTEE FOR THE REVIEW OF THE NATIONAL NANOTECHNOLOGY INITIATIVE, SMALL WONDERS, ENDLESS FRONTIERS: A RREVIEW OF THE NATIONAL NANOTECHNOLOGY INITIATIVE 11 (Nat'l. Acad. Press 2002) (hereinafter Committee For Review of NNI).
21. Id.
22. NNI: Leading to the Next Industrial Revolution. The Initiative and Its Implementation Plan, NSTC, July 2000, 38-40.
23. Committee For Review of NNI, *supra* note 20, at 12. NNI "nanotechnology" research activities are defined as: "research and technology development at the atomic, molecular, or macrmolecular levels, in the length scale of approximately 1-100 nanometer range, to provide a fundamental understanding of phenomena and materials properties at the nanoscale and to model, create, characterize, manipulate, and use structures, devices, and systems that have novel properties and functions because of their small or intermediate size. The novel and differentiating properties and function are developed at a critical length scale or matter typically under 100 nanometers. Nanotechnology research and development includes integration of nanoscale structure into larger material components, systems, and architectures. Within these larger scale assemblies, the control and construction of their structures and components devices remain at the nanometer scale." *Id.* at 12.
24. NNI: Leading to the Next Industrial Revolution, *supra* note 22, at 14-15.
25. *See* Committee for Review of NNI, *supra* note 20, at vi.
26. *Id.* at 17.
27. *Id.* at 18.
28. *Id.*
29. *Id.* at 20.
30. *Id.* at 21.
31. *Id.* at 23.
32. *Id.* at 25-26.
33. *Id.* at 28. "About one-third of the individual investigator activities under the NSF Functional Nanostructures programs have international collaborations. NSF has also sponsored young researchers for group travel to Japan, Europe, and other areas to present their work and visit centers of excellence in the field. Bilateral and international activities with the European Union, Japan, Korea, India, Switzerland, Germany, and Latin America have been under way since 2000. U.S. universities also have many foreign science and engineering graduate students." Committee Review of NNI, *supra* note 20, at 3-14.
34. *Id.* at 28-30.
35. The Nanoscale Systems in Information Technologies program at Cornell University has been partnering with industry to support a K—12 teachers' institute and a nanotechnology teaching laboratory. The Center for Science of Nanoscale Systems and their Device Applications at Harvard University has been fostering nano-focused public education activities in partnership with the Boston Science Museum. Finally, the Center for Directed Assembly Nanostructures at Rensselaer Polytechnic Institute has developed a partnership with industry and several smaller universities, some with large underrepresented populations. Outreach efforts have also been carried out through initiatives like the NanoManipulator at the University of North Carolina at Chapel Hill; Molecular Modeling and Simulation; the Web-based network at the University of Tennessee, and the Interactive Nano-Visualization in Science and Engineering Education program at Arizona State University. Traditional NSF outreach programs like Research Experiences for Teachers (RET) have also targeted nanoscale science and technology. *Id.* at 33.

36. For example, an exhibit titled "Making the Nanoworld Comprehensible" is on display at the University of Wisconsin and Discovery World Science Museum in Milwaukee. Other examples include "Internships for Creating Presentations on Nanotechnology Topics," at the Arizona Science Center; and "Small Wonders: Exploring the Vast Potential of Nanoscience," a traveling education program. *Id.* at 34.
37. *Id.* at 46-49.
38. 21st Century Nanotechnology Research and Development Act, S.189, 108th Cong. (2003).
39. For a more comprehensive discussion of the 21st Century Nanotechnology Research and Development Act, *see* Mike Honda, *Nanotechnology Legislation in the 108th Congress*, 1 NANOTECH. L&B 63 (2004).
40. President's Council of Advisors on Sci. & Tech. (PCAST), at http://www.ostp.gov/PCAST/pcast.html.
41. PCAST, *PCAST Drafted and Approved Four Reports in 2002*, at http://www.ostp.gov/PCAST/pcast2002rpt.html.
42. *See* Tom Kalil, *Next Steps For The National Nanotechnology Initiative*, 1 NANOTECH. L&B 55 (2004).

Chapter 10

1. Interview with Max Lagally, founder of nPoint, University of Wisconsin (Oct. 12, 2003).
2. Bayh-Dole Act, 35 U.S.C. §§ 200-212 (1994); 37 C.F.R. § 401 (2002).
3. Interview with Kathy Ku and Linda Chao, Stanford University Office of Technology Licensing, in Stanford, California (Jan. 16, 2003) [hereinafter "Stanford OTL Interview"]. *See also* Kenneth Sutherlin Dueker, *BioBusiness On Campus: Commercialization of University-Developed Biomedical Technologies*, 52 FOOD & DRUG L.J. 453, 498 (1997) ("There is, however, a significant problem of knowing how to structure exclusivity *a fortiori*. MIT's Lita Nelsen calls this situation 'one of the great theoretical terrors in the hearts of university technology transfer managers,' often expressed as 'what if the Cohen-Boyer patent[s] had arisen out of an industrially sponsored research agreement [or were otherwise exclusively licensed]?'").
4. *See generally* John C. Miller, *A Call To Legal Arms: Bringing Embryonic Stem Cell Therapies To Market*, 13 ALB. L.J. SCI. & TECH. 556 (2003).
5. Stanford OTL Interview, *supra* note 3.
6. For example, UC Berkeley includes mandatory sub-licensing provisions in their license agreements. Interview with Carol Mimura and Veronica Lanier, Berkeley Office of Technology Licensing, in Berkeley, California (Feb. 26, 2003).
7. 37 CFR § 401.15(k)(4) (2004).
8. Interview with Max Lagally, *supra* note 1.
9. Interview with Rebecca Goodman, UCLA Licensing Associate, in Los Angeles, California (Feb. 28, 2003).
10. Interview with Rich Wolf, Director of Caltech Office of Technology Transfer, in Pasadena, California (Feb. 28, 2003).
11. Fed. Lab. Consortium for Tech. Transfer Southeast Region, at http://www.southeastflc.org/ (last visited Apr. 12, 2004).
12. U.S. DEPT. OF ENERGY, OFFICE OF SCI., *National Laboratories*, at http://www.er.doe.gov/Sub/Organization/Map/national_labs_and_userfacilities.htm (last visited Apr. 12, 2004).

13. Fed. Lab. Consortium for Tech. Transfer Southeast Region, *Vision*, at http://www.federallabs.org/servlet/newContentObjServlet?LinkCoArID=2000-03-16-10-19-48-890-eportney&CoArRegion=National&parentID=1999-03-29-14-00-55-270-eportney (2004). *See also* Stevenson-Wydler Technology Innovation Act of 1980, Pub. L. No. 96-480, 94 Stat. 2311-20 (codified as amended at 15 U.S.C. §§ 3701-14 (2000)).
14. Under the Federal Technology Transfer Act of 1986, the FLC was chartered and funded to coordinate between the numerous individual agencies and laboratories in order to share information and "increase dialogue with state and local governments, businesses, academia, and other external participants." Fed. Lab. Consortium for Tech. Transfer, *supra* note 13.
15. Ronald L. Meeks, *Changing Composition of Federal Funding for Research and Development and R&D Plant Since 1990*, NSF InfoBrief (02-315), at http://www.nsf.gov/sbe/srs/infbrief/nsf02315/start.htm (Apr. 2002).
16. U.S. NAVAL RESEARCH LABORATORY, *Nanoscience and Technology*, at http://www.nanosra.nrl.navy.mil (last visited Apr. 12, 2004).
17. *See* DEPT. OF DEFENSE, OFF. OF TECH TRANSFER, *Tech Transit*, at http://www.dtic.mil/techtransit/ (last revised Feb 17, 2004).
18. U.S. ARMY SPACE AND MISSILE DEFENSE COMMAND, PUB. AFFAIRS OFF., at http://www.smdc.army.mil/TechCenter/TechTransfer/Tech_Transfer.html (last visited Apr. 12, 2004).
19. DEPT. OF DEFENSE, *Fiscal Year 2001 Budget*, at http://www.nanosra.nrl.navy.mil/muri.php (last visted Apr. 12, 2004).
20. *See* DEFENSE ADVANCED RESEARCH PROJECT AGENCY, at http://www.darpa.mil/body/dobdar.html (last updated Nov. 4, 2003).
21. Office of Basic Energy Sciences, at OFF. OF BASIC ENERGY SCI., *Nanoscale Science, Engineering, and Technology Research (NSET)*, at http://www.sc.doe.gov/bes/NNI.htm (last visited Apr. 12, 2004).
22. *See* DEPT. OF ENERGY, OFF. OF INDUSTRIAL TECHS., *Industrial Technologies Program*, at http://www.oit.doe.gov/redirects/oit.html (last visited Apr. 12, 2004).
23. DEPT OF ENERGY, *Laboratory Coordinating Council Memorandum of Cooperation*, May 22, 1995, at http://www.oit.doe.gov/lcc/lcc_moc.shtml (last updated Jan. 15, 2004).
24. NAT'L SCI. FOUND., *Federal Funds for Research and Development*, FY 2002.
25. *Id.*
26. Interview with Deepak Srivastava, Computational Nanotechnology, NASA Ames, in NANOTECH. BULL. Mar. 22, 2002, *available at* http://www.nanotech-bulletin.com/archive/3-22-02/nni.htm (last visited Apr. 12, 2004).
27. *See* NASA COMMERCIAL TECHNOLOGY NETWORK, at http://nctn.hq.nasa.gov/ (last visited Apr. 12, 2004).
28. *See* Srivastava, *supra* note 26.
29. NASA, COMMERCIALIZATION CTR. PROGRAM, at http://www.nasaincubator.csupomona.edu/program.htm (last visited Apr. 12, 2004).
30. NASA, CTR. FOR TECH. COMMERCIALIZATION, *Technology Solutions for New Threats*, at http://www.ctc.org/ (last visited Apr. 12, 2004).
31. HOWARD C. ANAWALT AND ELIZABETH ENAYATI POWERS, IP STRATEGY: COMPLETE INTELLECTUAL PROPERTY PLANNING, ACCESS, AND PROTECTION 3-25 to 3-26 (2002).
32. Robert P. Merges, *The Law and Economics of Employee Inventions*, 13 HARV. J. LAW & TEC. 1, 46 (Fall 1999).
33. *See, e.g.*, Koehring Co. v. Ed. Etnyre & Co., 254 F. Supp. 334 (N.D. Ill. 1966) (stating that employment agreement did not give an employer a "mortgage on all thoughts occurring to the employee" and holding that the employee's rough sketches and designs "were never developed [during] employment to the extent

that they constituted material subject to the agreement."); Jamesbury Corp. v. Worcester Valve Co., 318 F. Supp. 1 (D. Mass. 1970) ("Freeman virtually conceived patent '666 while employed . . . [Yet] it is impossible to find on the basis of the evidence that Freeman had completely conceived the entire invention at the time he left [the firm]. He had gotten to the point where no more than an additional few days or perhaps few hours of thinking was required for him to put his ideas on paper in a form substantially the same as his later patent application. The other key finding of fact is that Freeman deliberately refrained from reducing his ideas to drawings or written description until after his resignation.").

34. Merges, *supra* note 32, at 47.
35. *Id.* at 48 ("[I]t is a simple matter to leave trade secrets behind and still base a startup company on a concept developed at the old job.").
36. *See, e.g.,* PepsiCo. v. Redmond, 54 F.3d 1262 (7th Cir. 1995) (prohibiting former employee from working for a competitor for six months and permanently enjoining him from disclosing confidential information).
37. *See, e.g.,* DoubleClick, Inc. v. Henderson, No. 116914/97, 1997 WL 731413, 6 (N.Y. Sup. Ct. Nov. 7, 1997); Uncle B's Bakery, Inc. v. O'Rourke, 920 F. Supp. 1405, 1419 (N.D. Iowa 1996); Merck & Co. v. Lyon, 941 F. Supp. 1443, 1461 (M.D.N.C. 1996); Novell, Inc. v. Timpanogos Research Group, Inc., 46 U.S.P.Q.2d (BNA) 1197, 1204 (Utah D. Ct. 1998).
38. *Redmond,* 54 F.3d at 1267, 1270.
39. *See, e.g.,* Int'l Bus. Mach. Corp. v. Seagate Tech., Inc., 941 F. Supp. 98, 101 (D. Minn. 1992) (on remand from the 8th Circuit).
40. *See, e.g.,* Armorlite Lens Co. v. Campbell, 340 F. Supp. 273, 275 (S.D. Cal. 1972); Dorr-Oliver, Inc. v. United States, 432 F.2d 447, 451-52 (Ct. Cl. 1970); Ingersoll-Rand Co. v. Ciavatta, 524 A.2d 866, 870 (N.J. Super. Ct. 1987).
41. *See* Edward L. Raymond, Annotation, *Construction and Effect of Provision of Employment Contract Giving Employer Right To Inventions Made By Employee,* 66 A.L.R. 4th 1135, 1202-04 (1992).
42. Merges, *supra* note 32, at 53.
43. Cal. Bus. Prof. Code § 166000 (West 1997).
44. *See* Scott v. Snelling & Snelling, Inc., 732 F. Supp. 1034, 1042 (N.D. Cal. 1990).
45. Robert G. Bone, *A New Look At Trade Secret Law: Doctrine In Search of Justification,* 86 Calif. L. Rev. 241, 279 (March 1998).
46. Michael Krey, *Nanotech Investment Firm Sees Much More Than Hype,* Inv. Bus. Daily, Sept. 8, 2003, at LEXIS, Nexis Library (quoting Peter Hebert of Lux Capital).
47. Daniel Sorid, *Nano In Firm's Name Fuels Stock's Hefty Gain,* USA Today, Dec. 4, 2003, *available at* http://www.usatoday.com/tech/techinvestor/techcorporate news/2003-12-04-nano-nono_x.htm.
48. George Mannes, *The Five Dumbest Things On Wall Street in the Last Year or So,* THESTREET.COM, Jan. 1, 2004, at http://www.thestreet.com/_tscs/markets/dumbestgm/10134452.html.
49. We identified the following LLCs: General Nanotechnology LLC (software); Catalytic Materials LLC (carbon nanotubes); Integrated Nano-Technologies LLC (DNA electronics); KnowmTech LLC (neural networking); Metallicum LLC (nanostructured metals and alloys); Molecular Mechanisms LLC (actuators using molecular mechanisms); Moore Nanotechnology Systems LLC (nanomachine systems); Nano-C LLC (fullerenes); Nano-Tex LLC (textiles); Nanochron LLC (photonic logic devices); Nanomaterials Company LLC (nanopowders); Nanomaterials Research LLC (nanoelectronic devices); Nanomechanics LLC (software); Nanova LLC (nanopowders); Plasmonics LLC (nanocomposites); QuesTek Innovations LLC (nanostructured steels);

Robiobotics LLC (genome engineering); Sierra Pacific Research Company LLC (nano-silicate crystallization); Superior MicroPowders LLC (nanopowders); Technology Innovations LLC (microactuators capable of manipulating molecules and cells); Umech Technologies LLC (characterization and measurement tools); Versilant Nanotechnologies LLC (nano-materials fabrication). We did not attempt to identify any nanotech businesses organized as S Corps.

50. Legislation enacted in 2002 authorizes taxpayers to carry back five years a net operating loss arising during 2001 or 2002. *See* I.R.C. § 172(b)(1)(H) (2003).
51. I.R.C. § 382 (2003). The "change of ownership" rules are complicated and confusing.
52. A large percentage of the investment in start-ups is from tax-exempt institutional investors, such as pension funds and university endowments. Thus, they have no taxable income with which to offset losses. Further, these institutions are not exempt from taxes on income that would pass through to them from a business in the LLC form.
53. I.R.C. § 1361(b)(1)(A) (2003).
54. Interview with Robert Bradbury, CEO of Robobiotics (Oct. 1, 2003).
55. Interview with Terry Lowe, CEO of Metallicum (Oct. 1, 2003).
56. We agreed with the company we interviewed that we would not disclose certain information.
57. Interview with Max Lagally, *supra* note 1.
58. Larry Bock, *Following Mr. Robinson's Advice: The Story of Nanosys*, 1 NANOTECH. L&B 91 (2004).
59. The exchange is not taxable, provided that, immediately after the exchange, the transferor(s) of the property own stock representing 80% of the total combined voting power of all classes of stock entitled to vote and at least 80% of the total number of all classes of stock of the corporation. Further, the stock must not be issued at unduly low valuations. *See* DENNIS RICE, TAKING THE NANOTECHNOLOGY STARTUP FROM FORMATION TO ELECTRONIC PUBLIC OFFERING 5 (2002).
60. Under federal and state securities laws, the issuance of stock is considered the selling of a "security". The sale of a security must be registered with the U.S. Securities and Exchange Commission. However, under Regulation D, a startup company is exempt from registration. *Id.*
61. There are two primary restrictions on the use of ISOs. Initially, ISOs can only be given to employees and not directors or consultants. IRC §422(a)(2); Reg. §1.421-7(h)(2)-(3) (2004). Additionally, the exercise price of the options must be equal to the fair market value of the underlying stock on the day that the option is granted. IRC §422(b)(4).
62. If the company issues NQOs, employees are taxed upon the exercise of the option, and the company can deduct the difference between the option price and the value of the stock when the employee exercises the option. The employee is not taxed when receiving the ISO or exercising the option. IRC §421(a)(1), 422(a) (2004). (For high-income taxpayers, the ISO might trigger the alternative minimum tax.) Additionally, the employee can obtain capital gains treatment on the sale of the stock if she holds the stock two years from the date of grant of the option and one year from the date of exercise. The company cannot take any deductions in connection with the issuance of the ISO.
63. *See* Maryann Thompson, *Recruitment Spotlight: Hard Numbers on Net Executive Compensation*, THE INDUSTRY STANDARD, Nov. 1, 1999, at LEXIS, Nexis Library.
64. Most plans allow the company to negotiate accelerated vesting of unvested options with certain employees. Accelerated vesting can be triggered by death, disability, termination without cause, IPO, or sale of the company.

Some plans provide the corporation with the right to repurchase the stock at fair market value upon employment termination while other Plans allow employees to retain vested shares. Most plans restrict transfer of shares due to securities laws considerations and to prevent shares from being acquired by unknown people. For a more detailed discussion of these issues, *see* Rice, *supra* note 59.

65. In general, skilled counsel should be able to draft the Plan so that the issuance of options is exempt from registration under securities laws. The issuance of options is typically not subject to SEC registration under the "no sale rule." For a comprehensive discussion of issuance of options and securities laws, *see* Rice, *supra* note 59.
66. *See generally* ANNA LEE SAXENIAN, REGIONAL ADVANTAGE: CULTURE AND COMPETITION IN SILICON VALLEY AND ROUTE 128 (Harvard University Press, 1996).
67. *See* LARTA, *About Us, Overview*, at www.larta.org (last visited Apr. 12, 2004).
68. *See* MASS. TECH. COLLABORATIVE, *Massachusetts Nanotechnology Initiative*, at http://www.mtpc.org/nano/index.htm (last visited Apr. 12, 2004).
69. One such effort is known as the Texas Nanotechnology Initiative. *See* TEXAS NANOTECHNOLOGY INITIATIVE, *Frequently Asked Questions*, at www.texasnano.org/faqs.html (last visited Apr. 12, 2004).

Chapter 11

1. Interview with Peter Grubstein, Partner, NGEN Partners, in Santa Barbara, California (Feb. 27, 2004).
2. This phrase was coined by venture capitalist Peter Grubstein. *Id.*
3. For a more comprehensive discussion of business plan writing, see ERIC SIEGAL, BRIAN FORD, JAY BORNSTEIN, THE ERNST & YOUNG BUSINESS PLAN GUIDE (John Wiley & Sons, 1993); STANLEY R. RICH AND DAVID E. GUMPERT, BUSINESS PLANS THAT WIN MONEY LESSONS FROM THE MIT ENTERPRISE FORUM 25 (Harper & Row, New York 1985).
4. Interview with Peter Grubstein, *supra* note 1.
5. For a further description of the elements of business plans, we recommend the Small Business Association's website at www.sba.gov. The website business strategy and plan section is particularly helpful for start-ups. U.S. SMALL BUS. ADMIN., *Business Plan Basics*, at http://www.sba.gov/starting_business/planning/ basic.html.
6. This list has been compiled by analyzing many different business plans.
7. *See* Norman C. Bensley, *Legal Perspectives for the Nanotech Start-Up*, Presentation at Strategic Research Institute Conference on Nano-Financing (Mar. 2002).
8. *See* Catalytic Solutions, Inc., *MPC Technologies*, 2002, at http://www.catsolns.com/technology.htm.
9. *See, e.g.,* Cowen, et. al, *Which Types of Analyst Firms Make More Optimistic Forecasts?* (Harvard NOM Working Paper No. 03-46, Jul. 8, 2003), at http://ssrn.com/abstract=436686 (noting that market projections were inflated as the result of the incentives of analysts at investment banks).
10. *See generally* Lawrence Aragon, *KP's Khosla Pulls No Punches, Slams Plans for Nanosys IPO*, PRIVATE EQUITY WEEK, Jun. 7 2004, *available at* http://www.venture economics.com/vec/1070549978620.html.
11. *See generally* Marc Meyer, *The Strategic Integration of Markets and Competencies*, 17 INT'L J. TECH. MGMT. 677-95 (1999).
12. Carolina Braunschweig, *Nano Nonsense*, VENTURE CAP. J. 25 (Jan. 2003).

13. MICHAEL E. PORTER, COMPETITIVE ADVANTAGE 237 (New York, Simon & Schuster, 1985).
14. *Id.*
15. *See* Robert Cooper & Scott Edgett, *Overcoming The Crunch In Resources For New Product Development*, 46 RESEARCH TECH. MGMT 48 (May 2003).
16. Jeff Karoub, *Nanogate Puts the Nano Inside, And Let's Partners Do The Selling*, SMALL TIMES, Nov. 1, 2002, *available at* http://www.smalltimes.com/document_display.cfm?document_id=4942.
17. MICHAEL E. PORTER, COMPETITIVE STRATEGY 6 (New York, Simon & Schuster, 1998).
18. *Interview with NanoBio Corporation*, THE WALL STREET TRANSCRIPT, Aug. 2003, *available at* http://www.nanobio.com/downloads/TWSTinterview.pdf.
19. *See* Press Release, Nanosphere, *Nanosphere Signs Letter of Intent with TakaraBio Inc. to Form Development and Distribution Alliance* (Jul. 22, 2002) (on file with author), *available at* http://www.nanosphere-inc.com/3_media/1_pr/072202.html.
20. Richard Lockie, *Putting the Human Genome to Work*, Presentation for TM Bioscience at the Annual General Meeting (Jun. 4, 2003) (on file with author), *available at* http://www.tmbioscience.com/client/ files/AGMREPORTS/TMC%202003%20AGM%20Presentation%20June%203%2003%20v4.pdf.
21. Sherrie E. Zhan, *Choosing a Market Entry Strategy*, 12 WORLD TRADE 40-42, May 1999.
22. On June 10, 2004 Ruben Serrato addressed the City of Berkeley, Public Health Division in a talk entitled "Nanotechnology Health and Environmental Overview." It represented one of the first times, though unlikely the last, that a city held open public forums to address local concerns over health and environmental effects of nanotechnology. The transcript is *available at* http://www.ci.berkeley.ca.us/commissions/health/2004health/agenda/061004A16.htm.
23 Chapter 3 engages in a detailed discussion of how Michael Crichton's book, *Prey*, could give rise to a public backlash against nanorobot technologies.
24. For a discussion of the company's initial focus on developing drugs and drug delivery systems, *see* Allen Bernard, *CSixty Pioneers Drug Delivery Techniques Using Buckyballs*, Nanoelectronics Planet.com, NANOWATCH, Jan. 16, 2002, at http://www.nanoelectronicsplanet.com/features/article/0,4028,6571_956101,00.html. At the time of this writing, the company was in clinical trials for its dermatological and cosmetic applications. For an update on the company's efforts to these products, see C Sixty, Inc., *Research and Development, On-going Efforts*, at http://www. csixty.com/on-going.html.
25. For a lucid discussion and analysis of the rules governing information disclosure and fraud in securities markets see Steven Amchen, Jessica Cordova and Paul Cicero, *Securities Fraud*, 39 AM. CRIM. L. REV. 1037 (Spring 2002).
26. For a direct evaluation of different hardware products, see CVD Equipment Corporation, at http://www.cvdequipment.com; Nanonics Imaging Limited, at http://www.nanonics.co.il.
27. For a description of the typical sales and marketing expenses for small firms, see BUREAU OF LABOR STATISTICS, OCCUPATIONAL OUTLOOK HANDBOOK 27 (2003), *available at* http://www.bls.gov/oco/.
28. *See* 2002 10-K filings for the following companies:
Altair Nanotech: ALTI
NanoProprietary: NNPP
NanoPierce: NPCT
NanoScience Tech: NANSE
U.S. Global Nanospace: USGA

Nanophase: NANX
Nanogen: NGEN
Nanobac Pharmaceuticals: NNBP
Nanosignal: NNOS
Nanometrics: NANO

29. *See* 2002 10-K filing, *Management Discussion and Analysis*, at II-6.
30. Deborah House, *The Top Five Profit Drains and How to Plug Them*, 24 J. BUS. STRATEGY 32 (March 2003).
31. For a complete review of financial modeling and related concepts, see JERRY A. VISCIONE, FINANCIAL ANALYSIS: TOOLS AND CONCEPTS (1984).
32. STANLEY R. RICH AND DAVID E. GUMPERT, BUSINESS PLANS THAT WIN MONEY LES SONS FROM THE MIT ENTERPRISE FORUM 146-7 (Harper & Row, New York 1985).
33. *Id.*
34. *Id.*
35. The items usually included in an Income Statement are: Revenue/Sales, Cost of Goods Sold (material, labor), Operating Expenses (sales & marketing, research & development, general administrative items), Income or Expenses from Interest, and Income Taxes.
36. The items usually included in a Balance Sheet are: Current Assets (cash, value of investments, receivables, and inventory), Other Assets (property, plant and equipment, long term investments, intangibles), Current Liabilities (short term debt, accounts payable), Long Term Debt, and Stockholders' Equity (preferred stock, common stock, options/warrants, retained earnings, capital surplus).
37. The items usually included in a Statement of Cash Flows are: Cash at Beginning of Period, Net Income, Operating Adjustments (depreciation, changes in accounts receivable, changes in liabilities, changes in inventory), Investing Adjustments (capital expenditures, investments), Financing Adjustments (dividends paid, sale or purchase of stock, borrowings), and End Cash Balance.
38. For a thorough treatment of accounting issues, see James R. Hitchner, FINANCIAL VALUATION: APPLICATION AND MODELS (John Wiley & Sons, 2003).
39. We asked multiple people at the NSF, and we did not receive a clear defense of the figure or methodology used.
40. MARVIN CETRON, TECHNOLOGICAL FORECASTING 204 (Gordon & Breach, 1969).
41. For a complete description of sources and uses of cash in valuation analysis see generally Preston Estep, *Cash Flows, Asset Values, and Investment Returns: Tying Return Forecasting to Uses of Cash*, 29 J. PORTFOLIO MGMT. 17 (Mar. 2003).
42. William Weaver & Stuart Michelson, *A Practical Tool to Assist in Analyzing Risk Associated with Income Capitalization Approach Valuation or Investment Analysis*, 71 APPRAISAL J. 335 (2003).
43. JOSH LERNER, VENTURE CAPITAL AND PRIVATE EQUITY 181 (2000).
44. Anne Grimes, *V.C. Secret: Value Is in the Eye of the Beholder*, WALL ST. J., Nov. 4, 2002, at LEXIS, Nexis Library.
45. *Id.*
46. *See* Chapters 5, 10, and 13.
47. *See* Mark A. Lemley, *The Economics of Improvement in Intellectual Property Law*, 75 TEX. L. REV. 989, 1048-50, 1053, 1055 (1997) (discussing valuation problems for intellectual property); Gavin Clarkson, *Avoiding Suboptimal Behavior in Intellectual Asset Transactons: Economic and Organizational Perspectives on the Sale of Knowledge*, 14 HARV. J. L. & TECH 711 (2001) ("The difficulty of measuring and valuing innovation as embodied in a company's intellectual assets becomes particularly apparent when a company wants to acquire another company or the rights to exploit its intellectual assets."). For a comprehensive discussion of how

intellectual property and intangible assets are valued, see GORDON V. SMITH & RUSSELL L. PARR, VALUATION OF INTELLECTUAL PROPERTY & INTANGIBLE ASSETS 170 (3d ed. 2000).
48. For a comprehensive discussion of valuation techniques such as Comparables, Net Present Value, Adjusted Present Value, Venture Capital, and Options, see Lerner, *supra* note 43, at 181 -201.
49. UBS Warburg, *Global Valuation Group: Equity Analysis Seminars*, Presentation, Oct. 2000, at 6.
50. *Id.*
51. *Id.*
52. In financial terms, unless Nanogen's prospects for future profits were tripled as a result of their patent award, the associated tripling of value of the company was a mis-pricing.
53. UBS Warburg, *supra* note 49, at 9.
54. *Id.*
55. The trading price of GE's public shares implied a value for the company of approximately $312.7 billion while its net income in the last twelve months of operations was $14.1 billion.
56. These were valuation multiples commonly applied to growth companies recently in the Internet and biotech sectors.
57. For a complete review of discounted cash flow related valuation techniques we recommend Tom Copeland, et. al, DCF VALUATION 2000 MODEL (2000).
58. UBS Warburg, *supra* note 49, at 14.
59. For a comprehensive review of discount rate related issues we recommend JESSICA JAMES AND NICK WEBBER, INTEREST RATE MODELING: FINANCIAL ENGINEERING (John Wiley & Sons, 2000).
60. These estimates were prepared by evaluating actual transactions, examining published reports and discussing trends with venture capitalists. For a description of discount rates typically applied by venture capital firms see Lerner, *supra* note 43, at 181-201.
61. UBS Warburg, *supra* note 49.
62. For a continued discussion of terminal values for technology companies see Case R.H. and S. Shane, *Fostering Taking In Research and Development: The Importance of A Project's Terminal Value*, 29 DECISION SCIENCES 765-783 (1998).
63. UBS Warburg, *supra* note 49.

Chapter 12

1. Carolina Braunschweig, VCs *See Bottom, Expect To Do More Deals*, VENTURE CAP. J., July 1, 2003, at http://www.ventureeconomics.com/vcj/protected/1055428 676849.html.
2. Angel Investor News, *Nanotechnology Alliance Provides Funding for Nano Start-Ups*, Nov. 2003, at http://www.angel-investor-news.com/ART_nanotech2.htm.
3. *See* Central Coast Angel Network (CCAN), at http://www.ccangels.net.
4. Interview with Peter Grubstein, NGEN Partners, in Santa Barbara, California (Jan. 27, 2004).
5. Agencies do require companies to report intellectual property discovered using federal funds. Further, the federal government does retain a royalty free license to use the technology. Companies can be stripped of their intellectual property rights if they fail to follow reporting requirements. *See, e.g., General Terms and Conditions of SBIR Grants*, at http://www1.pr.doe.gov/sbirtoc.html.

6. Interview with Rachelle D. Hollander, National Science Foundation (Sept. 8, 2003).
7. Email from Sara B. Nerlove, SBIR Program Director at National Science Foundation, Sept. 11, 2003 (noting that "SBIR is highly competitive") (email on file with authors).
8. Interview with Max Lagally, founder nPoint, University of Wisconsin (Oct. 12, 2003).
9. In order to be eligible for grants under the SBIR/STTR program, the applicant must qualify as a small business. A small business is one that at the time of the Phase I and Phase II awards meets the following criteria:
   - It is independently owned and operated, is not dominant in the field of operation in which it is proposing, has its principal place of business located in the United States, and is organized for profit
   - It is at least 51 percent owned, or in the case of a publicly owned business, at least 51 percent of its voting stock is owned, by United States citizens or lawfully admitted permanent resident aliens.
   - It has, including its affiliates, a number of employees not exceeding 500, and meets the other regulatory requirements found in 13 CFR 121. Business concerns, other than licensed investment companies, or state development companies qualifying under Small Business Investment Act of 1938, 15 U.S.C. 661, et seq., are affiliates of one another when either directly or indirectly, (a) one concern controls or has the power to control the other; or (b) third parties (or party) control(s) or has the power to control both. Control can be exercised through common ownership, common management, and contractual relationships. The term "affiliates" is defined in great detail in 13 CFR 121.103. The term "number of employees" is defined in 13 CFR 121.106. Business concerns include, but are not limited to, any individual, partnership, corporation, joint venture, association, or cooperative.
10. *See* Defense Advanced Research Projects Agency, at http://www.darpa.mil/index.html.
11. *See* Defense Advanced Research Projects Agency, *Contracts Management Office Homepage*, at http://www.darpa.mil/cmo/.
12. *See* Nat'l Inst. of Standards & Tech., *Advanced Technology Program*, at http://www.atp.nist.gov/.
13. A list of the projects in nanotechnology that have been funded by ATP can be viewed at Nat'l Inst. of Standards & Tech. (NIST), *ATP's Project Portfolio in Nanotechnology*, Feb. 25, 2003, at http://www.nist.gov/public_affairs/atp_nanotech.htm (last updated Mar. 28, 2003).
14. *See* Zyvex, *Licensing and Alliances*, 2004, at http://www.zyvex.com/Alliances/government.html.
15. Ravi Chiruvolu, *VCs Turn The Screws*, BUSINESSWEEK, Aug. 2002, at 36 ("Building a venture-backed company should take five to seven years. If you spend any more time on a deal, your internal rate of return (IRR) takes a hit.").
16. Bob Zider, *How Venture Capital Works*, HARV. BUS. REV. 131, 133 (Nov.-Dec. 1998) ("Picking the wrong industry or betting on a technology risk in an unproven market segment is something VC's avoid.").
17. For a story of the poor investments in biotechnology made by one venture fund, see Ravi Chiruvolu, *Before You Do that 'Amazing' Biotech Deal, Read this Story*, VENTURE CAP. J., at 36 (Aug. 2002).
18. *See generally* GEORGE W. FENN ET AL., BOARD OF GOVERNORS OF THE FEDERAL RESERVE SYSTEM, THE ECONOMICS OF THE PRIVATE EQUITY MARKET (1995).
19. For example, Quantum Dot Corp., one of the first success stories in this burgeoning field, was forced to delay its product for several years due to mass production dilemmas. *See* Carolina Braunschweig, *Nano Nonsense*, 1120 VENTURE CAP. J. 18 (Jan. 2003).

20. Scott R. Burnell, *Private Funding Limited for Nanotech*, UPI, May 21, 2002, at LEXIS, Nexis Library (quoting Richard Shanley, a partner at the consulting firm Deloitte and Touche).
21. Braunschweig, *supra* note 19.
22. Mike McGehee, Stanford University, Winter 2003.
23. Steve Jurvetson, *Transcending Moore's Law With Molecular Electronics and Nanotechnology*, 1 NANOTECH. L&B 70 (2004).
24. Braunschweig, *supra* note 19.
25. David Rotman, *The Nanotube Computer*, 105 TECH. REV. 36, 45 (March 2002).
26. Braunschweig, *supra* note 19.
27. *Harnessing Innovation*, INDUSTRY WEEK, Dec. 2002.
28. Braunschweig, *supra* note 19.
29. We identified 59 nanotech companies, 17 of which had no publicly available information.
30. Carolina Braunschweig, *supra* note 19.
31. *Id.*
32. Interview with Peter Grubstein, *supra* note 4.
33. According to Mr. Grubstein, approximately 30% of NGEN's deals typically come from other nanotech investors such as Draper, Fisher, Jurvetson or Harris & Harris. These other firms refer deals to NGEN in order to obtain additional technology validation and co-investment on the deal. Approximately 25% of deals come directly from university referrals. The residual 45% of deals come directly from one of the three partners who actively seek out companies and who benefit from the reputation of NGEN. *Id.*
34. John Craig Venter, *John Craig Venter Unvarnished*, BIOIT WORLD, Nov. 12, 2002, at http://www.bio-itworld.com/archive/111202/horizons_venter.html.
35. DENIS RICE, TAKING THE NANOTECHNOLOGY STARTUP FROM FORMATION TO ELECTRONIC PUBLIC OFFERING 67 (2002).
36. *Id.*
37. *See* Jacquelin A. Daunt, *Venture Capital For High Technology Companies*, 17, at http://www.fenwick.com/pub/corp_pubs/vencap/Venture_Cap_Main_2002.htm.
38. *Id.*
39. *Id.*
40. Rice, *supra* note 35, at 71.

Chapter 13

1. Vivek Koppikar, Stephen Maebius, & J. Steven Rutt, *Current Trends In Nanotech Patents: A View From Inside The Patent Office*, 1 NANOTECH. L&B 24 (2004).
2. *See generally* Council on Gov't Relations: A Review of Industry-University Research Relationships (1996), *available* at http://www.cor.edu; George M. Low, *The Organization of Industrial Relationships in Universities*, *in* PARTNERS IN THE RESEARCH ENTERPRISE: UNIVERSITY-CORPORATE RELATIONS IN SCIENCE AND TECHNOLOGY 68, 71-74 (Thomas W. Langfitt et al. eds., 1983).
3. For a more detailed description of how these might be drafted and implemented, see John F. Hesselberth, *Technology Transfer From Academia: Prescription For Success and Failure*, in CONSORTIA AND STRATEGIC ALLANCES 151 (David V. Gibson & Raymond W. Smilor eds., 1992); Government-Industry-University Research Roundtable, *Simplified and Standardized Model Agreements for Industry-University Cooperative Research*, art 8.1 (1988).
4. For example, at MIT's Media Lab, "affiliate sponsorship" ($100,000 per year for three years) introduces sponsors to the overall work of the Lab and provides

limited access to intellectual property. "Consortium sponsorship" links a group of sponsors with a group of Laboratory faculty and research staff focused on common goals. Joining a consortium ($200,000 a year for a minimum of three years) entitles the sponsor to full intellectual property rights, license-fee free and royalty free, to all inventions in the Lab. *See* MIT Media Lab, *Sponsorship*, at http://www.media.mit.edu/sponsors/index.html.

5. NNIN facilities are located at: Cornell University, Howard University, Pennsylvania State University, Stanford University, University of California at Santa Barbara, Georgia Institute of Technology, Harvard University, North Carolina State University, University of Michigan, University of Minnesota, University of New Mexico, University of Texas at Austin, and University of Washington.
6. Nat'l Nanotechnology Initiative, *National Science Foundation User Centers and Facilities*, at http://www.nano.gov/html/centers/NSFcenters.html.
7. For example, the agreement form used by Stanford University for non-Stanford institutions using the facilities states that Stanford makes "no a-prior claim to inventions developed in the lab[.]"
8. Fed. Lab. Consortium, *Federal Technology Transfer Legislation and Policy* (Washington D.C.: Universal Technical Resources, 2002).
9. Los Alamos Nat'l Lab., *Technology Transfer Mechanisms* (Los Alamos, 2003).
10. *Id.*
11. *Id.*
12. *Id.*
13. *Id.*
14. Patent law provides patent holders with the right to exclude others from making, using, and selling the invention. § 35 U.S.C. 283 (2004).
15. We discussed difficulties associated with concluding multiple license agreements in chapter 5. Each patent holder attempts to extract the maximum royalty from the company, and the company is attempting to minimize the royalty it must pay. It is easy for patent holders to overlook the fact that, when a licensee must seek a number of licenses from different entities which are all trying to extract the maximum value, the business endeavor can quickly lose its value. Even holders of patents that contribute little value relative to the product can attempt to appropriate as much of the value of the improvement as possible.
16. *See* Chapter 5.
17. *See* Mark Modzelewski and Stephen Maebius, *Editorial*, 1 NANOTECH. L&B 7 (2004).
18. There are two exceptions to this rule. First, an inventor who first conceived the invention but the last to reduce it to practice will be denied priority if she did not exercise reasonable diligence in reducing to practice from a time just prior to when the first person who reduced to practice conceived the invention. Second, the later inventor will be given priority if the first inventor abandoned, suppressed, or concealed the invention.
19. In order to obtain priority back to the filing of the provisional, however, the disclosure in the provisional must enable the claims of the patent. New information obtained after the filing cannot be considered in determining disclosure sufficiency. Therefore, filing a provisional on an invention before it has been fully tested could increase the likelihood of non-enablement of the claims of the final patent.
20. Raj Bawa, *Nanotechnology Patenting In The U.S.*, 1 NANOTECH. L&B 31 (2004).
21. This estimate is based on interviews with several CEOs and patent attorneys.
22. *See* Chapter 5.
23. Bawa, *supra* note 20.

24. 35 U.S.C. § 282 (1994).
25. *Id.*
26. Such status can even be claimed for both provisional and conventional patent applications.
27. *See* CHISUM, ET. AL., PRINCIPLES OF PATENT LAW: CASES AND MATERIALS 97 (2001).
28. Roukes, et al., U.S. Pat. No. 6,593,731 *Displacement Transducer Utilizing Miniaturized Magnet and Hall Junction*, (issued July 15, 2003).
29. 35 U.S.C. § 113.
30. SRI Int'l v. Matsushita Elec. Corp. of Am., 775 F.2d 1107, 1121 (Fed. Cir. 1985).
31. *Compare* Scripps Clinic & Research Found. V. Genentech, Inc., 927 F.2d 1565 (Fed. Cir. 1991) (product-by-process claims not limited by process set forth in the claim) *with* Atlantic Thermoplastics Co. v. Faytex Corp., 970 F.2d 834 (Fed. Cir. 1992) (process set forth serves as limitation in determining infringement).
32. We provided detailed citations for laws governing infringement in Chapter 5.
33. *See* Festo Corp. v. Shoketsu Kinzohu Kogyo Kabushiki Co., 122 S. Ct. 1831 (2002).
34. Autogiro Co. of America v. United States, 384 F.2d 391, 397 (Ct. Cl. 1967).
35. Mycogen Plant Sci. , Inc. v. Monsanto Co., 243 F.3d 1316, 1327 (Fed. Cir. 2001).
36. Texas Digital Systems, Inc., v. Telegenix, Inc., 308 F.3d 1193, 1202 (2002); Brookhill-Wilk 1, LLC v. Intuitive Surgical, Inc., 334 F.3d 1294, 1300 (Fed. Cir. 2003).
37. Vitronic Corp. v. Conceptronic, Inc., 90 F.3d 1576 (Fed. Cir. 1996).
38. *See* John Molenda, *The Importance of Defining Novel Terms in Patenting Nanotechnology Inventions*, 1 NANOTECH. L&B (2004).
39. *Id.* (citing WEBSTER'S II NEW RIVERSIDE DICTIONARY 169 (Rev. ed. 1996)).
40. *Id.* (citing 4 MCGRAW-HILL ENCYCLOPEDIA OF SCIENCE & TECHNOLOGY 630 (8th ed. 1997) (emphasis added)).
41. Bruchez, Jr., et al., U.S. Pat. No. 6,500,622, *Methods of Using Semiconductor Nanocrystals In Bead-Based Nucleic Acids* (issued Dec. 31, 2002).
42. *See* Verdegaal Bros., Inc. v. Union Oil Co., 814 F.2d 628, 631, 2 U.S.P.Q.2d (BNA) 1051, 1053 (Fed. Cir. 1987) ("A claim is anticipated only if each and every element as set forth in the claim is found, either expressly or inherently described, in a single prior art reference.").
43. *In re* Rose, 220 F.2d 459, 105 USPQ 237 (CCPA 1955) (holding that claims to a lumber package "of appreciable size and weight requiring handling by a lift truck" were un-patentable over prior art lumber packages which could be lifted by hand because limitations relating to the size of the package were not sufficient to patentably distinguish over the prior art.); *In re* Rinehart, 531 F.2d 1048, 198 USPQ 143 (CCPA 1976) ("Mere scaling up of a prior art process capable of being scaled up, if such were the case, would not establish patentability in a claim to an old process so scaled."); Gardner v. TEC Systems, Inc., 725 F.2d 1338, 220 USPQ 777 (Fed. Cir. 1984), *cert. denied*, 469 U.S. 830, 225 USPQ 232 (1984) (holding that, where the only difference between the prior art and the claims was a recitation of relative dimensions of the claimed device and a device having the claimed relative dimensions would not perform differently than the prior art device, the claimed device was not patentably distinct from the prior art device).
44. *Id.* at 1346.
45. 158 USPQ 596 (CCPA 1968).
46. CHISUM, *supra* note 27, at 162. The written description doctrine could also be

invoked as a tool to limit the scope of these claims. Difficulties encountered in attempting to distinguish between the enablement and written description doctrines have generated judicial confusion and uncertainty. Because courts have generally used the enablement doctrine to narrow the scope of claims in biotechnology patents, this discussion will focus on the enablement doctrine.

47. 56 U.S. 62 (1854).
48. *Id.* at 112.
49. *Id.* at 113. ("[H]e claims an exclusive right to use a manner and process which he has not described and indeed had not invented, and therefore could not describe when he obtained his patent. The court is of opinion that the claim is too broad, not warranted by law.").
50. *Id.* at 113. ("For aught that we now know some future inventor, in the onward march of science, may discover a mode of writing or printing at a distance by means of the electric or galvanic current, without using any part of the process or combination set forth in the plaintiff's specification. His invention may be less complicated—less liable to get out of order—less expensive in construction, and in its operation. But yet if it is covered by this patent the inventor could not use it, nor the public have the benefit of it without the permission of this patentee.").
51. 11 F.3d 1046, 1052-53 (Fed. Cir. 1993).
52. *Id.* at 1048-49.
53. *Id.* at 1050.
54. 188 F.3d 1362 (Fed. Cir. 1999).
55. *Id.* at 1371-77. *See also* Amgen, Inc. v. Chugai Pharmaceutical Co., Ltd., 927 F.2d 1200, 1214 (Fed. Cir. 1991) (holding that a claim covering all possible DNA sequences encoding any polypeptide having an amino acid sequence "sufficiently duplicative" of EPO was not enabled, because of "the structural complexity of the EPO gene, the manifold possibilities for change in its structure, with attendant uncertainty as to what utility will be possessed by these analogs."); *In re* Vaeck, 947 F.2d 488, 495-96 (Fed. Cir. 1991) (holding that a claim to the production of the insecticidal Bacillus proteins within host cyanobacteria was not enabled when only one species of cyanobacteria was employed in the working examples of the specification.); *In re* Wright, 999 F.2d 1557 (Fed. Cir. 1993) (affirming the PTO's rejection of a claim pertaining to a live, non-pathogenic vaccine for a pathogenic RNA virus).
56. *See* Koppikar, Maebius, & Rutt, *supra* note 1.
57. *See* the U.S. PTO training guidelines for § 112 rejections in chemical and biotechnology inventions, which state as follows (emphasis supplied):

> To overcome a *prima facie* case of lack of enablement, applicant must demonstrate by argument and/or evidence that the disclosure, as filed, would have enabled the claimed invention for one skilled in the art at the time of filing. **This does not preclude applicant from providing a declaration after the filing date which demonstrates that the claimed invention works.** However, the examiner should carefully compare the steps, materials, and conditions used in the experiments of the declaration with those disclosed in the application to make sure that they are commensurate in scope, i.e., that the experiments used the guidance in the specification as filed and what was well known to one of skill in the art. Such a showing also must be commensurate with the scope of the claimed invention, i.e., must bear a reasonable correlation.

58. 559 F.2d 595 (CCPA 1977).
59. *Id.* at 600-601.

60. *Id.* at 606.
61. 126 F. Supp. 2d 69 (D. Mass. Jan. 19, 2001).
62. *Id.* at 160. In footnote 56, the court elaborated: "The reason for such a rule is clear. What would be the value in patenting a composition at all if, by making the slightest alteration in the method of making what is nonetheless the same product, a competitor were able to evade liability? A patent system that permitted such conduct would remove the carrot dangling in front of the inventor's nose. If inventors were so easily divested of their limited monopoly rights attendant to their novel, useful, and nonobvious contributions, they would likely abandon their pursuits and thereby inhibit progress. The law does not permit such an outcome." *Id.* at 160 n. 56. *See also* Chiron Corp. v. Genentech, 2002 U.S. Dist. LEXIS 19148 (E.D. Cal. June 24, 2002); United States Corp. v. Phillips Petroleum Co., 865 F.2d 1247, 1251 (Fed. Cir. 1989); *In re* Koller, 613 F.2d 819, 824-24 (C.C.P.A. 1980).
63. In determining who was first to invent, the USPTO Board of Patent Appeals and Interferences considers the "respective dates of conception and reduction to practice of the invention... [as well as] the reasonable diligence of one who was first to conceive and last to reduce to practice, from a time prior to conception by the other." 35 U.S.C. § 102(g). Priority of invention "goes to the first party to reduce an invention to practice unless the other party can show that it was the first to conceive the invention and that it exercised reasonable diligence in later reducing that invention to practice." Price v. Symsek, 988 F.2d 1187, 1190, 26 U.S.P.Q.2d (BNA) 1031, 1033 (Fed. Cir. 1993). If inventors cannot prove when the invention took place, the date of filing serves as the date that the invention took place.
64. 35 U.S.C. § 135.
65. 35 U.S.C. § 141; 35 U.S.C. § 145.
66. *See* 35 U.S.C. § 122(b)(2)(B) (2001); 37 C.F.R. 1.213(a) (2002).
67. *See* http://www.uspto.gov/web/offices/com/annual/2003/060404_table4.html.
68. A patent gives its owner the "right to exclude others from making, using, offering for sale, or selling" the invention "throughout the United States or importing the invention into the United States . . . ." 35 U.S.C. § 154(a)(1)-(2).
69. *See* WORLD INTELL. PROP. ORG. (WIPO), Patent Cooperation Treaty ("PCT") (1970), at http://www.wipo.int/pct/en/treaty/about.htm.
70. An applicant can wait until the end of the 20th month after the filing of the international application to commence filing in specific countries. This 20-month period is extended by a further 10 months where the applicant chooses to ask for an "international preliminary examination report", a report which is prepared by one of the major patent offices and which gives a preliminary non-binding opinion on the patentability of the claimed invention. *See id.*
71. *See* Chapter 5 for a detailed discussion of damages.
72. *See, e.g.,* Ortho Pharm. Corp. v. Smith, 959 F.2d 936, 944, 22 U.S.P.Q.2d (BNA) 1119, 1126 (Fed. Cir. 1992) (explaining that an opinion letter by counsel may be an important factor in a determination of willful infringement).
73. *See* Michael J. Meurer, *Controlling Opportunistic and Anti-Competitive Intellectual Property Litigation*, 44 B.C. L. REV. 509, 510 (March 2003) ("Some IP owners value their property rights chiefly as "tickets" into court that give them a credible threat to sue vulnerable IP users. Socially harmful IP litigation is common because the rights are easy to get and potentially apply quite broadly, and the problem is growing worse because of the expansion of the scope and strength of IP law.").
74. Prior to 1999, the challenging party could not actively participate in the reexamination proceeding. In 1999, Congress enacted the Intellectual Property and

Communications Omnibus Reform Act of 1999, permitting *inter partes* reexamination wherein the challenging party would have an expanded role in the reexamination proceeding. Subtitle F of the American Inventors Protection Act of 1999 is entitled the Optional Inter Partes Reexamination Procedure Act of 1999, Pub. L. No. 106-113, 113 Stat. 1501, 1501A-567 to 1501A-570 (1999). The new inter partes procedure is only applicable to patents that were filed on or after the effective date of the legislation, November 9, 1999.
75. *See* 35 U.S.C. § 306.
76. The moving party must prove its right to a preliminary injunction in light of four factors: (1) a reasonable likelihood of success on the merits; (2) irreparable harm if the injunction were not granted; (3) balance of the hardships; and (4) the impact of the injunction on the public interest. *See generally* Reebok Int'l Ltd. v. J. Baker, Inc., 32 F.3d 1552 (Fed. Cir. 1994).
77. *See* Mark A. Lemley and Colleen V. Chien, *Are the U.S. Patent Priority Rules Really Necessary*, 54 HASTINGS L.J. 1299, n. 99 (July 2003).
78. *See* Markman v. Westview Instruments, Inc., 517 U.S. 370, 372 (1996). The court's decision on the construction of patent claims is known as a "Markman" ruling.
79. ANTHONY B. ASKEW AND ELIZABETH C. JACOBS, 2000 WILEY INTELLECTUAL PROPERTY LAW UPDATE 3 (Aspen Law & Business 2000).
80. DONALD A. GREGORY, ET AL., INTRODUCTION TO INTELLECTUAL PROPERTY LAW 212-13 (1994).
81. Unif. Trade Secrets Act, § 1(4) (1996).
82. *See, e.g.*, Zoecon Industries v. American Stockman Tag Co., 713 F.2d 1174 (5th Cir. 1983); Metallurgical Industries, Inc. v. Fourtek Inc., 790 F.2d 1195, 1201 (5th Cir. 1986).
83. Unif. Trade Secrets Act § 1(2) (1996).
84. 18 U.S.C. § 1839 (2000).
85. *See* RESTAMENT (THIRD) OF UNFAIR COMPETITION § 39 (1995).
86. *See* RESTAMENT (THIRD) OF UNFAIR COMPETITION § 39 (1995).
87. *See, e.g.*, Trailer Leasing Co. v. Assocs. Commercial Corp., No. 96 C2305, 1996 WL 392135, at 6 (N.D. Ill. July 10, 1996) (holding that agreement barring employee from disclosing "any methods or manners" of business was unenforceable); Nasco, Inc. v. Gimbert, 238 S.E.2d 368, 369-70 (Ga. 1977) (holding that non-disclosure covenant barring use or disclosure or "any information concerning any matters affecting or relating to the business of employer" was overbroad and unenforceable).
88. *See, e.g.*, Warner & Co. v. Solberg, 634 N.W.2d 65, 73 (N.D. 2001) (upholding employee non-solicitation covenant which was "narrowly drawn" to penalize only active solicitation by departed employee).
89. Specifically, a litigant cannot argue that (1) the mark is not inherently distinctive and lacks secondary meaning; and (2) the mark is confusingly similar to a mark or trade name that someone else used prior to the registrant and continues to use. MARGRETH BARRETT, PATENTS, TRADEMARKS, & COPYRIGHTS 115 (1998).
90. *See* Worthington Foods, Inc. v. Kellogg Co., 732 F. Supp. 1417, 1435 (S.D. Ohio 1990).
91. *In re* Quick-PrintCopy Shops, Inc., 616 F.2d 523, 525, 205 U.S.P.Q. 505, 507 (C.C.P.A. 1980).
92. *See* 15 U.S.C. § 1052(f) (1996).
93. *See* 15 U.S.C. § 1052(d) (1996) (The PTO may refuse to register a trademark if it so resembles a previously registered mark "as to be likely, when used on or in connection with the goods of the applicant, to cause confusion, or to cause mistake, or to deceive.").

94. Denis T. Rice, Taking the Nanotechnology Startup From Formation to Electronic Public Offering 59 (2002).
95. *See, e.g.*, Haughton Elevator Co. v. Seeberger, 85 U.S.P.Q. 80 (Dec. Com. Pat. 1950) (canceling "escalator" trademark as becoming generic).
96. Copyright Act of 1976, 17 U.S.C. §§ 101-1101 (2002).
97. The copyrightability of software was codified in the 1976 Copyright Act and the corresponding amendments in 1980. *See* Peter S. Menell, *An Analysis of the Scope of Copyright Protection for Application Programs*, 41 Stan. L. Rev. 1045, 1046-48 (1989). *See also* Apple Computer, Inc. v. Franklin Computer Corp., 714 F.2d 1240, 1249 (3d Cir. 1983) (holding programs in machine-readable form are appropriate subject matter for copyright).
98. *See* 17 U.S.C. § 106 (2002).
99. *See* Entertainment Research Group, Inc. v. Genesis Creative Group, Inc., 122 F.3d 1211, 1217 (9th Cir. 1997).
100. *See* Campbell v. Acuff-Rose Music, Inc., 510 U.S. 569, 577 (1994). Courts evaluate the following factors in determining fair use: (1) the purpose and character of the use, including whether the use is of a commercial nature; (2) the nature of the copyrighted work; (3) the amount and substantiality of the portion used in relation to the copyrighted work as a whole; and (4) the effect of the use upon the potential market for or value of the copyrighted work. 17 U.S.C § 107 (2002).
101. *See* Atari Games Corp. v. Nintendo of America, Inc., 975 F.2d 832 (Fed. Cir. 1992) (holding copying of software to be fair use when it was necessary to obtain access to the software's functional elements).
102. 17 U.S.C. § 408(a) (2002).
103. 17 U.S.C. § 411 (2002).

Chapter 14

1. Interview with Randy Bell, CEO of Nanotechnologies Inc. (Nov. 20, 2003) [hereinafter Bell Inverview].
2. Fenwick & West, LLP, Jacqueline A. Daunt and George Von Gehn, *Corporate Partnering for High Technology Companies 2*, 2000, available at http://www.fenwick.com/docstore/publications/corporate/Corporate_Partnering.pdf.
3. Jeff Coburn, *All for One: Strategic Alliances between Firms Are Good for Clients*, Business, 17 Legal Mgmt., Sept.-Oct. 1998, at 46-47.
4. Stephen Fraidin & Radu Leletiu, *The Role of Lawyers in Strategic Alliances: Strategic Alliances and Corporate Control*, 53 Case W. Res. L. Rev. 865 (2003).
5. *See generally* Cynthia Robbins-Roth, From Alchemy to IPO (Perseus Publishing, 2001).
6. L.J. Sellers, *Lilly's International Family*, 21 Pharmaceutical Executive, Mar. 2001, at 40-54.
7. Thomas Villeneuve, et al., Corporate Partnering: Structuring and Negotiating Domestic and International Strategic Alliances 1-1 (1997 Supplement).
8. Kathryn Rudie Harrington, Managing For Joint Venture Success 16 (1986).
9. Henry W. Chesbrough, *Making Sense of Corporate Venture Capital*, 80 Harvard Business Review 90 (March 2002).
10. Bell Interview, *supra* note 1.
11. Interview with Peter Grubstein, NGEN Partners (Jan. 27, 2004).
12. According to one investment banker: "[T]here are far fewer 'mini-Mercks' now in existence. Mini-Mercks are companies that announce on the day of their

formation that they expect to be a fully integrated pharmaceutical company with manufacturing and marketing capabilities. Some companies have achieved this, but it is very difficult and expensive. Instead, there are companies today that will help to purify proteins, and other companies, called contract research organizations ("CROs"), that will help to design and carry out clinical trials. In addition, there are contract manufacturers to reduce investment in manufacturing facilities." *Symposium: Financing the Biotech Industry: Can the Costs Be Reduced?*, 4 B.U. J. SCI. & TECH. L. 1, ¶ 42 (1997).
13. *See generally* Jeffrey Dyer, Prashant Kale, & Harbir Singh, *How to Make Strategic Alliances Work*, MIT SLOAN MGMT. REV., Summer 2001, at 37.
14. Villeneuve, et al., *supra* note 7, at 1-13.
15. For a more detailed discussion of how nanotech start-ups should go about seeking a corporate partner, *see* Daunt & Gehn, *supra* note 2. These materials were primarily relied on in writing this section.
16. Bell Interview, *supra* note 1.
17. Daunt & Gehn, *supra* note 2, at 3.
18. *Shaping Winning Business Strategies with Game Theory*, 19 FIN. EXECUTIVE 70 (March 2003).
19. *Id.* (noting that preparation is necessary to fully understand bargaining power, define negotiating position, and gain insight into effective negotiating tactics).
20. *See* Thomas O. Davenport, *The Integration Challenge*, 87 MGMT. REV. 25-28, Jan. 1998.
21. Villeneuve, et al., *supra* note 7, at 1.
22. *See, e.g., id.*
23. Daunt & Gehn, *supra* note 2, at 15.
24. For a more detailed discussion of factors likely to contribute to successful partnerships, *see* Dyer, Kale, and Singh, *supra* note 13.
25. Bell Interview, *supra* note 1.
26. For a description of issues in dispute resolution *see* Robert M. Howard, *Pre-Trial Bargaining & Litigation: The Search for Fairness and Efficiency*, 34 L. & SOC'Y REV. 431 (2000).
27. *See, e.g.*, STEPHEN S. COHEN & GAVIN BOYD, CORPORATE GOVERNANCE AND GLOBALIZATION: LONG RANGE PLANNING ISSUES (2000).
28. *See, e.g.*, DAVID J. BENDANIEL, ARTHUR ROSENBLOOM, AND JAMES HANKS, INTERNATIONAL M&A, JOINT VENTURES AND BEYOND: DOING THE DEAL (Wiley & Sons, March 2002).
29. Companies that hire foreign scientists may have to wade through the export control laws. *See* chapter 7.
30. *See* Richard Read, *High-Tech Companies Stake Claims in China's Hinterland*, NEWHOUSE NEWS SERVICE, Jan. 7, 2004, at LEXIS, Nexis Library.
31. *See generally* JOON K. PARK, INTELLECTUAL PROPERTY LAWS OF EAST ASIA (1997).
32. *See* WTO, *Overview: The TRIPS Agreement*, at http://www.wto.org/english/tratop_e/trips_e/intel2_e.htm.

Chapter 15

1. Edward Rashba, Daniel Gamota, Doug Jamison, John Miller, and Kirk Hermann, *Standards in Nanotechnology*, 2 NANOTECH. L&B (2004). A substantial part of the article was reprinted in this chapter with permission of *Nanotechnology Law & Business*.
2. *See, e.g.*, Josh Lerner, *Boom and Bust In The Venture Capital Industry and the Impact on Innovation*, 87 ECON. REV. 25 (Oct. 2002), at LEXIS, Nexis Library

("[A]ll too often these periods find venture capitalists funding firms that are too similar to one another."). As Lerner notes, this pattern of behavior is supported by theoretical works in "herding" by investment managers. When the success of investment managers is based on a comparison with their peers, they tend to make similar investments. *See, e.g.*, Andrea Devenow and Ivo Welch, *Rational Herding in Financial Economics*, 40 EUROPEAN ECON. REVIEW 603 (April 1996).
3. *See* Lerner, *supra* note 2.
4. *Id. See also* Bob Tedeschi, *The Pet Supply Business Is Finding That A Site May Serve Mostly To Guide Shoppers To Stores and Catalogs*, N.Y. TIMES, Oct. 28, 2002, at LEXIS, Nexis Library ("Companies like Pets.com and others helped put a face on the e-commerce lunacy. . . . [I]t is difficult to conjure the former runaway enthusiasm for the online pet supplies category, which, after books, CD's and toys, was one of the darlings of the e-commerce venture capital community. Hundreds of millions of dollars were poured into companies like Pets .com, Petopia.com and Petstore.com in 1998 and 1999.").
5. *See* Lerner, *supra* note 2.
6. *Id.* ("In many cases, the firms were liquidated when further financing could not be arranged. In others, the firms shifted their efforts into other, less competitive areas, largely abandoning the initial research efforts. In yet others, the companies remained mired with their peers for years in costly patent litigation.").
7. *Id.*
8. Edmund Kitch advanced this argument in the context of the desirable breadth of intellectual property. He argued that broad rights on upstream research are desirable, because they provide incentives for commercial development of the technology and allow the owner of the prospect to coordinate research efforts to prevent wasteful duplication. Edmund Kitch, *The Nature and Function of the Patent System*, 20 J. L. & ECON. 265, 276 (1977).
9. This argument has also been advanced to support consolidation in biotechnology. *See, e.g.*, Alvin R. Chin, *The Misapplication of Innovation Market Analysis to Biotechnology Mergers*, 3 B.U. J. SCI. & TECH. L. 6 (1997) ("[E]xpensive R&D programs, and the uncertainty of innovation present formidable obstacles for the mostly smaller firms that comprise the biotechnology industry. Biotechnology mergers are an important way for firms to pool financial resources, lower operating costs, and enhance R&D efficiency by bringing together complementary pieces of a larger R&D solution.").
10. *See* Lerner, *supra* note 2 (noting that in many cases, further financing of similar biotechnology start-ups could not be arranged).
11. *See, e.g.*, SYMPOSIUM REPORT: INTELLECTUAL PROPERTY LAW AND THE VENTURE CAPITAL PROCESS 18 (1989) ("A TRO can knock out the start-up by damaging its reputation and drying up venture capital funding.").
12. IBM holds patent 5,424,054, which claims a "hollow carbon fiber having a wall consisting essentially of a single layer of carbon atoms." Bethune, et. al., Pat. No. 5,424,054, *Carbon Fibers and Method for Their Production* (June 13, 1995).
13. T. ALLEN AND REINIER KRAAKMAN, COMMENTARY AND CASES ON CORPORATE LAW 11-18 (2001) (Preliminary Version); WILLIAM BRATTON, CORPORATE FINANCE 720-21 (2003).
14. Allen and Kraakman, *supra* note 13, at 11-1.
15. *Id.*; Jacqueline A. Daunt, *Mergers & Acquisitions For High Technology Companies*, available at http://www.fenwick.com/pub/corp_pubs/Mergers_&Acquisitions/M&A_2002_main.htm.
16. In a stock acquisition, the acquirer must contract separately with each shareholder of the target company. Bratton, *supra* note 13, at 725. This transaction is not used because not only are the transaction costs exorbitant, but there is a risk

that not all shareholders will agree to the deal. Asset sales are not used because target's avoid these transactions. In a stock for stock merger, the shareholders of the target company must approve the merger under DCL § 251(c), and dissenting shareholders can be "squeezed out" with appraisal rights under § 262(a).
17. See, e.g., Tom Stein & Matthew Debellis, *Desperately Seeking Partners: VCs Try To Line Up Mates For Struggling Start-ups*, INVESTMENT DEALER'S DIGEST, Aug. 5, 2002, at LEXIS, Nexis Library.
18. Two analysts observe, "Pulling off such deals is difficult. In particular, mergers between two private companies, known as private-to-privates, can be extremely tough." Stein & Debellis, *supra* note 17. Adds Bart Schachter, a general partner at Blueprint Ventures: "Everybody wants to do private-to-private deals, but very few actually succeed. It takes a tremendous amount of intestinal fortitude, because valuation, team composition and who comes out on top are all very subjective. I've been holed up in these tough discussions all day in a hotel room trying to decide who comes out on top." *Id.*
19. For example, assume two nanotechnology companies engage in a stock-for-stock merger. Assume Company A is worth $10 million, Company B is worth $20 million, and one million shares of stock have been issued by each company. The stock of Company A is worth $10 per share, and the stock of Company B is worth $20 per share. A rational, market-oriented exchange would involve Company B issuing one share of stock in exchange for every two shares held by the shareholders of Company A. Thus, each side has an incentive to maximize their respective valuations to obtain a greater percentage of the combined company.
20. For a good discussion of mergers and acquisitions of publicly traded companies, see Alfred Rappaport and Mark L. Sirower, *Stock or Cash? The Trade-Offs For Buyers and Sellers in Mergers and Acquisitions*, HARV. BUS. REV. (Nov.-Dec. 1999).
21. *Id.*
22. Arti Rai, *Fostering Cumulative Innovation in the Biopharmaceutical Industry: The Role of Patents and Antitrust*, 16 BERKELEY TECH. L.J. 813, 834 (Spring 2001) (noting that companies are reluctant to disclose proprietary information in licensing negotiations because such information is only protected by trade secret law, which is a "rather weak form of protection.").
23. Chapter 12 provides a more detailed explanation of anti-dilution protections.
24. Douglas Cogen, lawyer experienced in private-private mergers, San Francisco, California (March 2003).
25. Generally, discussions of how the value of convertible preferred stock is determined by the rights attached to such stock take place in the context of discounting common stock from the value of preferred shares. *See, e.g.,* Gerard Boyce, *E-Finance: Understanding Convertible Preferred Stock*, 224 NYLJ 5 (Feb. 2000) ("While there is no set formula for such a discount, it has not been uncommon to *see* the value of common shares discounted as steep as 75 percent to 90 percent from the value of the preferred shares. The greater the disparity in rights between the common and the preferred shares, the greater the support for a larger discount in value.").
26. As Cogen explains, there are a "million different permutations of how the cards get shuffled." Cogen, *supra* note 24.
27. *Id.*
28. *See, e.g.,* D.G.C.L. § 251(c).
29. *See, e.g.,* Manuel A. Utset, *Reciprocal Fairness, Strategic Behavior & Venture Survival: A Theory of Venture Capital-Financed Firms*, 2002 WIS. L. REV. 45, 101 (2002) ("A number of studies, however, have shown that entrepreneurs tend to be more over-optimistic than professional managers and non-entrepreneurs."); Gaylen N. Chandler & Erik Jansen, *The Founder's Self-Assessed Competence and*

Venture Performance, 7 J. Bus. Venturing 223 (1992); Arnold C. Cooper et al., *Entrepreneurs' Perceived Chances for Success*, 3 J. Bus. Venturing 97 (1988); Norris Krueger, *The Impact of Prior Entrepreneurial Exposure on Perceptions of New Venture Feasibility and Desirability*, 18 Entrepreneurship Theory & Prac. 5, 13 (1993).

30. For a detailed discussion of how such plans are set up, see Jacquelin A. Daunt, *Mergers & Acquisitions: A Strategy For High Technology Companies: 2002 Update*, available at http://www.fenwick.com/pub/corp_pubs/Mergers_&_Acquisitions/M&A_2002_main.htm (noting that a cash retention bonus plan is "easy to implement, easily understood by the participants, cost effective and generally does not require shareholder approval or securities law compliance.").
31. 15 USCS § 18 (2003).
32. *Id.*
33. D.O.J. & F.T.C., Antitrust Guidelines For the Licensing of Intellectual Property 3.2.3 (1995), http://www.usdoj.gov/atr/public/guidelines/ipguide.htm [hereinafter Licensing Guidelines] (defining innovation markets).
34. *Id.* Unlike antitrust analysis of product markets, there are no geographic limitations on innovation markets analysis.
35. Mary L. Azcuenaga, *Recent Issues In Antitrust and Intellectual Property*, 7 B.U.J. Sci. & Tech. L. 1, 14 (2001); FTC Hearings on Competition in Innovation (1995) (statement of D.J. Schafer), at http://www.ftc.gov/opp/global/schafer.htm (noting that "such mergers [in the pharmaceutical industry] do raise real antitrust concerns, including those based on new product pipelines as well as existing products.").
36. *In re* Ciba-Geigy Ltd., No. 961-0055, 1996 F.T.C. LEXIS 701, at 5 (F.T.C. Dec. 5, 1996).
37. *Id.* at 90-95.
38. *See, e.g.*, Thomas A. Piraino, *A Proposed Antitrust Analysis of Telecommunications Joint Ventures*, 1997 Wis. L. Rev. 639, 679-80 (1997) ("Most upstream joint ventures are integrated enough to produce real efficiencies, and thus the courts should be able to conclude after minimal inquiry and without a market power analysis that their efficiencies outweigh their anticompetitive effects. Research joint ventures allow firms to combine their resources to achieve synergies in developing new telecommunications services. In addition to cost savings from the elimination of redundant research efforts, such ventures allow firms to share the complementary assets which are often necessary for the successful commercialization of a new technology.").
39. *See* International Standards Organization, *The ISO Standards Bookshop*, at http://www.iso-standards-international.com/.
40. Edward Rashba and Daniel Gamota, *Anticipatory Standards and the Commercialization of Nanotechnology*, 5 Journal of Nanoparticle Research 401, 402 (2003) [hereinafter Rashba & Gamota].
41. *See* American National Standards Institute, *About ANSI*, at http://www.ansi.org/about_ansi/introduction/introduction.aspx?menuid=1 (last visited May 31, 2004).
42. Mark A. Lemley, *Intellectual Property Rights and Standards Setting Organizations*, 90 Cal. L. Rev. 1889, 1903, 1973-80 (2002).
43. The TBT Agreement does not designate specific organizations as the developers and promulgators of international standards. Rather, the agreement sets out principles for the standards development process, including openness, consensus, transparency, and due process.
44. *See* NIST, *Standards*, at http://www.nist.gov/public<uscore>affairs/standards.htm#Documentary.

45. For a more comprehensive discussion of the role standards played in the development of Wi-Fi, see Rashba & Gamota, *supra* note 40.
46. Jay Lyman, *MEMS Standards, While Small, May Mean Much For Industry*, SMALL TIMES, Aug. 26, 2003, at http://www.smalltimes.com/document_display.cfm?document_id=6556 (quoting Eric Novak).
47. M2 PRESSWIRE, Nov. 4, 2003.
48. *See* CHINA ECONOMIC NET, *National Nanotech Center Names Principal Investigator*, at http://en.ce.cn/main/index.shtml (last visited May 21, 2004).
49. *See* IEEE, *IEEE Nanotechnology Standards*, at http://grouper.ieee.org/groups/nano/.
50. *Id.*
51. *See* IEEE, *Nanotechnology Standards Working Group*, at http://grouper.ieee.org/groups/1650/.
52. NIST, *Nanotechnology Is BIG at NIST*, at http://www.nist.gov/public_affairs/nanotech.htm.
53. The assurance must be either: (1) a general disclaimer that the patentee will not enforce any of its present or future patents whose use would be required to implement the proposed IEEE standard against any person or entity using the patent(s) to comply with the standard; or (2) a statement that a license will be made available without compensation or under reasonable rates, with reasonable terms and conditions that are demonstrably free of any unfair discrimination.

Chapter 16

1. James E. Schrager, *Editorial*, 6 J. PRIVATE EQUITY 1 (June 22, 2003).
2. Douglas J. Cummings & Jeffrey G. MacIntosh, *Venture-Capital Exits in Canada and the United States*, 53 U. TORONTO L.J. 101 (2003).
3. For a more general description of corporate finance, see STEPHEN A. ROSS, RANDOLPH W. WESTERFIELD, & JEFFREY JAFFE, CORPORATE FINANCE (McGraw Hill 2002).
4. *Id.* at 164.
5. For a complete discussion of these advantages see John R. Graham, *How Big Are the Tax Benefits of Debt?*, 55 J. FIN. 1900-1942 (2000).
6. *See* Robert Hendershott, *Net Value: Wealth Creation (and Destruction) During the Internet Boom*, Dec. 2001, at http://lsb.scu.edu/finance/faculty/Hendershott/working_papers/hendershott%20net%20value.pdf.
7. *See generally* Robert C. Merton, *On the Pricing of Corporate Debt: The Risk Structure of Interest Rates*, 29 J. FIN. 449-470 (1974).
8. For a good discussion of convertible debt, see Mark Davis & Fabian Lischka, *Convertible Bonds with Market Risk and Credit Risk*, Working Paper, Tokyo-Mitsubishi Int'l, at 1 (1999).
9. Timothy J. Harris, *Bridge Financing Over Troubled Waters*, 6 JOURNAL OF PRIVATE EQUITY 59 (2002).
10. For more detailed comparisons of IPOs versus acquisitions for companies, see JAMES C. BRAU, BILL FRANCIS, NINON KOHERS, THE CHOICE OF IPO VERSUS TAKEOVER: EMPIRICAL EVIDENCE (2001).
11. Eli Okef & Matthew Richardson, *The IPO Lock-Up Period: Implications for Market Efficiency and Downward Sloping Demand Curves*, New York University, Leonard N. Stern School Finance Department Working Paper Series 99-054, at 3 (Jan. 2000).
12. Brau, *supra* note 10, at 30.
13. *Id.* at 4.

14. *Id.* at 7.
15. Okef & Richardson, *supra* note 11, at 4.
16. For a discussion of the rigorous requirements for going public, *see* Stephen I. Glover, *The Offerings that Precede and Initial Public Offering—How to Preserve Exemptions and Avoid Integration*, 24 SECS. REG. L.J. 3-37 (1996).
17. *Id.* at 4.
18. *See* Jeffrey A. Brill, *"Testing the Waters"—The SEC's Feet Go from Wet to Cold*, 83 CORNELL L. REV. 525, 545 (1998).
19. Alix Nyberg, *Sticker Shock*, CFO MAGAZINE, Sept. 2003, *available at* http://www.cfo.com/article/1,5309,10546l1Ml667,00.html.
20. Company press releases.
21. *See* Theodor Kohers, *Takeovers of Technology Firms: Expectations Vs. Reality*, 30 FIN. MGMT. 35, 39 (2001).
22. *See* James Dukart, *U.S. Israeli Nanoparticle Makers Merge their Metals and Markets*, SMALL TIMES, Oct. 30, 2002.
23. For a description of investment banks and the process of working with them, see ROBERT G. ECCLES & DWIGHT B. CRANE, DOING DEAL: INVESTMENT BANKS AT WORK (1998).
24. For a detailed guide to implementing a deal, see FENWICK & WEST, LLP, Jacqueline A. Daunt, *Mergers & Acquisitions For High Technology Companies*, 2002, at http://www.fenwick.com/docstore/publications/corporate/MA.pdf .
25. For a detailed discussion of issues involved in negotiating a letter of intent, see generally Andrew R. Klein, Comment, *Devil's Advocate: Salvaging the Letter of Intent*, 37 EMORY L.J. 139 (1998).
26. Mergers require a shareholder vote on the part of both the target and acquiring company, except when the acquiring company is much larger than the target. D.G.C.L. § 251(c). Dissenting shareholders can be "squeezed out" with appraisal rights. *See id.* at § 262(a). The acquiring company assumes all of the liabilities of the target company.
27. In an asset purchase agreement, the buyer must identify the assets to be acquired, conduct due diligence with respect to those assets, and contract to purchase the assets. Generally, there are high transaction costs associated with purchasing an entire operating business through the purchase of the assets of the business. A sale of substantially all assets is a fundamental transaction requiring shareholder approval, but purchases of assets do not require a shareholder vote. *See, e.g.*, D.G.C.L. § 271. The acquiring company does not assume all liabilities of the target company.
28. In a stock purchase agreement, the acquiring company purchases all, or a majority of, the company's outstanding stock. A stock purchase is not feasible if a target has a large number of shareholders or even one shareholder with substantial holdings who is opposed to the acquisition. After the acquiring company obtains all of the target's stock, the target becomes a subsidiary, with all of its assets and liabilities intact.
29. *See generally* Steven A. Bank, *Mergers, Taxes & Historical Realism*, 75 TUL. L. REV. 536 (Nov. 2000).
30. In many cases, parties will negotiate a break-up fee in case the deal fails to close. A break-up fee provides liquidated damages to be paid by the party responsible for disrupting the deal. The break-up fee should reasonably approximate the likely damages. Possible damages include the value of services provided by lawyers, investment bankers, accountants, and lending institutions as well as foregone profit from a potentially successful alternative acquisition.
31. A definitive agreement will contain detailed representations and warranties about the company's business. The aquiree must disclose any information that

might make the representations and warranties inaccurate in an "exception schedule." The acquiree must also provide detailed lists of its assets, contracts, and liabilities to the definitive agreement as part of the "disclosure schedule." As a general rule, the nanotech company should disclose as much information as possible to avoid disputes over indemnity later. The disclosure requirement should be limited to information that is "material" and the acquiree has knowledge of. The acquiree should insist that the acquiror is not entitled to indemnification unless a certain damages threshold is reached. The acquiror will want to place a portion of the merger consideration in escrow as security for such indemnity obligations.

32. An acquisition can cause a great deal of employee stress and instability. As such, acquisition negotiations should be kept confidential until a definitive agreement is concluded. Meetings should be held away from business premises and only key employees should be informed. Before the definitive agreement is announced, the merging companies must decide which employees will be retained and which will be released. For employees that the company wishes to retain, compensation agreements should be designed and steps should be taken to integrate these employees into the new corporate culture. Golden parachutes might be used, which are arrangements that provide a key employee with benefits equal to three or more times the employee's average annual compensation over the last five years. The merged entity must flesh out employee health plans, profit sharing plans, bonus plans, employee loans, stock options, and other employee benefits. Termination of employees must comply with the Worker Adjustment and Retraining Notification Act. 29 U.S.C. §§ 2101-2109 (2000). WARN Act requires that terminated employees receive either 60 days termination notice or 60 days severance pay. Employees that will ultimately be released may be needed for a period of time. "Transition" packages can be put together that motivate employees to remain with the company during the transition period. Such packages might include special option vesting, severance, bonus payments, and assistance in finding new jobs. The merging parties must meet with each employee to discuss their particular situation and future role in the merged entity.

33. If the transaction involves the acquiring company issuing stock in exchange for the stock or assets of the nanotech company, it either must register the shares under the Securities Act of 1933 or find an applicable exception. Three devices can be used to exempt registration in such a transaction: (1) issuance of shares in a private placement in reliance upon §4(2) of the 1933 Act; (2) registration of the transaction with the SEC on a Form S-4 Registration Statement; (3) use of a state fairness hearing and reliance on §3(a)(10) of the 1933 Act.

34. *See* Phyllis Diamond et al., *Broker Dealers: Wall St. Agrees to $ 1.4 Billion Payment, Broad Reforms, Resolving Conflict Changes*, 34 SEC. REG. & L. REP. (BNA) 2037 (Dec. 23, 2002).

35. Sarbanes-Oxley Act of 2002, Pub. L. No. 107-204, 501, 116 Stat. 745, 791-93 (2002).

36. *See* Lawrence Aragon, *Khosla Apologizes To Nanosys About Published Remarks*, PRIVATE EQUITY WEEK, June 21, 2004, at http://www.ventureeconomics.com/vec/1070550015536.html.

37. Brau, *supra* note 10.

38. Interview with Douglas Jamison, Harris & Harris Group, Inc., Feb. 17, 2004.

39. For a more detailed discussion of initial public offerings, see ROSS GEDDES, IPO AND EQUITY OFFERINGS (2002).

40. 15 U.S.C. § 77e(c) (2002).

41. Under Rule 135, the issuer can publish its name, the basic terms of the securities, the amount of the offering, the anticipated timing of the offering, a brief

statement of the manner and purpose, whether the issuer is directing its offering to a particular class purchasers, and any statements required by law. 17 C.F.R. § 230.135 (2002).
42. 15 U.S.C. § 77e (2002).
43. 15 U.S.C. § 77j(b) (2002).
44. Memorandum of the Statutory Revision Committee, SEC Release No. 33,224 (June 6, 1947), at 1947 WL 6715.
45. 15 U.S.C. § 77e(b)(2) (2002).
46. *See* SEC Act Release No. 33-7233 (Oct. 6, 1995); Exchange Act Release No. 34-36345 (Oct. 6, 1995); SEC Release Nos. 33-7234, 34-36346 (Oct. 6, 1995); SEC Release Nos. 33-7314, 34-37480 (July 25, 1996); Securities Act Release Nos. 33-7288 (May 9, 1996); SEC Release Nos. 33-7516, 34-39779, IA-1710, IC-23071 (Mar. 23, 1998); SEC Release Nos. 33-7855, 34-42712 (Apr. 24, 2000); Secs. Act Release No. 33-7856 (Apr. 28, 2000).
47. SEC Release No. 33-7233, 34-36345 (Oct. 6, 1995) (stating that it views "information distributed through electronic means as satisfying the delivery or transmission requirements of the federal securities laws if such distribution results in the delivery to the intended recipients of substantially equivalent information as these recipients would have had if the information were delivered to them in paper form.").
48. *See generally* Jack A. Rosenbloom, *Direct Public Offerings on the Internet: A Viable Means of Obtaining Capital?*, 2000 COMPUTER L. REV. & TECH. J. 85 (2000).
49. *See* DENIS T. RICE, TAKING THE NANOTECHNOLOGY STARTUP FROM FORMATION TO ELECTRONIC PUBLIC OFFERING 101 (2002).

Chapter 17

1. Soc'y of the Plastics Industry, *Size and Impact of the U.S. Plastics Industry*, Plastics Data Source, Dec. 2003, *available at* http://www.plasticsdatasource.org/impact.htm#summary.

About the Authors

John C. Miller

John is vice president, Intellectual Property at Arrowhead Research Corporation, a publicly-traded nanotech company. His duties at Arrowhead include monitoring the intellectual property landscape and licensing and enforcing patents held by Arrowhead and its subsidiaries. John is also a managing editor of *Nanotechnology Law & Business*, a peer-reviewed, quarterly journal.

He has published various articles on legal and policy issues in information technologies, biotechnology, and nanotechnology and recently spoke to California lawmakers at the 2004 California Nanotechnology Policy Briefing.

John is a member of the California bar and federal courts in the Northern District of California. He graduated *Order of the Coif* from Stanford Law School.

Ruben Serrato

Ruben is a business professional with extensive experience in private equity financing, mergers and acquisitions, and corporate strategy. His finance background includes investment banking, venture capital and strategy development work with Lehman Brothers, Liberty Media and, currently, Canon USA.

Ruben is Canon's Limited Partner representative in NGEN Partners, a venture capital firm focused on nanotechnology investments. In his duties with NGEN, Ruben reviews nanotech business plans and evaluates potential nanotech investments. As part of Canon's Research and Development group, Ruben wrote the initial business plan for Canon's first American investment in nanotech, and he has since been involved in negotiating joint ventures and technology acquisitions for the new company. He also helped to conceive and launch Canon U.S. Life Sciences, a nanobio subsidiary developing advanced molecular diagnostics.

Ruben graduated with degrees in Economics and Political Science from Stanford University and he is currently completing his master's degree at Stanford.

Jose Miguel Represas-Cardenas

Jose Miguel is the recipient of a Department Fellowship from the Department of Electrical Engineering at Stanford University, where he is currently a graduate student (on leave 2003–2004). His research interests are quantum transport phenomena and atomically-precise fabrication technology. He received a BS in Mechanical and Electrical Engineering from ITESM in Monterrey, Mexico.

Griffith A. Kundahl

Griff serves as general counsel for the NanoBusiness Alliance, the national nonprofit trade association for the nanotechnology industry. He also acts as NanoBusiness Alliance vice president for the Western Region. His duties include advising the NanoBusiness Alliance on day-to-day legal matters and developing business partnerships and liaisons in the western United States. Griff has also served as a member of the Colorado Technology Alliance Nanotechnology Council, a body that advises Colorado's Governor and Secretary of Technology on issues relating to nanotechnology. Additionally, Griff is a founder and advisor to the Colorado Nanotechnology Initiative. Griff is an Associate Editor of *Nanotechnology Law & Business*.

Griff is a member of the Colorado bar and federal courts in Colorado. He is a graduate of the University of Pennsylvania (BA), the University of Alabama (MA), and the University of Denver College of Law (JD).

Griff currently practices law in Denver, Colorado, with the firm of Campbell Bohn Killin Brittan & Ray, LLC. He focuses in the areas of technology, general civil litigation, trials and appeals, business law, and employment law.

Index

Abbott Laboratories, 170
Abbreviated new drug application (ANDA), 93
Accelyrs, 280
Acquisitions
  deal process, 274–275
  examples of, 272–273
  as financing mechanism, 268, 270–271
  preparation for, 274
  regulatory issues, 274
  targets of, 274
Action Group on Erosion, Technology, and Concentration (ETC), 51–54, 58, 60, 63, 133
Acusphere, 203, 276
Adenosine triphosphate (ATP), 92
Advanced biotechnology, 45
Advanced Imaging Systems, 272
Advanced Measurement Laboratory (AML), NIST, 263–265
Advanced Technology Materials, Inc., 70, 216
Advanced Technology Program (ATP), as funding resource, 124, 176, 191, 194, 283
Advertising, 173
Advion Biosciences, 203
Advisory panels, 99
Aerospace applications, 28–29
Age of Spiritual Machines, The (Kurzweil), 44
Agile Materials & Technologies, Inc., 201–202
Agreement on Technical Barriers to Trade, 62
Agreement on the Application of Sanitary and Phytosanitary Measures, 62
Agricultural biotechnology, 43
Air pollution control, 80
Air Products, 202, 238, 243
Alien Technology, 203
Alivisatos, Paul, Dr., 156
Allergenic extracts, 88
Aluminum nanoparticles, 22
Amberwave Systems, 200
American National Standards Institute (ANSI), 261
American Society for Mechanical Engineers (ASME), 262
American Supersonic Transport, 120
Ames Center for Nanotechnology, 148
Amyotrophic lateral sclerosis, 171
Angel investors, 189–191

Angstrovision, 200
Angular momentum, 24
Animal studies, 54, 56
Antidilution protection, 207, 256
Antitrust, 81
Apax Globis Partners & Co., 247
Apparatus claims, patent applications, 216
Apple, 158
Applied Epi, 272
Applied Nanotechnologies, 253
Arbitrary and fanciful trademarks, 231
ARCH Venture Partners, 20
Ardesta, 200, 205
Argonne National Laboratory, 159
Ariadne Capital, 247
Arms Control and Disarmament Agency, 107
Arryx, 201
Articles of incorporation, 154–155
Artificial brains, 24–25
Artificial intelligence, 98
Artificial Muscle, 202
Asbestos, 52, 62
Asbestos Ban and Phaseout, 62
Asian market, 170
Asian research institutions, 248
ASML Holding NV, 238, 244
Assembler, 29, 31
Asset purchase agreements, 275
Athenson, Bill, 170
Atlantic Nano Forum, 79
Atomic energy, 80
Atomic force microscope, 16, 21, 172
AtomWorks, 159
Aveka, 273–274

Baker, James, Dr., 96, 156, 169
Balance sheets, 175
BASF, 202, 239
Bawendi, Moungi, Dr., 156
Bayer, 170, 202
Bayh-Dole Act, 140, 143–144
Bell, Randy, 237
Best Efforts agreement, 278
Betanapthylamine, 56
Big 5 Consulting Firms, 187
BinOptics, 201
Biological detection applications, 23–24

357

Biological products, clinical testing of, 88
Biotech derived therapeutics, 88
Biotechnology
 innovations in, 47–48, 197
 patents, 68
 products, 97
Block, Steven M., 13–14
Block copolymers, 20
Blocking patents, 79–80
Blonder, Greg, 201–202
Blood and blood components, 88
Blood collection and processing, 90
Board of directors, functions of, 155, 275
Bock, Larry, 38, 156
Boeing, 202
Bond offering, 268
Bottom-up fabrication, 15–16, 122, 179
Bridge financing, 269
Bronchoalveolar inflammation, 55
Brus, Louis, Dr., 156
Buckministerfullerene, 17
Buckyballs, 27
Bucky USA, 253
Building nanotechnology
 large, 29–31, 167–168
 small, 23–29, 167–168
Bulge bracket investment banks, 187
Bureaucracy, 112, 128–129
Bureau of Industry And Security (BIS), 105–108, 110
Business entities, nano start-ups, 153–154
Business plan
 Executive Summary, 163–165
 financial modeling, 174–175, 179
 importance of, 161–162
 information sources for, 176–178
 key elements of, 163–164
 market selection, 166–171
 operations, 172–174
 outsourcing, 186–187
 purpose of, 162–163
 risk disclosure, 171–172
 sources and uses, 179–181
 valuation, 181–186
Bylaws, 55

Cabosil amorphous silica, 55
California Molecular Electronics Corporation, 280
California Nanosystems Institute (CNSI), 117, 159
Cancer/carcinogenecity, 57, 60
Canon, 202
Canon nanotubes, 252–253
Capstone Turbine, 203
CarboLex, 253
Carbon, 218
Carbon Nanotechnologies, 76, 143, 238, 253
Carbon nanotubes, 17–18, 38, 52–53, 55–56, 64, 69–70, 72, 79, 215–216, 224, 259
Carbon Solutions, 253
CAS, 213
Cash flow, 174–175

Catalytic Solutions, 37, 165, 202
C corporations (C Corps), 153–154, 157, 282
Celera Genomics, 206
Cellular therapies, 28, 90, 95
Center for Biological and Environmental Nanotechnology (CBEN), 64
Center for Biologics Evaluation and Research, (CBER), 88–92, 99–100
Center for Devices and Radiological Health (CDRH), 87–91, 98–100
Center for Drug Evaluation and Research (CDER), 86–93, 100
Center for Functional Nanomaterials, 146
Center for Integrated Nanotechnologies, 146
Center for Nanophase Materials, 146
Center for Nanoscale Materials, 146
Central Coast Angel Network, 190
Chemical interactions, 91
Chemical patents, 68
Chemical self-assembly, 16
Chemical sensors, 23, 168
Chemical vapor deposition (CVD), 77, 80, 172
Chlorogen, Inc., 201
CibaoGeigy, 259
CIENA, 203
Citizen advisory panels, 48
Citrix, 203
Classification, FDA regulation, 86–93
Clayton Act, 257–258, 275
Clean rooms, 172
Clinch River Breeder Reactor, 120
Clinical testing, importance of, 98
Clinical trials, significance of, 94–96, 171
Clinton Administration, 115–116
Cloning, 19
Coal, conversion to liquid fuel, 22
Coatings, 22, 169
Coatue, 201, 273
Cohen, Linda, 120
Colloidal crystals, 218
Colvin, Vicki, Dr., 58, 96
Combined research and curriculum development (CRCD), 124
Commerce Control List (CCL), 106–108, 110, 113
Commercialization, 130, 135, 239–240, 262, 283–284
Committee on Foreign Investment (CFIUS), 111
Common law doctrines, 229
Common stock, 206, 256
Communications devices, 26
Compact Model Council (CMC), 265
Compaq, 203
Comparable Company Analysis, 182–184
Competition, as influential factor, 168
Composite patents, 68
Compulsory licensing, 79–80
Computational nanotechnology, 21
Computer chips, 24
Computer Retrieval of Information on Scientific Projects (CRISP), 177
Confidential information, 227–228

Confinement, 17
Consolidation
  benefits of, 251–252, 254–255, 284
  merger agreements, 255–258
  regulatory concerns, 258–259
Consultants, 99, 187
Consumer electronic industry, 34
Continuation-in-part patent application, 29–220
Continuum Photonics, Inc., 201
Contract law, 150
Contractual tools, 229–230
Conversion rights, venture capital criteria, 207
Convertible debt, 268–269
Cooperation, benefits of, 284
Cooperative licensing agreements, 75–77
Cooperative Research and Development Agreements (CRADAs), 99, 124, 210
Copyrights, 233–234
Corporate claims, patent applications, 215–216
Corporate governance
  articles of incorporation, 154–155
  board of directors, 155
  management, 156–157
  scientific advisory board (SAB), 155–156
Corporate partnering
  benefits of, 284
  strategic fit, 274
Corporate spinoff, 149–151
Corporation investment, 118
Co-sale agreement, 208
Cost-benefit analysis, 63
Credibility, sources of, 156, 174, 191, 239
Cross-licensing agreements, 76, 211
C Sixty, 38, 171
Customer service, 173
CVC, 272
Cypress, 203

Databases, of inventions, 213
Data storage, 25
Debt, as financing mechanism, 269
Decision-making skills, significance of, 282
Deemed export rule, 106
Defense Advanced Research Projects Agency (DARPA)
  Contracts Management Office (CMO), 193
  as funding resource, 177, 191, 193–194, 283
Defense Security Service (DSS), 111
Defense systems, 29
Dehumanization, 44
Dekker, Cees, 24
Dendrimers, 27, 218
Deoxyribonucleic acid (DNA), 19–20, 27, 43
Department of Defense (DOD)
  Comptroller, 177
  grant funding, 193
  security issues, 103, 111
  technology dissemination, 146–147
Department of Energy (DOE), 112, 146
Department of Justice Reauthorization Act, 68
Descriptive trademarks, 231–232
Device system, patent applications, 216

Diabetes, 97, 179
Diagnostic applications, 27–28, 34, 167
Diagnostic devices, 96
Diagnostic tests, 170
Dialog, 213
Digital Instruments, 273
Dip-pen lithography, 36
Dip pen nanolithography, 16
Direct exposure, health effects, 57
Direct mail, 173
Direct public offerings (DPOs), 279–280
Direct sales force, 173
Discera, 200
Disclosure
  business plan, 171–172
  consolidations, 256
  corporate spin-offs, 150
Discounted cash flow (DCF), 182, 184–186
Discount Rate, 185
Disease screening, 28
Disposal systems, commercial, 172
Dispute resolution, 144, 247
Distribution, corporate partnering, 243, 245
Dividend preferences
  corporate partnering, 243
  venture financing criteria, 207
DNA sequence analysis, 169
Draper Fisher Jurvetosn (DFJ), 200–201
Drexler, K. Eric, 29–31
Drug, defined, 86
Drug delivery devices, 171
Drug delivery systems, 27, 34, 91
Drug development, 27, 60, 171
Due diligence, in venture financing process, 203, 205
DuPont, 34, 56, 202, 277
DuPont Electronic Technologies, 238, 246
D-Wave Systems, 199, 201
Dynamic random-access memory (DRAM), 25, 265

Eagle Picher, Inc., 245
EaglePicher Technologies, 238
Ear99, 106
Early stage financing
  financing strategy, 190
  government funding, 191–194
  seed round, 190–191
  venture capital, 194–208
Early-stage companies, valuation metrics, 183, 186
E-beam lithography, 15
Echo Technologies, 192
Economic Espionage Act (EEA), 229
Edgar Online, 178
Education programs, 48
Egelhoff, William, Jr., 262
EiComendex, 213
Eisenberg, Rebecca, 77
Electromagnetic radiation, 15
Electronic applications, 24–26, 168–169
Electronic devices, 38
Eli Lilly, 236

Embossing, 15
Embryonic stem cells, 143
Emerging markets, growth in, 179
Emerging technologies, 43, 83, 97–99, 100, 236
Employee(s)
    employment contracts, 150
    stock option plans, 157–158
Enablement doctrine, 221–222
End-product development, 76
Energy conversion and storage, 26
Energy industry, 34
Enforcement process, patents, 226
Engineered nanomaterials, 55
Enplas, 238, 245
Environmental Protection Agency (EPA), regulation by, 51, 53, 60–64, 133, 171, 173
Environmental regulation
    data analysis, 56–58
    data collection, 54–56
    increasing environmental concern, 52
    international law, 62–63
    legal analysis, 60–63
    normative analysis, 58–60
    recommendations for, 63–64
    regulatory action, timing of, 53–54
    regulatory review, 53
    toxicity, 54
    U.S. law, 61–62
Environmental risks, 170–171
Epidemiological studies, 54, 56
Equity investment, 243–244
Espin Technologies, 192
Essilor, 238
Estoppel, 217
Ethical implications
    First Amendment protection, 46–47
    immediate action, 42–43
    legal analysis, 46
    overview of, 41–42
    policy analysis, 43–46
European Committee for Standardization (CEN), 262
European research institutions, 248
Exclusive licensing, 143
Executive Summary, business plan, 163–165
Exit opportunities
    acquisitions, 271–275
    financing mechanisms, 268–271
    initial public offerings, 267–269, 275–280
Export Administration Regulations (EAR), Department of Commerce
    applications to nanotechnology, 108–110
    defined, 105
    summary of, 105–108
Export Control Classification Number (ECCN), 106–107
Export control laws
    Department of Commerce, 105–110
    future directions for, 113
    importance of, 112
    regulation of, 104–105
    significance of, 171
    State Department, 110–111
    state of, 104
Exposure, toxicity research, 57

Factica, 213
Fair pricing, 82
Family, as financial resource, 189–190
FDAMA, 89
FedBizOpps, 177
Federal funding
    future directions for, 128–130
    historical perspectives, 115–116
    management of, 120–128
    overview of, 116
    reasons for, 119
Federal Laboratory Consortium, 177
Federal Laboratory Consortium for Technology Transfer (FLC), 146
Federal Trade Commission (FTC), 259
Fees, licensing, 144
FeRx, 192
Feynman, Richard, 14–15, 281
Field emission devices, 38, 168
Films, macroscopic, 22–23
Final Prospectus, 279
Financial modeling, business plan components
    independent estimates, 179
    market estimates, 175
    methodology, 174
    third party sources, 175
    time frame, 174–175
Financial projections, importance of, 174–175, 179–180
Financial resources
    federal funding, 115–116
    foreign public spending, 117–118
    private investment, 118
    state funding, 117
Financing
    early stage, see Early stage financing
    future rounds, 208
    stages of, 283
Firm Commitment, 278
First Amendment protection, 46–47
FirstGov, 177
Five Forces Model, 169
505(b)(2) application process, 93, 95, 102
Five Star Technologies, Inc., 199, 202
Flash technology, 25
Flat panel displays, 34
Flexics, 201
Food and Drug Administration (FDA) regulation
    agency effectiveness strategies, 98–99
    challenges of, 86–101
    classification problems, 86–93
    product approval, 93–96
    scientific expertise issues, 96–101
    significance of, 83–86, 171, 173
    state of the FDA, 84
Food, Drug, and Cosmetic Act, 88

Foreign investment, as financial resource, 195
Foreign patent applications, 223–224
Foreign public spending, 117–118
Foresight Guidelines, 48–49
Founding technologists, managerial functions, 156
Fqubed, 202
Freitas, Robert, 92
Friends, as financial resource, 189–190
Frontier Carbon Corporation, 202, 253
Fullerene antioxidants, 171
Fullerene International Corporation, 202
Fullerenes nanotubes, 17, 21
Full ratchet formula, 207

General Agreement on Tariffs and Trade (GATT)
 provisions, 62
 TRIPS Agreement, 249
Generic trademarks, 231–232
Gene therapies, 88, 98–99
Genetic antisense technology, 222
Genetic engineering, 42
Genetic testing, 99
Genomics, 98
Globalization
 foreign corporations and inveestors, 247–248
 foreign ideas/talent, 247–248
 intellectual property, protection of, 247, 249
 manufacturing abroad, 247, 249
Goddard Center, 148
Good Manufacturing Practices (GMPs), 87
Google, 158
Government agencies, as information resource, 176
Government funding
 Advanced Technology Program (ATP), 194, 283
 criterion for, 191
 Defense Advanced Research Projects Agency (DARPA), 193–194, 283
 SBIR/STTR grants, 193, 283
Government interest, 47
Government interventions, patent process, 79–80
Government laboratory spinoff
 characteristics of, 146–149
 intellectual property, generation of, 210–211
Grubstein, Peter, 163, 203
Guidance Document, FDA regulation process, 95

Handylab, 200
Harrigan, Kathryn, 236
Harris & Harris Group, Inc., 201
Health risks, 170–171
Heat shock protein 60, 19
Heath, James, Dr., 156
Heeger, Alan, 202

Heller, Michael, 77
Henney, Jane, 97–98
Hewlett-Packard (HP), 24, 118, 158
High-dose studies, 57
High-growth companies, valuation metrics, 183, 185–186
HIV, 171
Holdover clauses, 150, 229
Holmenkol (Loba), 238
Holmenkol Sport-Technologies GmbH & Co., 244
Homopolymers, 222
Honjo Chemical Corporation, 202
Hotmail, 200
House Government Operations Committee, 105, 108
HP Labs, 197
Hyperion Catalysis International, 69, 71, 216, 253
HyperNex, 201

IBM, 24–25, 34, 70–71, 118, 216, 224–225, 253
IBM SXM, 272
IEEE 802, 261, 264–265
IFI CLAIMS, 213
Iijima, Sumio, 17
Iljin Nanotech, 253
Imago, 201
Immunicon, 199
Impinj, 200
Incentive stock options (ISOs), 157
Income statements, 175
Indirect exposure, health effects, 57
Industrial production, 59
Industrial structure
 companies commercializing nanotechnology, 34, 37–38
 established companies integrating nanotechnology, 34, 36
 model of, 35, 38
Industry journals, as information resource, 176
Inevitable disclosure doctrine, 229
Infringement
 copyright, 233
 corporate partnering and, 245
 patent, 211, 217, 224–226
Inhalation exposure, 62
Inhalation toxicity, 55, 57
Inherent anticipation doctrine, 220–221
Initial public offering (IPO)
 benefits of, 267–269, 275
 candidates for, 277–278
 examples of, 276
 future directions for, 279–280
 mergers and acquisitions (M&A), 270
 process overview, 278–279
 regulation of, 276
 SEC registration, 275–276
InMat, 199, 202
InnovaLight, 200, 203
Innovation Engine, 247

362 INDEX

Inorganic nanomaterials, 17–18
INPADOC, 213
INSPEC, 213
Institute of Electrical and Electronics Engineers (IEEE), 261–265, 284
Insulin, 97
Insurance companies, as financial resource, 195
Intel, 34, 239, 277
Intellectual property
   commercialization and, 283
   copyrights, 233–234
   corporate spin-offs, 149
   generation of, 210–211
   initial, 140–151
   market segmentation and, 170
   patents, 211–227, 283
   protection strategies, 71, 74, 77, 82, 209–210
   regulation of, 134
   trademarks, 230–233
   trade secrets/confidential information, 227–230
   in venture financing process, 203, 205
Intematix, 201
Intercenter Agreement, 90, 92
Interference proceedings, pending patents, 222–223
Interferon, 97
International agreements, 45
International Electrotechnical Committee (IEC), 261
International law, 62–63
International Organization for Standardization (ISO), 261, 265
International Traffic in Arms Regulations (ITAR), State Department
   application to nanotechnology, 110–111
   defined, 110
   summary of, 110
Internet stocks, 162
Investigation Device Exception (IDE), 87
Investigational New Device (IND), 87
Investment banks
   acquisition preparation, 274
   as financial resource, 187, 195
   initial public offerings (IPOs) process, 276, 278
In vivo replication, 96
Ion Optics, 200
Ion Tech Inc., 272
Iron nanocrystals, 64

Jacobson, Elizabeth, 86
Japanese nanotechnology programs, 117–118
JAPIO, 213
JEDEC, 265
Joy, Bill, 42
Juniper Capital Ventures Ptd. Ltd., 247
Jurisdictional designation, FDA regulation process, 89–91
Jurvetson, Steve, 196, 200

Kahan, Jonathan, 95
Key management, 174, 242
Khosla, Vinod, 166, 181, 276
Konarka Technologies, 38, 200–202
Kurzweil, Ray, 44

Labeling, investigational drug, device, or biological products, 88
Lagally, Max, 37, 144
Langer, Robert, 86
Langley Research Center, 148
Language, in patent application, 217–218
Lanham Act, 231
Laser, Zia, 38
Lasers, 38
Legal analysis, 46
Letter of intent (LOI), 274–275
Leukemia, 97
LexisNexis, 178
Liability
   corporate partnering and, 244
   mergers, 275
Licensing
   agreements, *see* Licensing agreements
   components of, 79–82
   consolidations and, 254
   corporate partnering, 243, 246
   patents, 226
   standardization and, 265
Licensing agreements
   government laboratory spinoffs, 149
   patent infringement, 225–226
   university spin-offs, 140, 143–145
Licensing negotiations, 144
Lieber, Charlie, Dr., 24, 156
Life sciences, 167, 170
LightConnect, Inc., 202–203
Limited liability companies (LLCs), 153–154, 282
Limited liability partnerships, 195
Liquidation preferences
   corporate partnering, 243
   venture financing criteria, 206–207
Lithography, 15
Litigation
   patent infringement, 54, 79, 211, 214, 217, 225–227, 246, 254
   trademark, 233
Location, significance of
   Massachusetts, 159
   New York, 159
   Northern Illinois, 159
   Silicon Valley, 158–159
   Southern California, 159
   Texas, 159
   Washington, D.C., 160
Logos, 230
Lotus, 203
Loyalty, 229
Lucent, 34
Luminus Devices, 201

INDEX

Lung cancer, 62
Luxtera, 203

McGehee, Mike, 196
Magnesium oxide particles, 55
Magnetic resonance (MR) scanners, 94
Managerial functions, 156–157, 163. *See also* Key management
Manufacturing
  corporate partnering, 243–245
  in foreign countries, 249
  operations, 172–173
March-in rights, 210
Marketing, *see* Sales and marketing
marketresearch.com, 178
Market segmentation, 167
Market selection, business plan guidelines
  customer identification, 166
  influential factors, 167–171
  market focus, 166
  market segmentation, 167
Marlowe, Donald, 99
Mass production, 59
Massachusetts Technology Collaborative, 159
Materials and Electrochemical Research Corporation, 202, 253
Matsushita Electric, 238, 277
Maturation stage, in business life cycle, 174
Mechanical interaction, 91
Mechanosynthesis, 29
Medical device, regulation of, 87–88, 90
Mergers, as financing mechanism, 268, 270–271
Mergers and acquisitions (M&A), *see* Acquisitions; Mergers
Merging companies
  agreements between, 255–256
  agreement within company, 257–258
MesoSystems, 200
Mesothelioma, 62
Metal nanoparticles, 216
Meyyappan, Meyya, 29
Microbivores, 92
Microelectromechanical systems (MEMs), 200–201, 203, 261
Microelectronics, 24
Microfabrica, 201
Micronics, 200
MicroOptical Devices, 200
Microtechnano, 253
Microtechnology, 197
Mirkin, Chad, 16, 37, 156
Mission statement, 163
MIT, 144, 217
Mitsubishi Chemical Corporation, 202
Mitsubishi Corporation, 202
Mitsubishi Gas Chemical, 238
Mobil, 34
Modzelewski, Mark, 117
Molding, 15
Molecular Foundry, 146

Molecular Imprints, 199, 201
Molecular manufacturing, 29–31, 110, 113
Molecular Nanosystems, 38, 77, 143, 253
Molecular Solutions, 201
Monarch, 272
Monoclonal antibodies, 91
Monopolies, 170, 275
Moore's law, 24
Morgenthaler Ventures, 201–202
MPEG-2 compression, 76
MRC Process Equipment, 273
Multex, 178
Multi-walled nanotubes (MWNTs), 17
Munitions List, export control laws, 110–111

Naming a company, 151–152
Nano Age, 132
Nano Lab, 253
NanoBio Corporation, 27, 38, 156, 169–170, 192
Nanobio tools, 179
NanoBusiness Alliance, 72, 196, 200
NanoBusiness Angels, 190
Nanocarblab, 253
Nanocompany start-up process
  business plan, 161–187
  components of, 139–160
  consolidation, 251–259
  corporate partnering, 235–247
  early stage financing, 189–208
  exit opportunities, 267–280
  globalization, 247–249
  intellectual property, 209–234
  standardization, 259–265
NanoCoolers, 201
Nanocor, 238
Nanocyl, 253
NanoDevices, 273
Nanodots, 218
Nanogate Technologies, 37, 238, 244
Nanogen, 183, 211, 276, 285
Nanogram Devices Corporation, 170, 199, 201, 238, 245
NanoInk, Inc., 37, 156, 192, 199, 230
NanoInvestorNews, 33
Nanoledge, 253
Nanolithographic printing, 230
Nanomed Pharmaceuticals, 60
Nanomedical products, 83
Nanomedical research, 28
Nanometrics, 151, 173, 285
Nanomix, 38, 151–152, 192, 203, 238, 246, 253
Nanomuscle, 199
NanoNexus, 199
NanoOpto Corporation, 166, 199, 201–202, 238, 245
Nanoparticles, 21, 218
Nanopharma Corporation, 201
Nanophase Technologies Corporation, 37, 200–201, 276–277

Nanopowders Industries, 274
Nanorobots, self-replicating, 43, 45, 48, 171.
    See also Nanorobot technology
Nanorobot technology, 30, 92, 96–97
Nanoscale Exploratory Research, 124
Nanoscale Science, Engineering, and
    Technology (NSET), 121
Nanoscale Science Research Centers (NSRC),
    146–147
Nanoscience
  building blocks of, 15–19
  historical perspectives, 14–15
  organic nanomaterials, 19–20
  tools, 21–22
Nanosolar, 170, 199
Nanosphere, 38, 156, 170, 192, 199, 202, 245
Nanostream, 199
Nanosys and Quantum Dot Corp., 70–71
Nanosys, Inc., 74, 156–157, 170, 181, 192,
    194, 197, 199–201, 203, 212, 238, 271,
    275–277, 285
Nanotech Partners Limited, 202
NanoTech Resources Inc., 160
Nanotechnologies, Inc., 201, 238, 243
Nanotechnology
  applications, overview of, 22–31
  defined, 13–14
  future growth of, 277, 281–282, 285
  potential for, 281
*Nanotechnology and Homeland Security New
    Weapons for New Wars* (Ratner/Ratner),
    103
Nanotechnology business, start-ups, *see*
    Nanocompany start-up process
Nanotechnology Business Alliance, 117
Nanotechnology center, 78
Nanotechnology Customer Partnership
    Meeting, 79
Nanotechnology research
  banned, 46–47
  regulation of, 44–45
  university list, 141–143
Nanotechnology Research and Teaching
    Facility, University of Texas, 125
Nanotechnology Standards Initiative (IEEE),
    284
Nanotex, 37
Nanotubes, 17–18, 21, 23. *See also* Carbon
    nanotubes
Nanotweezers, 21–22
Nanowires, 17–18, 21, 23, 38, 215
Nanox, 199
Nantero, Inc., 38, 76, 199, 201, 238, 244
National Accreditation Board of Laboratories,
    262
National Aeronautics and Space
    Administration (NASA)
  Commercial Technology Division (NCTD),
    148
  grant funding, 193
  nanotech research, 146–148
  product development and, 19, 28–29, 55, 64

National Formulary, 87
National Industrial Security Program (NISP),
    111
National Institute for Nanotechnology, 118
National Institute of Standards and
    Technology (NIST), 177, 194, 262–264
National Institutes of Health (NIH), 48, 99,
    193
National Nanofabrication Infrastructure
    Network, 210
National Nanotechnology Initiative (NNI)
  development of, 115–116, 119–120
  fund allocation, 140, 146
  as information resource, 176
  review of, 123–126
  structure of, 121–123
  themes of, 122–123
National Research Council (NRC), 123, 128
National Science and Engineering
    Nanotechnology, 176
National Science and Technology Council
    (NSTC), 121, 128
National Science Foundation (NSF), 14, 64,
    121, 123–125, 130, 166, 175, 193
National security, 101
NEC, 118
Negotiations
  consolidations, 255–257
  corporate partnering, 242
Nelson, Lita, 82
Neophotonics Corp., 201
Neurodegenerative diseases, 60
New Drug Application (NDA), 87, 93, 95
Newspapers/newsletters, as information
    resource, 178
NGEN Partners, 202–204
Niche investment banks, 187
Noll, Roger, 120
Non-compete covenants, 150–151
Noncompetition agreement, 230
Nondisclosure agreements, 229–230
Nonfinancial corporations, as financial
    resource, 195
Non-merger techniques, 255
Nonqualified options (NQOs), stock option
    plans, 157
North Carolina Citizens Technology Forum,
    48–49
Novaled, 199
Novartis, 259
Npoint, 37, 192
Nuclear power, 43, 48
Nuclear Regulatory Commission, 112
Nugen, 199
NVE Corp., 211

Obviousness doctrine, 220–221
Office of Biologics Research and Review
    (OBRR), 97–98
Office of Biotechnology, FDA, 98, 101
Office of Combination Products, FDA, 90, 101
Office of Defense Trade Controls (DTC), 110

INDEX

Office of Foreign Assets Control (OFAC), 111
Office of Homeland Security, 104
Office of Nanotechnology, FDA, 101, 134
Office of Science and Technology Policy (OSTP), 121, 125
1-D nanomaterials, 17
Operations
   customer service, 173
   key management, 174
   manufacturing, 172–173
   research and development, 172
   sales and marketing, 173
   venture financing process, 208
Optical detectors, 169
Optical tweezers, 22
Optimag, 272
Optimism, 133
Optiva, Inc., 199, 201–202
Optobionics, 203
Organic light-emitting diodes (OLEDs), 26
Organic nanomaterials, 19–20
Organizational philosophy, 163
Organizational structure
   C corporations (C Corps), 153–154, 157, 282
   limited liability companies (LLCs), 153–154, 282
   S corporations (S Corps), 153–154
Outsourcing, business plan, 186–187
Overture, 200
Ownership issues, 237, 256–257, 270
Oxonica Ltd, 202

Parametric Technology, 200
Partnering arrangements
   benefits of, 235–236
   distribution, 243, 245
   equity investment, 243–244
   examples of, 238
   licensing, 243, 246
   manufacturing, 243–245
   motives for, 236–237
   risk of, 239–240
   search for partner, 240–243
   start-up contributions, 241–242
   success factors, 246–247
   technology development, 244
Partnerships, 170, 195
Patent and Trademark Office (PTO)
   Customer Partnership Meeting, 213, 220
   foreign patent applications, 223
   functions of, 211, 213
   infringement litigation, 225
   patent law, 66–68
   patent system goals, 66
   reform, 78–79
   rejection by, 219–223
   review of nanotechnology patents, 68–77
   trademark registration, 231–233
Patent applications
   claims, 215–219
   written description, 214–215

Patent Cooperation Treaty (PCT), 223–224
Patent law, summary of, 66–68
Patent pools, 76, 80–82
Patents
   applications, *see* Patent applications
   enforcement of, 226
   filing process, 212–214, 223–224
   foreign applications, 223–224
   importance of, 211–213
   infringement of, 67, 211, 217, 224–226
   issuance of, 223
   licensing, 226
   litigation, 211, 214, 226–227
   market segmentation and, 170
   PTO rejection, response to, 219–223
   PTO review, *see* Patents, PTO review
   standardization and, 264–265
Patents, PTO review
   broad and overlapping claims, 69–71
   claims rejection, 69
   judicial action, 79
   licensing, government intervention, 79–82
   number of patents, 71–74
   process overview of, 68–69
   transaction costs, 75, 77
Pay to play provisions, 207
Peer review process, 194
Penn State University, 194
Performance Plastics Products Inc. (3P), 238
Personnel Exchange Agreement (PEX), 211
Pesticides, 52, 64
Pharmacy-in-a-cell, 92–93
Phoenix Bioscience, 200
Photolithography, 15–16, 45
Photonic crystals, 26
Photovoltaics, 26, 38, 169–170
Physical Sciences, 192
Pionetics, 202
Plutonium, 112
Polaris Ventures, 203
Policy analysis, 43–46
Policy and regulation
   desirability of nanotechnology, 132–133
   environmental regulation, 51–64, 133
   export control, 103–113, 134
   FDA regulation, 83–102, 134
   federal funding, 115–130
   fundamental organization of, 133–134
   government support and, 135
   national security, 103–113
   patent and trademark office (PTO), 65–82, 133–134
   policymakers, functions of, 132
   recommendations, 47–49
   societal and ethical implications, 41–49
Polychlorinated biphenyls (PCBs), 52, 64
Polymers, 20–21
PolyOne, 238
Porter, Michael, 167, 169
Positional assembly, 16
Powerspan, 202
Power systems, backup, 172

INDEX

Precautionary principle, 58–60
Preferred stock, 206–207, 243, 256
Premanufacturing notice (PMNs), 53
Premarket Approval Application (PMA), 87, 94
Pre-money valuation, 206
Present value (PV), 184
President's Council of Advisors on Science and Technology (PCAST), 128–129
Price-to-earning (P/E) multiple, 183
Printing, 15
Private companies, valuation of, 181
Private equity investors, 197–198
Private investments, 118
Private-private mergers, 255
Process/method claims, patent applications, 216
Product approval, FDA regulation problems
 generally, 95–96
 new versions of existing products, 93–94
 new versions of other technologies, 94–95
Product-by-process claims, patent applications, 216–217
Product liability, 54
Professional journals, as information resource, 178
Professional societies, 99
Promotions, 173
PROMT, 213
Prospectus, 279
Protein-based nanostructures, 19
Proteomics, 98
PsiloQuest, 202
Public health, 62, 83
Public Health Service Act, 88
Public health service agencies, 99
Public offerings, 207
Public perception, as influential factor, 170–171
Public relations, 173
Pyrograf Products, 253

Quantomix, 199
Quantum Dot Corporation, 74, 230
Quantum dots, 18, 38, 218
Quantum Solar Energy, 273
Questech Corp., 201

Random-access memory (RAM), 25
Rat studies, 64
Ratner, Daniel and Mark, 103
Raytheon, 34
r-DNA experimentation, 48
Recombinant DNA Advisory Committee (RAC), 48
Recombinant DNA technologies, 97, 143
Recordkeeping, corporate spinoffs, 151
Redemption preferences, corporate partnering, 243
Redemption provisions, venture financing criteria, 207
Referrals, in venture financing, 203

Regional Center for Nanofabrication Manufacturing Education, Pennsylvania State University, 125
Registration rights, venture capital criteria, 207
Regulation, as influential factor, 171
Replication, 19, 30–31, 96
Research & Development (R&D)
 federal funding, 120–128, 194
 high-risk projects, 194
 local resistance to, 170–171
 operations, in business plan, 172
Research Corporation Technologies, 202
Research experience for undergraduates (REU), 124
Residual value, 185–186
Revenue growth, 183–184
Rice University, 143
Right of first refusal, 208
Risk disclosure, 171–172
Roche, 34, 170
Rocket propellants, 22
Roco, Mihail, 58
Rosseter Holdings, 253
Royalties, licensing, 144

SAIC, 277
Sales and marketing, 173
Sales force, 173
Sandoz, 259
Santur, 199
Sarbanes-Oxley Act of 2002, 271, 276
Sayre, Phil, Dr., 57
Scanning electron microscope (SEM), 52
Scanning tunneling microscope (STM), 21
Scientific advisory board (SAB), 155–156
Scientific expertise, FDA regulation
 acquisition problems, 100–101
 emerging technologies, 97–99
 internal training, 102
 nanomaterials, 96–97
 significance of, 97
Sciperio Inc., 194
SCISEARCH, 213
S corporations (S Corps), 153–154
Securities and Exchange Act of 1933, 270, 278
Securities and Exchange Commission (SEC), 207, 269–271, 275, 278–280
Seed round, 189–191
Self-assembly, 18–19
Self-replication technology, 45, 48–49
Semiconductor industry, 24, 34
Semiconductor manufacturing, 197
Semiconductor nanocrystals, 18–19, 21, 55, 218
Semiconductor patents, 68
Sensicast, 200
Sensicore, 200, 202, 205
Sensors, 38, 42, 168
Service mark protection, 230
Sevin Rosen Funds, 203
Sewing Machine Combination, 76

Shapiro, Carl, 77
Significant New Use Notice (SNUN), 53
Significant New Use Rule (SNUR), 53, 63
Silicon nanocrystals, 64
Silicon Valley, 196, 200, 255
Simple nanotechnology
  defined, 22–23
  market segmentation, 167–168
  societal and ethical implications of, 41–42
Single-walled nanotubes (SWNTs), 17–18
Sionex, 199
SiWave, 201
Small Business Administration (SBA), 177
Small business innovative research (SBIR) grants, 124, 130, 191, 193, 183
Small business technology transfer (STTR) grants, 124, 191, 193, 283
Smalley, Richard, 17, 31, 43
Small Times, 200
Societal implications
  First Amendment protection, 46–47
  immediate action, 42–43
  legal analysis, 46
  overview of, 41–42
  policy analysis, 43–46
Solar energy, 26
Solicore, 201
Solubest, 199
Somatic cell, 88
Sources and uses, business plan components
Southwest NanoTechnologies, 253
Spectroscopy, 23–24
Spin-offs
  corporate, 149–151
  government laboratory, 146–149, 210–211
  university, 140–145, 210
Spintronic devices, 24
Spiritual machines, 42, 44, 132
Sponsorship, 210
Standardization
  anticipatory standards, 261–262
  defining standards, 259–261
  efforts toward, 262–264
  importance of, 259–260
  setting standards, 264–265
Standard MEMS, Inc., 194
Standards developing organizations (SDOs), 260–261
Stanford University, 144, 216
Starfire Electronic Development and Marketing, Inc., 70
Starfire Electronics, 216
Starting a nanotech company
  corporate governance, 154–157
  corporate spinoff, 149–151
  employee stock options, 157–158
  initial intellectual property, 140–151
  government laboratory spinoff, 146–149, 210–211
  initial intellectual property, 140–151
  location, 158–161

naming the company, 151–152
organizational structure, 152–154
stock issuance, 157
university spin-off, 140–145, 210
Start-up companies
  components, of, 35–37, 77
  funding for, 118–11
  licensing, 80–82
  product development, regulation of, 60
State Department, 110–111
State funding, 117
Stevenson-Wydler Technology Innovation Act of 1980, 146
Stock-for-stock mergers, 255
Stock issuance, 157
Stock purchase agreements, 275
Strategic fit, 274
Strategic partners, selection process, 240–243
Strategic Partnership for Research in Nanotechnology (SPRING), 160
Sublicensing, 143
Suggestive trademarks, 231
Sumitomo, 238
Sun Nanotech, 253
Superior MicroPowders, 273
Synergistic effects, 56
Synfuels Program, 130

TakaraBio Inc., 170, 245
Target market, selection factors
  competition, 168
  Five Forces Model, 169
  partnerships, 170
  public perception, 170–171
  regulation, 171
  technological development speed, 168
Technical Barriers to Trade Agreement (TBT), 260–261
Technological advances, impact of, 94
Technological progress, impact of, 44–45
Technology dissemination, 147–148
Technostart, 247
Telecommunications industry, 34
Telesales force, 173
Terminal values, 185–186
Terrorism, 104
Texas Nanotechnology Initiative, 117, 160
Therafuse, 200
Thermo Microscopes, 272
Thin-films, 168
Thomason Derwent, 213
Thomson Delphion, 213
Thomson Research, 178
3i, 247
Tissue engineering, 28, 88, 95, 98–99
Tissue Proteomics Progra, 99
Tissue Reference Group, 90
Titanium dioxide, 55
Tombstone ad, 279
Top-down approach, 15, 18
Toxicity, 54–58

Toxic Substance and Control Act (TSCA), 53, 61–64
Toxic tort litigation, 54
Trademark
  characteristics of, 230
  litigation, 233
  ownership establishment, 230–232
  utilization of, 232–233
Trade secrets
  components of, 227–228
  consolidations and, 256
  corporate spinoffs, 150–151
Trade shows, 173
Tragedy of the anticommons, 77, 79
Trailer clauses, 150, 229
Translume, 200
Tree diagrams, patent review, 72–74
TRIPS Agreement, 249
TSCA Inventory, 53
Tufts Center for the Study of Drug Development, 99
21st Century Nanotechnology Research and Development, 120, 126–128
21st Century Nanotechnology Research and Development Act, 47
2-D nanomaterials, 17

UC Berkeley, 24, 217
UCLA, 24
Ultrafine particles, health hazards, 54–56, 58
Underwriters, IPOs, 278
Unfair competition, 229
Uniform Trade Secrets Act (UTSA), 228
United Online, 200
U.S. Census Bureau, as information resource, 177
United States Copyright Office, 234
U.S. Customs Service, 231
U.S. Foreign Trade Zone, 105
U.S. law, 61–62
United States Pharmacopoeia, 87
United States Treasury Department, 111
University license agreements
  price negotiations, 144
  standard provisions, 145
University licensing office (OTL), 140, 143
University of Texas-Dallas, 194
University spin-offs
  components of, 140–145
  intellectual property, generation of, 210
University technology transfer (OTT) office, 140, 144
Uranium, 112
User Facility Agreement (UFA), 211
User fee, new device applications, 88

Vaccines, 60, 88
Valuation
  consolidations, 256
  discounted cash flow (DCF), 184–186
  Discount Rates, 185
  exogenous variables in, 186
  implications of, 181–182
  multiples, 182–184
  terminal values, 185–186
  in venture financing process, 206
VEECO, 37, 272–273
Venter, Craig, 206
Venture capital
  consolidations, 258
  corporate partnering and, 239, 241
  investing, 195–199
  sources of, 77, 149, 189, 203–206
  top nanotech venture firms, 200–203
Venture capitalists (VCs)
  control issues, 237
  financing criteria, 195
  functions of, 156, 166, 190
  industry perceptions, 196
  voting rights, 208
Venture Capital Journal, 196
Venture financing
  future rounds of, 208
  terms and conditions, 206–208
  sources of, 153, 156
  top nanotech venture firms, 200–203
  in 2003, 199
Venture investment, 196–195
Vesting schedule, stock option plans, 57–158
Viruses, 19–20
Voting rights, 208, 243

Water systems, 172
Water treatment plants, 64
Weighted average formula, 207
Weinbaum, Barry, 166
Wildavsky, Aaron, 59
Williams, Stan, 197
Work For Others (WFO) agreement, 211
World Trade Organization (WTO), 63
Wyko Optical Metrology, 273

Xenotransplantation, 88

Yahoo!, 158
Yang, Peidong, Dr., 156

0-D nanomaterials, 17
Zettacore, 201
Zyvex Corporation, 194